14th International Conference on Electrostatics 2015

AA002899

Journal of Physics: Conference Series Volume 646

Southampton, United Kingdom
12 – 16 April 2015

Editors:

James Matthews

ISBN: 978-1-5108-1473-8
ISSN: 1742-6588

Printed from e-media with permission by:

Curran Associates, Inc.
57 Morehouse Lane
Red Hook, NY 12571

Some format issues inherent in the e-media version may also appear in this print version.

Printed by Curran Associates, Inc. (2016)

For permission requests, please contact the Institute of Physics
at the address below.

Institute of Physics
Dirac House, Temple Back
Bristol BS1 6BE UK

Phone: 44 1 17 929 7481
Fax: 44 1 17 920 0979

techtracking@iop.org

Additional copies of this publication are available from:

Curran Associates, Inc.
57 Morehouse Lane
Red Hook, NY 12571 USA
Phone: 845-758-0400
Fax: 845-758-2634
Email: curran@proceedings.com
Web: www.proceedings.com

Table of contents

Volume 646

Electrostatics 2015
12–16 April 2015, Southampton, UK

Papers

Bill Bright Memorial Lecture

Biological electrostatics

Electrodynamics

Electrospin and electrospray

Energy

Environmental and space applications

Hazards

Materials, powders and fluids

Measurement and instrumentation

Modelling and simulation

Non-thermal plasmas

Preface

Electrostatics 2015, supported by the Institute of Physics, was held in the Sir James Matthews building at Southampton Solent University, UK between 12th and 16th April 2015. Southampton is a historic city on the South Coast of England with a strong military and maritime history. Southampton is home to two Universities: Solent University, which hosted the conference, and the University of Southampton, where much work is undertaken related to electrostatics. 37 oral and 44 poster presentations were accepted for the conference, and 60 papers were submitted and accepted for the proceedings.

The Bill Bright Memorial Lecture was delivered this year by Professor Mark Horenstein from Boston University who was, until recently, Editor-in-Chief of the Journal of Electrostatics. He spoke on *The contribution of surface potential to diverse problems in electrostatics* and his thorough knowledge of the subject of electrostatics was evident in the presentation. The first session was chaired by the Conference Chair, Dr Keith Davies, whose experience in the field showed through his frequent contributions to the discussions throughout the conference.

Hazards and Electrostatic Discharge have formed a strong core to Electrostatics conferences for many years, and this conference contained sessions on both Hazards and on ESD, including an invited talk from Dr Jeremy Smallwood on *ESD in Industry – Present and Future*.

Another strong theme to emerge from this year's programme was Non-Thermal Plasmas, which was covered in two sessions. There were two invited talks on this subject: Professor Masaaki Okubo gave a talk on *Development of super-clean diesel engine and combustor using nonthermal plasma hybrid after treatment* and Dr David Go presented a talk on *Atmospheric-pressure ionization processes: New approaches and applications for plasmas in contact with liquids*.

A new innovation to the conference this year was the opportunity for conference sponsors to present to the delegates a technical presentation related to their work. Chilworth Technology and Infolytica both took advantage of this opportunity. David Firth from Chilworth Technology delivered some case studies related to process safety and Chris Emson from Infolytica compared the different types of modelling software used in industry and academia. For two days of the conference, an exhibition was held for delegates to meet and discuss their work with interested companies.

Sessions on Modelling and Simulation and on Measurement and Instrumentation were included. Recent successful IOP meetings on Electrospinning and Electrospray prove that this is an important topic, and were the subject of a session in the conference, including an invited talk by Dr Horst von Recum on *Electrospun materials for affinity based drug delivery*. The conference finished with a session on Environmental and Space Applications.

The Southampton Yacht Club provided a fitting venue for the conference dinner on the Wednesday evening. Meal times, and conference dinners in particular, are always a great opportunity to meet with other workers in related fields, and there were many conversations started in question and answer sessions that continued over a plate of food. Within the conference dinner, prizes were awarded for the best student work. Ladislav Konopka's talk in the modelling and simulation session discussed how different particle sizes can be shown to transfer charge in a modelled system. Matthias Perez's poster presented early work on the use of a small-scale wind turbine to generate wind power.

The discussions both within the lecture theatre and the ongoing discussions that occur over coffee and tea in between sessions are often a place where new ideas are shared. In fact, the presentation submitted by Dr Atsushi Ohsawa, *Charge neutralisation from the side surface of an insulating plate*, acknowledged an inspiration from a question raised at a previous Electrostatics conference in Budapest in 2013. In these proceedings the conference committee took the decision to transcribe the questions and answers to leave a lasting record of the conversations that took place after each oral presentation and they are included in the printed proceedings.

I am very grateful to Keith Davies and the conference committee, and Joanne Hemstock and other IOP staff, who provided advice and assistance throughout the whole process. A peer reviewed proceedings is not possible without willing expert reviewers who are able to provide reviews on abstracts and submitted papers. I am grateful to all who undertook reviews throughout the process.

I hope to see many of this year's delegates in four years time for Electrostatics 2019.

James Matthews
Conference Programme Chair and Proceedings Editor, Electrostatics 2015.

Peer review statement

All papers published in this volume of *Journal of Physics: Conference Series* have been peer reviewed through processes administered by the proceedings Editors. Reviews were conducted by expert referees to the professional and scientific standards expected of a proceedings journal published by IOP Publishing.

Journal of Physics: Conference Series **646** (2015) 012001 doi:10.1088/1742-6596/646/1/012001

The Contribution of Surface Potential to Diverse Problems in Electrostatics

M. Horenstein[1]

Professor, Dept. of Electrical and Computer Engineering
Boston University, 8 Saint Mary's St., Boston, MA USA 02215 (mnh@bu.edu)

Abstract. Electrostatics spans many different subject areas. Some comprise "good electrostatics," where charge is used for desirable purposes. Such areas include industrial manufacturing, electrophotography, surface modification, precipitators, aerosol control, and MEMS. Other areas comprise "bad electrostatics," where charge is undesirable. Such areas include hazardous discharges, ESD, health effects, nuisance triboelectrification, particle contamination, and lightning. Conference proceedings such as this one inevitably include papers grouped around these topics. One common thread throughout is the surface potential developed when charge resides on an insulator surface. Often, the charged insulator will be in intimate contact with a ground plane. At other times, the charged insulator will be isolated. In either case, the resulting surface potential is important to such processes as propagating brush discharges, charge along a moving web, electrostatic biasing effects in MEMS, non-contacting voltmeters, field-effect transistor sensors, and the maximum possible charge on a woven fabric.

1. Introduction

Surface potential is a well-known phenomenon. While the surface of a conductor at fixed voltage is an equipotential with field lines normal to the surface, the surface potential of a charged insulator, as in figure 1, can vary with position, and the field lines need not be surface normal. When an insulating sheet charged to ρ_s covers a ground plane, as in figure 2, a surface potential $V_S = (\rho_s/\varepsilon)d$ develops with capacitance ε/d per m^2. Unlike a "real" capacitor charged by current, the surface capacitance instead requires on charge deposition. This paper examines the role of such surface potentials in electrostatics..

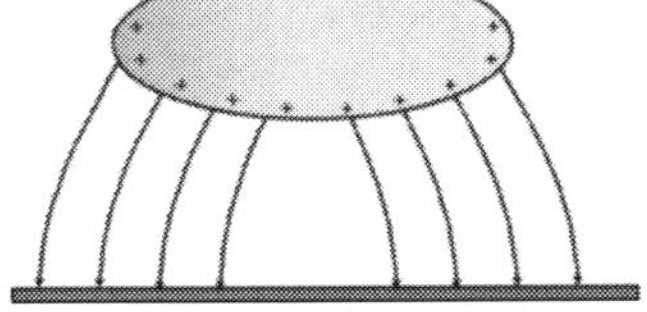

Figure 1 – Charged, isolated insulator

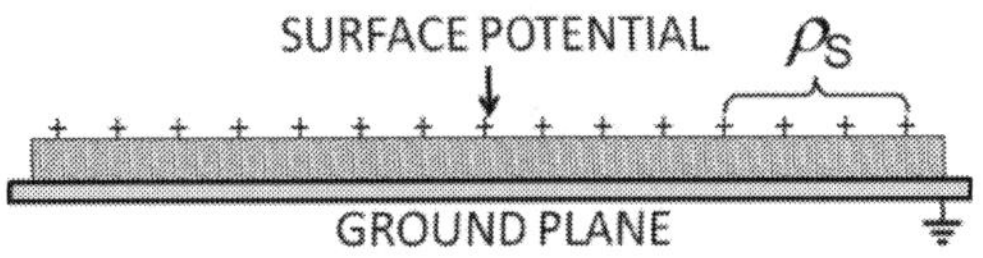

Figure 2 – Charged sheet over a ground plane

2. Equilibrium charge distribution along a moving web

When an insulating web passes over a metal roller, triboelectrification and charge accumulation at the peel point can cause arcing, disruptive forces, material damage, and dust precipitation. Understanding how surface charge distributes in equilibrium is important to preventing these nuisances. In the system of figure 3, the charge distribution depends on surface potential, conductivity, capacitance, and transport speed. The discrete model of figure 4 provides insight into the related role of surface potential [1].

1 To whom any correspondence should be addressed.

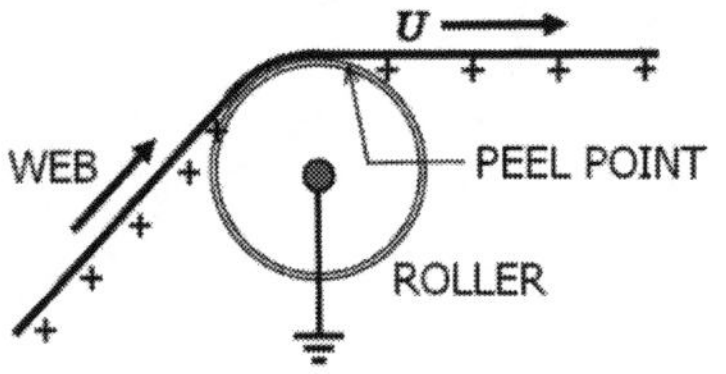

Figure 3 – Web charged by roller

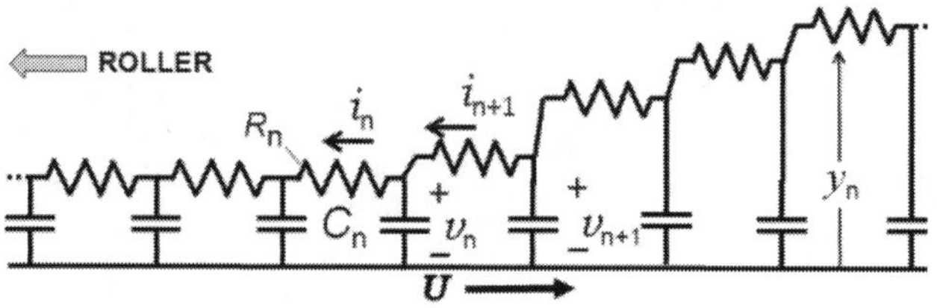

Figure 4 – Model for peel point between web and roller

The web, moving at speed U, is divided into small segments of area A having height y_n and capacitance $C_n = \varepsilon_0 A/y_n$ to the roller, and surface resistance R to adjacent segments. A uniform surface charge at $t = 0$ is assumed. An iterative calculation then proceeds in small time steps. As a segment leaves the roller at the peel point, its y_n increases and its C_n decreases. The segment voltage $V_n = Q_n/C_n$ thus rises, causing charge to flow back toward the roller via the surface potential difference between segments. After each Δt, the charge flow between segments is found via Ohm's law: $i_n = (v_n - v_{n-1})/R$, and charges are updated by $\Delta Q_n = (i_{n+1} - i_n)\Delta t$. The web is then advanced by $\Delta x = U\Delta t$. After many time steps, the charge distribution reaches equilibrium. A plot of charge per meter web width for $r_s = 10^{10}\ \Omega$/sq, $Q_0 = 0.2\ \mu$C/m, and $U = 1$ m/s is shown in figure 5. The back flow of charge toward the peel point is evident. The results compare favourably with the analytical solution of Clum [2] in figure 6.

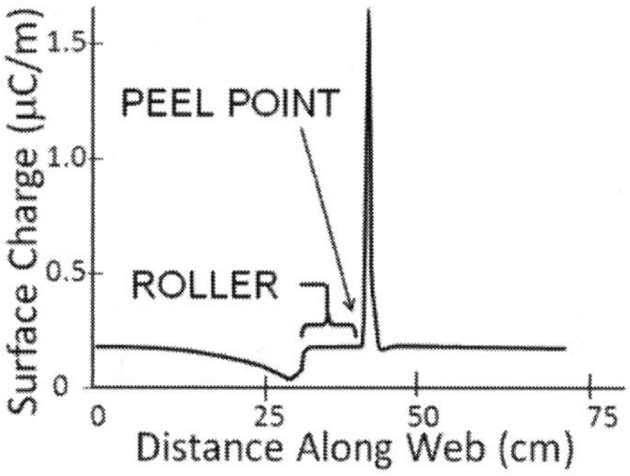

Figure 5 – Results of charge-backflow model from [1]

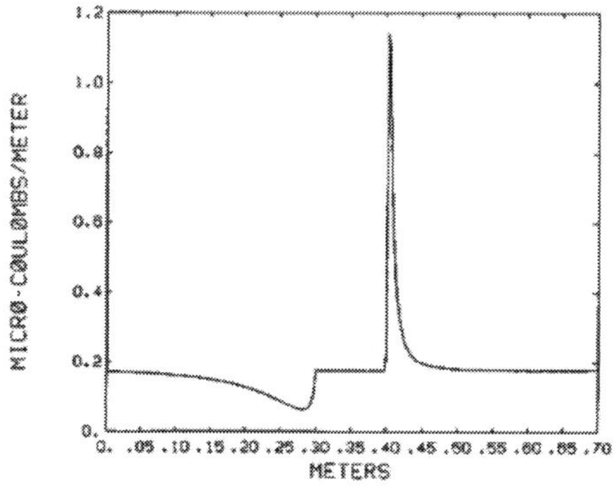

Figure 6 - Result obtained by Clum [2]

3. Direct gate field-effect transistor

Measuring DC fields has always been a challenge in electrostatics. Field mills and surface vibrators, which rely on a variable capacitance ($i = v\ dC/dt$) are reliable but require moving parts. One attempt to build a solid-state field sensor [3] is shown in figure 7. A depletion-mode (internally biased) field-effect transistor (FET) was fabricated with no gate electrode. Its oxide layer was exposed to an incident E-field which provided an equivalent gate field for modulating the channel current i_D.

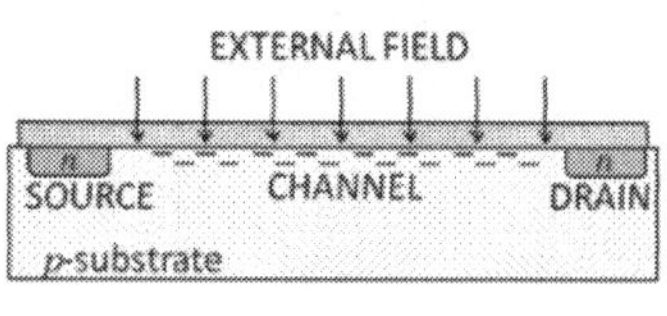

Figure 7 – FET with no gate

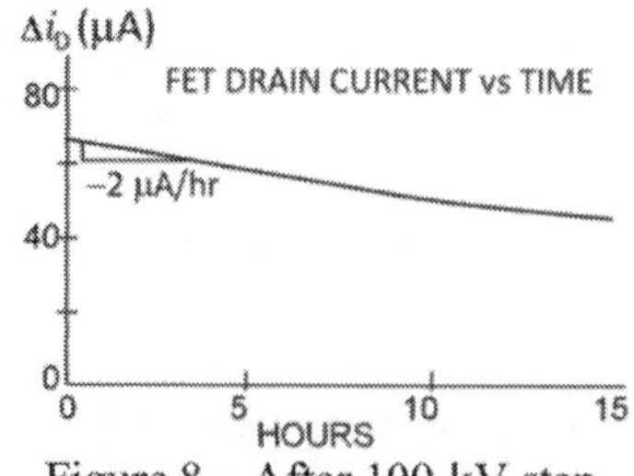

Figure 8 – After 100-kV step

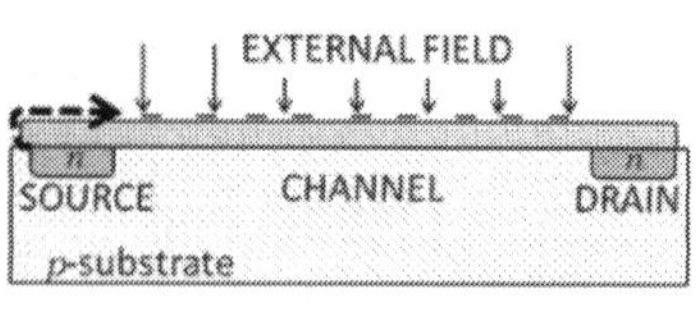

Figure 9 – Surface charge migration

Figure 8 shows the i_D from a 100-kV/m step function, yielding an initial current of 60 μA. With the field maintained, the current fell −2 μA per hour, rather than holding steady. This drift can be explained by *surface potential*. Later experiments showed that electrons or negative ions from the source region slowly flowed via the surface resistance of the oxide layer, as in figure 9. This charge altered the surface potential of the oxide, in turn reducing the modulating field in the FET channel.

4. Measuring surface charge with a non-contacting voltmeter

Non-contacting voltmeters, such as those made by Monroe™ and Trek™, measure surface potential with a variable capacitance probe, detector, and HV amplifier. The probe voltage V_{PROBE} is raised until its detected field nulls to zero. The surface potential $\rho_s d/\varepsilon$ of a charged sheet covering a ground plane is easily measured. When the charged insulator is *not* on top of a ground plane, as in figure10, the surface potential will vary with the position of both sheet and probe. The NCV, however, will still adjust V_{PROBE} until the probe sees zero field.

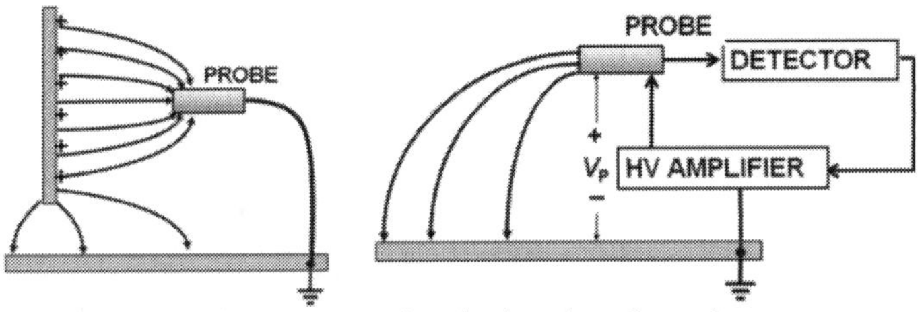

Figure 10 – Measuring isolated surface charge

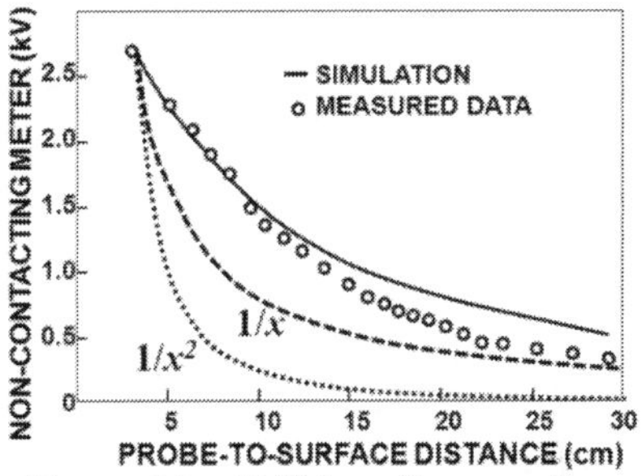

Figure 11 – NCV reading vs. distance

The resulting V_{PROBE} is found by superposition [4]. One field component is from the surface charge with the probe grounded (figure 10, left). The other component (figure 10, right) is due to just V_{PROBE} with no surface charge present. Both superimposed field components depend on geometry, hence the V_{PROBE} enforced by the NCV will depend on probe position. The position-dependent NCV reading can be predicted by field modelling. First the surface-charge field is found by finite-element analysis or charge simulation. The probe-field pattern is similarly found, but made proportional to the unknown V_{PROBE}. The unknown V_{PROBE} is then chosen so as to null the E-field at the probe. Figure 11 shows the predicted and measured NCV readings as a function of axial probe position for a 10-cm disc charged to 10 μC/m^2. As the distance increases, V_{PROBE} doesn't fall as $1/x$ or $1/x^2$, as one might expect for a charge distribution somewhere between a point and a plane, but rather as something less steep than either. This result suggests that field calculations are indeed needed to interpret the NCV reading.

5. Maximum surface charge on a woven fabric

Many textiles exhibit a maximum surface charge limit related to composition and construction. In one experiment [5], cloth samples on a ground plane were corona "sprayed" with positive ions at 4 kV to 9 kV, as in figure 12. From SEM photos, each sample was characterized by a weave pattern, like those in figure 13, and assigned a cloth "pore size" based on the width of interweave air channels and the overall weave pattern. The highest obtainable V_S values under corona charging were measured. Though widely scattered, the results of figure 14 show a trend whereby the maximum possible surface potential V_S occurs on cloth samples having the smallest pores. This trend can be explained by back ionization, well known in the precipitator industry. As ions collect on a highly resistive, porous layer over a ground plane, V_S increases until breakdown occurs in the pores. Ions of both polarities are generated, causing local regions of high conductivity which limit any further charge accumulation.

Pore breakdown will follow the classic Paschen discharge curve [6] which can be described (for parallel electrodes) by the curve-fit equation $E_{BK} = (6.2 + 312/g)$ V/m [7], where the gap g is in μm. The maximum sustainable V_S for a tightly woven fabric can thus be estimated by finding E_{BK} for the fabric and computing the resultant $V_{SMAX} = \frac{\varepsilon_0}{\varepsilon_f} T(6.2 + \frac{312}{d})$. Here T is the cloth thickness in μm. Figure 14 compares this curve with measured values of V_{Smax} (white circles) and their averages (black circles.) The near correlation between the measured, charge-limited surface potentials and those predicted by modelling suggests that the Paschen breakdown model for fabric pores is plausible.

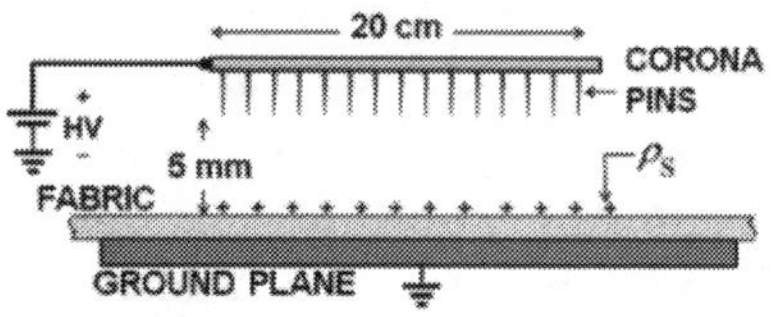

Figure 12 – Corona charging fabric

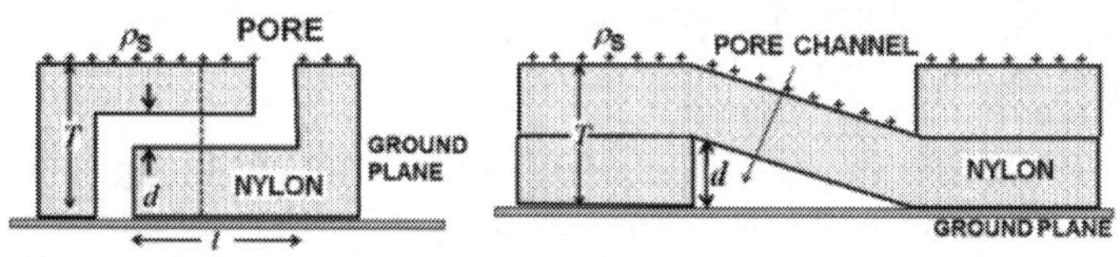

Figure 13 – Models for the pores of two woven samples.

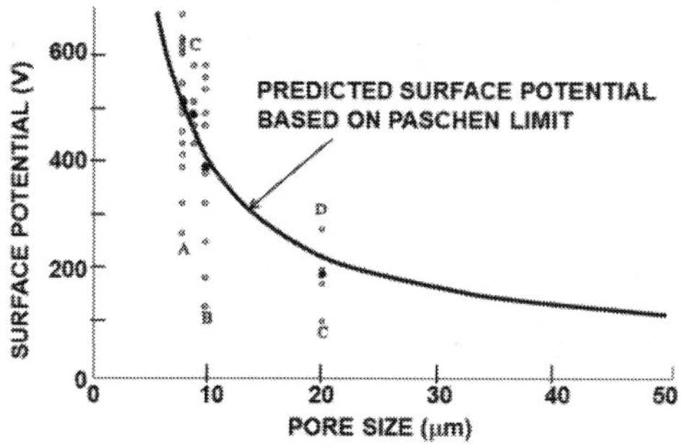

Figure 14 – Maximum surface potential attainable with corona charging

6. Electrostatic effects in MEMS devices

A classic dual-cantilever MEMS actuator in shown in figure 15 (upper). Applying 30-300 V causes the polysilicon membrane to deflect downward due the electrostatic force that is counteracted by the mechanical restoring force of the membrane. The field in the 2 – 5 μm gap g falls below the Paschen breakdown minimum. If the structure is modelled as a simple capacitor/spring system, the downward electrostatic force is $F_{elec} \approx -V^2\varepsilon A/2y^2$, and the mechanical restoring force is $F_{mech} = k(g-y)$. The displacement vs. voltage solution for these equations is shown figure 16 (black and white circles to be explained shortly.) The deflection curve is more-or-less parabolic with increasing voltage until the "snap through" point, where the restoring force can no longer hold back the electrostatic force. Above snap-through, the membrane collapses onto the nitride layer. This is a well-known effect in MEMS.

The plot of figure 17 shows the deflection of a MEMS membrane actuator energized at +28 V DC for several minutes [8]. Over time, deflection increases, even though the voltage is constant. This increase in deflection is caused by a surface potential effect. Specifically, over time, negative charges migrate onto the nitride layer, causing its surface potential to acquire a negative bias, as in figure 15 (lower), thereby increasing the net field in the gap and, in turn, the membrane deflection. Given enough time, the field will be screened entirely out of the nitride layer, effectively reducing the gap distance and increasing deflection for a given applied voltage. The white dot in figure 16 shows the initial operating point of the actuator with +28 V applied. The black dot shows the deflection for the same voltage with the effective gap spacing reduced by the thickness of the nitride layer. Figure 18 demonstrates a different type of surface-potential effect. The plot shows the membrane height with 35 V applied. At 25 seconds, the voltage is increased to 50 V for 10 sec, causing snap-through. Contact charging leaves residual positive charge on the nitride layer, thereby raising its surface potential and reducing the gap field, leading to a smaller deflection for the same voltage after the snap-through event.

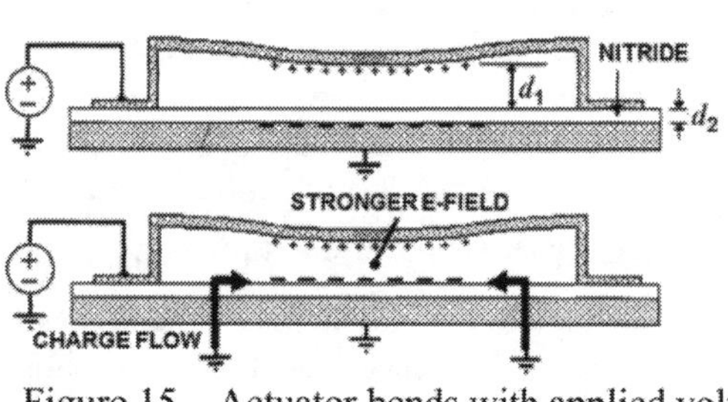

Figure 15 – Actuator bends with applied voltage

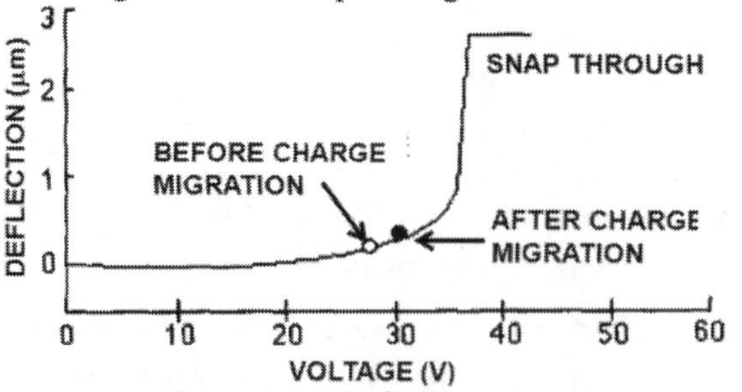

Figure 16 – Deflection vs. voltage

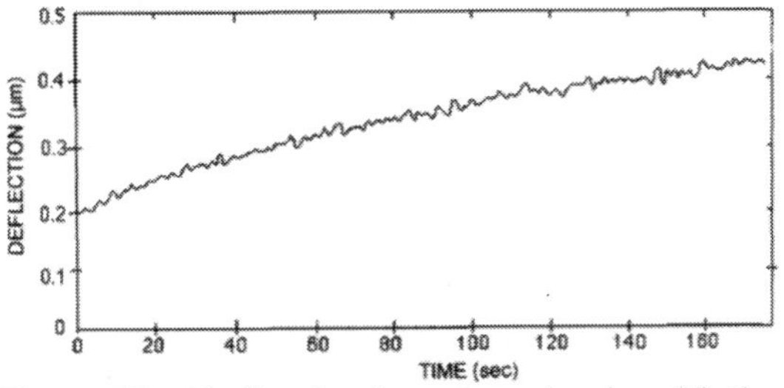

Figure 17 – Deflection increases slowly with time

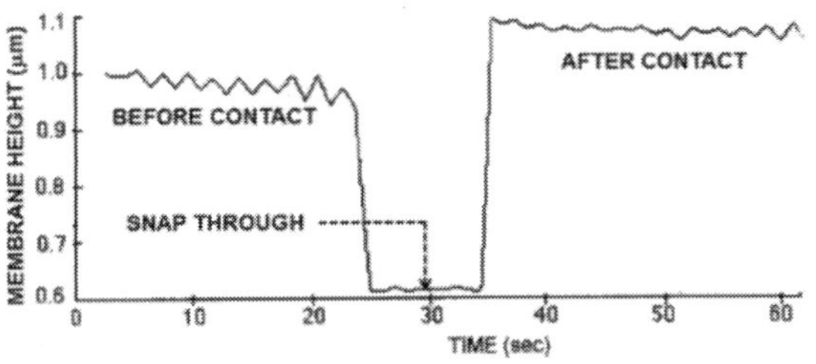

Figure 18 – Contact charging after snap-through

7. Modelling propagating brush discharges

Propagating brush discharges can release a lot of energy, making them the most destructive of electrical discharges. They occur when a charged, insulating layer on a ground plane is point-discharged from above. The energy content of such a discharge can be predicted by a model that accounts for surface potential, capacitance, and resistance. The charged surface is divided into segmented capacitor cells (figure 19) bridged by spark gaps to adjacent cells, as shown in figure 20.

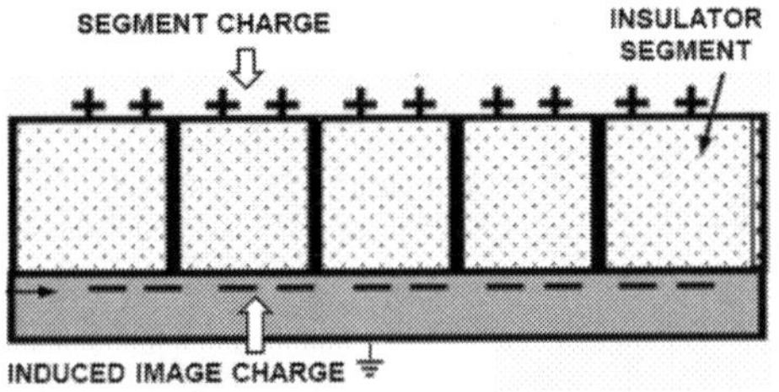

Figure 19 – Segmented model of charged surface

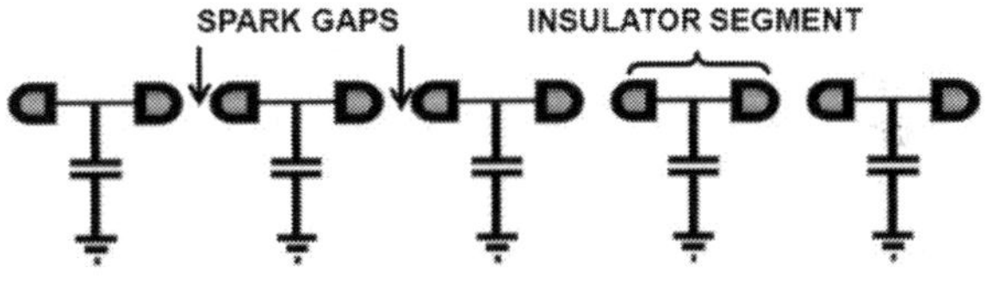

Figure 20 – Spark-gap model between segments

Each segment of charge ρ_s is tightly coupled to the ground-plane. The nth segment's potential is $(\rho_s/\varepsilon)d$. Charge flows between cells if the mutual spark gap is bridged by a tangential E-field in excess of the 3-kV/mm breakdown of air. Charge can flow by surface breakdown between cells and is what drives the "propagation" of the discharge. Once bridged, a spark gap is modelled as a resistor (0.1 Ω /mm) until the gap field falls to 75 V/mm for negative charge and 300 V/mm for positive [10]. The dynamic progression of the discharge is found by time iteration. After Δt, cell voltages are found, spark gaps are bridged if appropriate, and inter-cell currents computed from Ohm's law. The charge redistributes with each Δt. The steps continue until all spark gaps are below breakdown. Figures 21 - 24 compare calculated post-discharge ρ_S distributions with measured results from the literature [11-14].

8. Summary

This paper has addressed the role of surface potential in diverse areas of electrostatics. The effects of surface potential show up in many different situations, from field sensing, to webs, textiles, discharges and MEMS. The charge on insulator surfaces and the resulting fields and forces are often responsible for observed phenomena. At other times, surface potential causes behaviour to differ significantly from the ideal or expected.

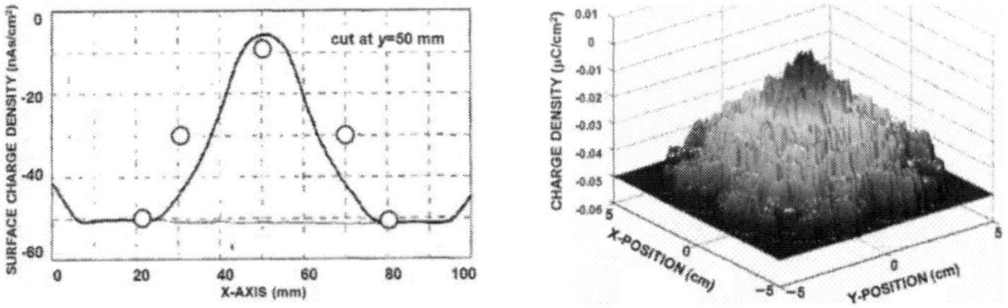

Figure 21 – Predictions of Crager [10] (circles) vs Muller et al [11]; Right: ρ_S predicted by Crager [10]

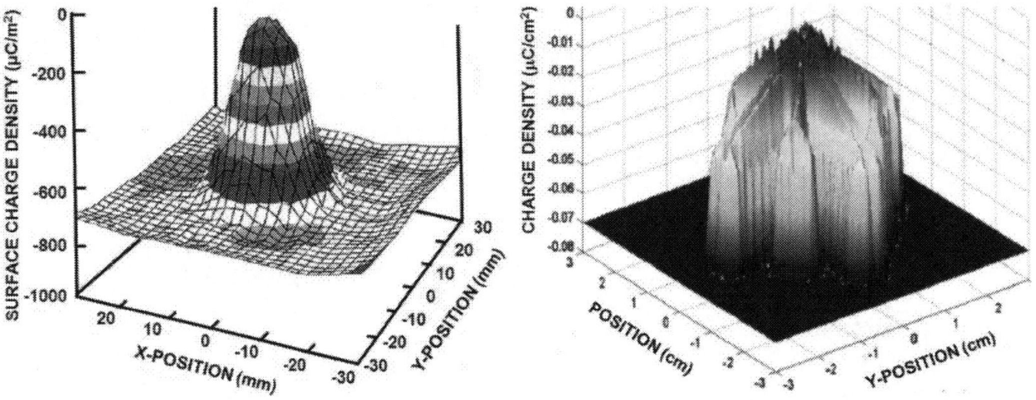

Figure 22 – Left: Measured ρ_S from Davidson et al [13]; right: ρ_S predicted by Crager [10]

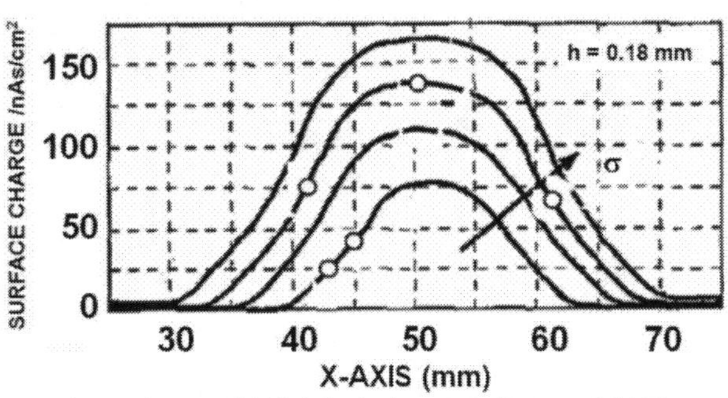

Figure 23 – Crager [10] (circles) vs Muller et al [12]

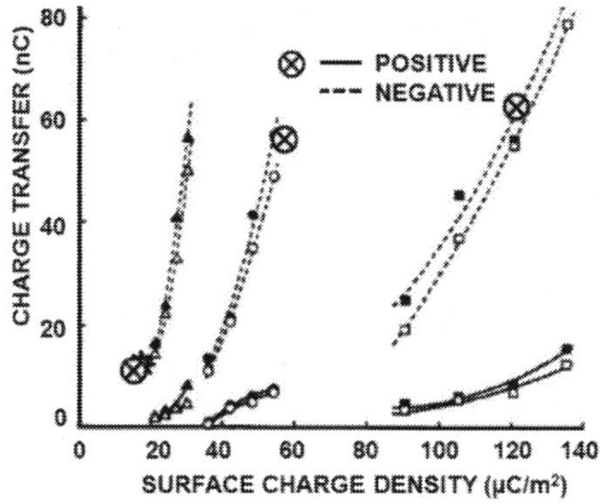

Figure 24 – Crager ⊗ [10] vs Matsui et al [14]

9. References

[1] Horenstein M 2000 *IEEE Trans. Ind. Appl.* **36** 1
[2] Clum KL 1989 IEEE Trans. Ind. Appl. **25** 1 36-40
[3] Horenstein M 1985 *IEEE Trans. Electron Dev.* **ED-2** 3 44-47
[4] Horenstein M 1993 *Proc. IEEE IAS Annual Mtg.* **3** 1811-1816
[5] Horenstein M 1995 *J. Electrostatics* **35** 31-40
[6] Cobine JD 1958 *Gaseous Conductors* New York: Dover
[7] Crowley J, private communication
[8] Horenstein M 1997 *J. Electrostatics* **42** 69-81
[9] Horenstein M 2011 *IEEE Trans. Ind. Appl.***47** 1 344 – 349
[10] Crager JC and Horenstein, 2005 *J. Electrostat.* **63**, no. 3/4, 239–247
[11] Mueller L, Feser K et al 2001 *Proc. Annu. Rep. Conf. Elect. Insul. Dielectr. Phenom.* **110–113**
[12] Mueller L and Feser K 2002 *Proc. Annu. Rep.Conf. Elect. Insul. Dielectr. Phenom.***122–125**
[13] Davidson JL, Williams TJ and Bailey AG 2002 *J. Electrostat.* **56** 1 29-42
[14] Matsui M and Murasaki N, Fujibayashi K, Wei X, 1995 *J. Electrostat.* **35** 1 73–82

Safe housing ensured by an electric field screen that excludes insect-net permeating haematophagous mosquitoes carrying human pathogens

Y Matsuda[1], K Kakutani[2], T Nonomura[1], J Kimbara[3], K Osamura[4], S Kusakar[1] and H Toyoda[1]

[1]Faculty of Agriculture, Kinki University, Nara, Japan,

[2]Pharmaceutical Research and Technology Institute, Kinki University, Osaka, Japan,

[3]Research and Development Division, Kagome Co., Ltd., Tochigi, Japan,

[4]Technical Development Unit, Panasonic Environmental Systems and Engineering Co., Ltd., Osaka, Japan

E-mail: ymatsuda@nara.kindai.ac.jp

Abstract. An electric field screen can be used to keep mosquitoes out of houses with open windows. In this study, doubly charged dipolar electric field screens (DD-screens) were used to capture mosquitoes entering through a window. The screen had two components: three layers of insulated conductor iron wires (ICWs) in parallel arrays and two electrostatic direct current (DC) voltage generators that supplied negative or positive voltages to the ICWs. Within each layer, the ICWs were parallel at 5-mm intervals, and connected to each other and to a negative or positive voltage generator. The negatively and positively charged ICWs are represented as ICW(–) and ICW(+), respectively. The screen consisted of one ICW(+) layer with an ICW(–) layer on either side. The Asian tiger mosquito (*Aedes albopictus*) and house mosquito (*Culex pipiens*) were used as models of vectors carrying viral pathogens. Adult mosquitoes were blown into the space between the ICWs by sending compressed air through the tip of an insect aspirator to determine the voltage range that captured all of the test insects. Wind speed was measured at the surface of the ICW using a sensitive anemometer. The result showed that at ≥1.2 kV, the force was strong enough that the ICWs captured all of the mosquitoes, despite a wind speed of 7 m/s. Therefore, the DD-screen could serve as a physical barrier to prevent noxious mosquitoes from entering houses with good air penetration.

1. Introduction

Electric field screens have been used to create spore-free and pest-free spaces for crop production and preservation [1]. The first electrostatic spore precipitator that we reported was a screen that created a non-uniform electric field around insulated copper conductor wires arranged in parallel [2]. The electric field generated an electrostatic force that could be harnessed to attract fungal conidia entering the field. Unfortunately, the spore precipitator was ineffective at trapping major insects that fly into greenhouses. The second device used to solve this problem was a three-layered version of the electric field screen, in which grounded metal mesh was placed on both sides of the original spore precipitator to make dielectric poles; it captured insects that were blown into the inner space of the screen [3]. In this system, a negative charge released from the insects to the grounded net is detected immediately after they enter the electric field [4,5]. We postulated that the strong attraction between the insects and insulated conductor iron wire (ICW) was a result of this release. In fact, this postulate held for different insect pest species, including mosquitoes [6]. Successful insect capture depends on the formation of an electrostatic barrier with no spaces through which the insects can pass. However, parallel ICWs with negative charges create a static electric field with no electricity transfers between the ICWs or insects, which is not effective for attracting insects to enter the field. In a survey of screens installed on greenhouse windows, we observed that insects rarely entered such a static electric field. Therefore, we constructed a doubly charged dielectric electric field screen (DD-screen) in which the ICWs were oppositely charged with equal voltages using two voltage generators to provide an insect exclusion screen with multiple gap-free electric fields [1].

We are especially interested in ways to exclude hematophagous mosquitoes from houses. In tropical and subtropical countries, these noxious mosquitoes transmit various viral and bacterial pathogens and malarial plasmodia that cause severe diseases in humans. The current standard countermeasure is the use of an insecticide-impregnated woven net with larger mesh sizes (mesh diameter 5 mm) for increased air-permeability [7]. However, this method has some potential risks, including the development of insecticide-resistant pests [8] and continuous contact with exuded neurotoxic insecticides by the users [9]. Our electrostatic technique should create pest-free living spaces in houses by installing the screens on windows because its strong electrostatic force captures insects while allowing good air penetration.

2. Materials and methods

2.1. Electric field screen

A three-layered DD-screen [10] was used in this study (Fig. 1). An iron conductor wire was insulated by passing it through a transparent insulator vinyl sleeve and was used to construct the electric field screen. The screen had two components: three layers of ICWs in parallel arrays and two electrostatic DC voltage generators that supplied negative and positive voltages to the ICWs. Within each layer, the ICWs were parallel at 5-mm intervals and connected to each other and to a negative or positive voltage generator. The negatively and positively charged ICWs are represented as ICW(–) and ICW(+), respectively. The screen consisted of one ICW(+) layer with an ICW(–) layer on either side. The layers were parallel and 2 mm apart, and the ICWs in the different layers were offset from each other. The generators were linked to each other to make an electric circuit producing electric fields between ICW(–) and ICW(+). Both generators were operated with 12 V storage batteries, with power supplied by a 15-watt solar panel, to supply equal negative and positive voltages (0.4–1.4 kV) to the ICWs. In this system, free electrons from ICW(+) were pushed out to ICW(–), and the opposite surface charges on the ICWs acted as dipoles that form an electric field between them.

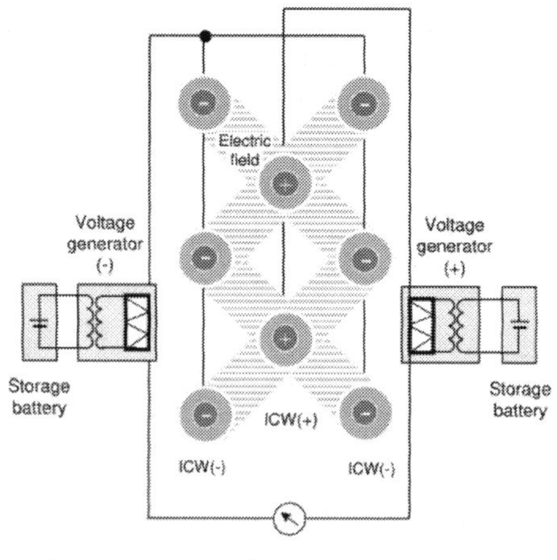

Fig. 1. Construction of a three-layered electric field screen.

2.3. Test mosquitoes and insect-capturing assay

The Asian tiger mosquito (*Aedes albopictus*) and house mosquito (*Culex pipiens*) were used as model vectors of viral pathogens. Adults of both mosquitoes were purchased from Sumika Technoservice (Hyogo, Japan) and maintained in a growth chamber (25.0 ± 0.5°C, 12-h photoperiod of 4000 lux), using our standard method.

Adults of the test insects were collected with an insect aspirator and blown into the space between the ICWs by sending compressed air through the tip of an insect aspirator to determine the voltage range that captured all test insects. Wind speed was measured at the surface of the ICW using a sensitive anemometer. To confirm the successful capture of mosquitoes with the ICW, we directed a blower (max. wind speed 7 m/s at the ICW) at the captured insects for 10 minutes. Twenty adults were used for each voltage tested. The experiments were repeated three times, and the data are presented as the mean ± SD, with a statistical analysis.

3. Results and discussion

Females of both the Asian tiger mosquito and house mosquito invade houses nocturnally and diurnally, respectively, to suck blood from humans [11]. The prevention of entry by these mosquitoes is a fundamental approach to protect humans from the diseases transmitted by pest mosquitoes. For this purpose, the ability of the electric field screen to trap the mosquitoes blown into the field at different wind speeds was examined. Different wind speeds were achieved by changing the distance between the tip of an insect aspirator and the surface of ICWs; wind speeds of 1, 2, and 3 m/s were created at distances of 5, 10, and 15 cm, respectively. **Table 1** lists

Table 1. Percentage of Asian tiger mosquito and house mosquito adults captured by ICWs of three-layered DD-screen

Pest insects tested	Wind speed (m/sec)	Negative and positive voltages (kV) applied to ICWs						
		0	0.4	0.6	0.8	1	1.2	1.4
Asian tiger mosquito	1	0	0	42.1 ± 5.3 a	76.4 ± 4.7 a	100 a	100	100
	2	0	0	41.4 ± 2.3 a	75.9 ± 3.4 a	100 a	100	100
	3	0	0	45.0 ± 3.1 a	77.2 ± 5.6 a	100 a	100	100
House mosquito	1	0	0	25.7 ± 3.1 b	56.1 ± 1.4 b	99.2 ± 0.6 a	100	100
	2	0	0	27.1 ± 3.9 b	55.6 ± 2.4 b	98.6 ± 0.5 a	100	100
	3	0	0	35.8 ± 4.5 b	50.8 ± 2.2 b	99.8 ± 0.1 a	100	100

Twenty adults were used for each voltage and wind speed, and the means and standard deviations were calculated from five replicates. The different letters on the mean values in each vertical column indicate significant differences ($p < 0.05$) according to Tukey's method.

the percentages of the two species of mosquito captured by the screens at different voltages (0.4–1.5 kV) for each wind speed. The mechanism of insect capture by the DD-screen has been described in our previous paper [1]. Stronger forces of ICWs were necessary to capture mosquitoes carried at higher wind speeds, and the force increased with the voltage applied to the ICWs. In addition, there was no significant difference in the capture rate between the two mosquito species for any combination of wind speed and voltage. At ≥1.2 kV, the force was strong enough that the ICWs captured all of the mosquitoes, regardless of the wind speed. The mosquitoes were attracted to the near ICW. Both ICW(–) and ICW(+) were similarly effective at capturing approaching mosquitoes; in fact, the force at both sites was strong enough to capture the mosquitoes despite a wind speed of 7 m/s. At lower voltages, however, the force was insufficient to capture the mosquitoes permanently; the captured mosquitoes waved their legs, twisted their bodies, and then flew away from the ICWs; otherwise, they were blown away from the ICW by the blower. Therefore, this study provides an experimental basis for the practical application an electric field screen in house windows.

In addition to its ability to capture mosquitoes, the low electric power consumption of our screen system is important for practical use. In the electric field screen, the voltage generator is the only driving part requiring an electric power supply, and its electric power is 0.5W, equivalent to that of a small LED pilot lamp. This enabled the use of a photovoltaic power generation method for supplying electric power to the voltage generators. Using this system, we continuously operated four DD-screens attached to a house window for 1 month.

4. Conclusion

The offset arrangement of ICWs in the DD-screen results in gap-free electric fields to ensure the complete prevention of entry by mosquitoes and wider spaces for good air penetration. More importantly, it is possible to place major parts (the voltage generator, current flow-controlled storage battery, and electric wires connecting the ICWs to the generator) inside the aluminium window frame and seal them with silicon resin to waterproof the screen. In addition, the solar panel used for photovoltaic power generation can be attached on the frame surface. This self-sufficient electric structure of the DD-screen enables installation of the screen on house windows in areas with no electrical infrastructure. Therefore, this study strongly supports the possibility that safe housing that eliminates pathogen-carrying insect pests can be realised with the installation of DD-screens on house windows.

Acknowledgement

This work was supported in part by JSPS KAKENHI Grant Number 25450375 from the Ministry of Education, Culture, Sports, Science and Technology, Japan.

References

[1] Matsuda Y, Kakutani K, Nonomura T, Kimbara J, Kusakari S, Osamura K and Toyoda H 2012 *Journal of Applied Physics* **112** 116103

[2] Matsuda Y, Ikeda H, Moriura N, Tanaka N, Shimizu K, Oichi W, Nonomura T, Kakutani K, Kusakari S, Higashi K and Toyoda H 2006 *Phytopathology* **96** 967

[3] Matsuda Y, Nonomura T, Kakutani K, Takikawa Y, Kimbara J, Kasaishi Y, Osamura K, Kusakari S and Toyoda H 2011 *Crop Protection* **30** 155

[4] Kakutani K, Matsuda Y, Haneda K, Sekoguchi D, Nonomura T, Kimbara J, Osamura K, Kusakari S and Toyoda H 2012 *Journal of Electrostatics* **70** 207

[5] Nonomura T, Matsuda Y, Kakutani K, Kimbara J, Osamura K, Kusakari S and Toyoda H 2014 *Journal of Electrostatics* **72** 1

[6] Kakutani K, Matsuda Y, Haneda K, Nonomura T, Kimbara J, Kusakari S, Osamura K and Toyoda H 2012 *Annals of Applied Biology* **160** 250

[7] Bobanga T, Ayieko W, Zanga M, Umesumbu S, Landela A, Fataki O, Mandoko AS, Tshibamba J and Nyabola L 2013 *Journal of Vector Borne Disease* **50** 206

[8] N'Guessan R, Corbel V, Akogbéto M and Rowland M 2007 *Infectious Diseases* **13** 199

[9] Sharma SK, Tyagi PK, Upadhyay AK, Haque MA, Mohanty SS, Raghavendra K and Dash AP 2009 *Acta Tropica* **112** 181

[10] Takikawa Y, Matsuda Y, Nonomura T, Kakutani K, Kimbara J, Osamura K, Kusakari S and Toyoda H 2014 *Aerobiologia* **30** 435

[11] Bowen M F 1991 *Annual Review of Entomology* 36 139

Avoidance of an electric field by insects: Fundamental biological phenomenon for an electrostatic pest-exclusion strategy

Y Matsuda[1], T Nonomura[1], K Kakutani[2], J Kimbara[3], K Osamura[4], S Kusakari[1] and H Toyoda[1]

[1]Faculty of Agriculture, Kinki University, Nara, Japan,

[2]Pharmaceutical Research and Technology Institute, Kinki University, Osaka, Japan,

[3]Research Institute, Kagome Co., Ltd., Tochigi, Japan,

[4]Technical Development Unit, Panasonic Environmental Systems and Engineering Co., Ltd., Osaka, Japan

E-mail: ymatsuda@nara.kindai.ac.jp

Abstract. An electric field screen is a physical device used to exclude pest insects from greenhouses and warehouses to protect crop production and storage. The screen consists of iron insulated conductor wires (ICWs) arrayed in parallel and linked to each other, an electrostatic DC voltage generator used to supply a negative charge to the ICWs, and an earthed stainless net placed on one side of the ICW layer. The ICW was negatively charged to polarize the earthed net to create a positive charge on the ICW side surface, and an electric field formed between the opposite charges of the ICW and earthed net. The current study focused on the ability of the screen to repel insects reaching the screen net. This repulsion was a result of the insect's behaviour, *i.e.*, the insects were deterred from entering the electric field of the screen. In fact, when the screen was negatively charged with the appropriate voltages, the insects placed their antennae inside the screen and then flew away without entering. Obviously, the insects recognized the electric field using their antennae and thereby avoided entering. Using a wide range of insects and spiders belonging to different taxonomic groups, we confirmed that the avoidance response to the electric field was common in these animals.

1. Introduction

An electrostatic-based crop protection method was previously developed as a spore precipitation screen for fungal pathogens [1] and as an insect exclusion screen for glasshouse and warehouse pests [2]. This structure consisted of a single-charged dipolar (SCD) screen in which earthed metal nets were placed on both sides of a spore precipitator to create dielectric poles. This screen was able to capture all insect pests that passed the through the screen net. In addition to insect-capture, we found that the SCD screen repels insects reaching the screen net [2,3]. The insects on the charged screen net placed their antennae inside the screen, then flew away without entering the screen. Insects apparently detected an electric field with their antennae and avoided entry. Nevertheless, this finding applied to a limited number of insect species: whiteflies (glasshouse pest) [3], and cigarette beetles and vinegar flies (warehouse and food processing factory pests) [2]. In this study, we clarify whether the insect-repelling functionality of the SCD screen is effective across insect species. We used a wide range of insects and spiders belonging to different taxonomic groups (13 orders, 45 families, 62 genera and species) and confirmed that the avoidance response to the electric field was common across these animals.

2. Materials and methods

2.1. Electric field screen

A copper conductor wire (2 mm diameter, 0.9 m length), insulated by passing through a transparent insulator vinyl sleeve (1 mm thickness, 1 × 10^9 Ω), was used to construct the SCD-screen. The insulated conductor wires (ICWs) were parallel and spaced at 5-mm intervals; they were connected to each other and to a negative voltage generator. Two earthed stainless nets were placed on one side of the ICW layer, with a separation of 3 mm (**Fig. 1**). The ICWs were negatively charged to dielectrically polarise the ICW insulator sleeve. The negative surface charge of the ICWs causes electrostatic induction in the earthed nets (conductor), creating an opposing surface charge on the ICW-side surface of the nets. An electric field forms between the opposing charges of the ICW layer and the earthed nets [2].

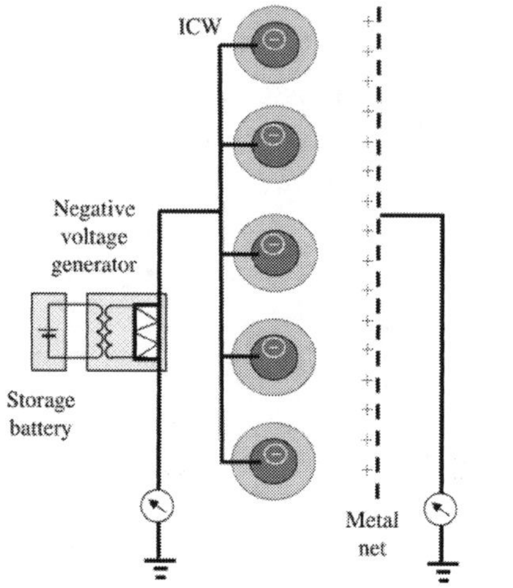

Fig. 1. Construction of an electric field screen.

2.2. Insect avoidance assay

The avoidance assay was conducted using two pieces of apparatus: a transparent acrylic cylinder (30 cm diameter, 40 cm length) partitioned into two parts with the SCD-screen (screen-cylinder) placed horizontally, and a screen-cylinder with a straw pole placed upright on the bottom (**Video supplement 1**). The ICWs were negatively charged with a range of voltages (0.1–8.0 kV). Test insects were released at the bottom of the screen cylinder to observe their actions as they flew up (for flies) or climbed an erect straw (for ladybird beetles) or cylinder wall (for others) to the earthed net of the screen. The assay involved 57 insect species and 5 spider species (**Table 1**). These organisms were collected on the university campus throughout the year. Screens with different mesh sizes (1.6–5.0-mm mesh) were used depending on the body size of the test insects and spiders. Twenty adults were used for each insect (and spider) and per voltage tested, and the experiments were performed five times.

3. Results and discussion

The avoidance assay showed that all test insects and spiders were deterred from passing through the screen net. The initial voltage at which the insects avoided the screen net varied among species (**Table 1**). Insects reaching the net placed their antennae inside the screen and subsequently refused to enter. The videos show examples of these movements for a leaf beetle scrambling up a wall and a ladybird beetle climbing a straw pole (**Video supplement 1**). In contrast, spiders (with no antennae) inserted their legs and then moved from the screen net without entering (**Video supplement 2**). In all cases, this avoidance behaviour became conspicuous when increased voltage was applied to the ICW. As may be expected, smaller insects or insects with longer antennae showed avoidance behaviour for lower applied voltage. However, at >2.0 kV, all insects and spiders moved from the net immediately after they reached it, regardless of body size. Thus, the present study demonstrates that an electric field across the screen acts as an electrostatic pest exclusion barrier. In previous papers, the electric field screen was shown to capture insects blown inside the space between the ICWs and the screen net [2,4,5]. From these results, we conclude that the insect-capturing capability of the singly charged dipolar electric field screen complements unsuccessful insect repulsion.

4. Conclusion

The aim of the present work was to generalise the insect-repellent function of an electric field screen for pest exclusion to ensure safe production and preservation of crops. A broad range of insects and spiders were used for this purpose, and all were deterred from entering the screen. At a particular voltage (2.0 kV) applied to the ICWs of the screen, the screen was effective in repelling all targets approaching the screen net. These results demonstrate the importance of pest repulsion as a primary function in physical pest management.

Table 1. Insects avoiding a dipolar electric field

Order	Family	Genus and species	Common name	Voltage (kV) of avoidance
Coleoptera	Anobiidae	*Lasioderma serricorne*	Cigarette beetle	0.8
	Attelabidae	*Euops splendidus*	Leaf-rolling weevil	1.8
	Bruchidae	*Callosobruchus chinensis*	Azuki bean weevil	1.2
	Cerambycidae	*Chlorophorus annularis*	Bamboo longicorn beetle	2.8
		Phytoecia rufiventris	Chrysanthemum longicorn beetle	1.9
	Chrysomelidae	*Argopistes coccinelliformis*	Ladybug mimicking leaf beetle	3.2
		Aulacophora femoralis	Cucurbit leaf beetle	1.3
		Chrysolina aurichalcea	Mugwort leaf beetle	1.9
		Gallerucida bifasciata	Dioscorea leaf beetle	2.8
		Gastrophysa atrocyanea	Japanese green duck leaf beetle	1.2
		Gonioctena rubripennis	Wisteria leaf beetle	2.8
		Lema cirsicola	Leaf beetle	2.8
		Ophraella communa	Ragweed leaf beetle	1.2
	Coccinellidae	*Coccinella septempunctata*	Seven-spotted ladybird beetle	2.4
		Aiolocaria hexaspilota	Ladybird beetle	1.2
		Epilachna vigintioctopunctata	Twentyeight-spotted ladybird beetle	2.4
		Harmonia axyridis	Asian ladybird beetle	2.1
	Curculionidae	*Anosimus decoratus*	Weevil	2.5
		Episomus turritus	Weevil	4.3
		Eugnathus distinctus	Weevil	3.2
		Nesalcidodes trifidus	Snout weevil	4.3
	Elateridae	*Pectocera fortunei*	Click beetle	2.1
	Meloidae	*Epicauta gorhami*	Blister beetle	0.4
	Mordellidae	*Mordella brachyura*	Tumbling flower beetle	3.2
	Oedemeridae	*Xanthochroa atriceps*	False blister beetle	4.5
	Rhynchophoridae	*Sitophilus oryzae*	Rice weevil	4.5
	Tenebrionidae	*Plesiophthalmus nigrocyaneus*	Mimawari beetle	4.5
		Tribolium castaneum	Red flour beetle	2.4
		Uloma latimanus	Black fungus beetle	0.5
Hemiptera	Aleyrodidae	*Bemisia tabaci*	Sweet potato whitefly	0.9
	Aphididae	*Myzus persicae*	Green peach aphid	1.5
	Cicadellidae	*Nephotettix cincticeps*	Green rice leafhopper	0.3
	Tettigellidae	*Bothrogonia ferruginea*	Black tipped leafhopper	0.5
	Lygaeidae	*Geocoris varius*	Large white-spotted seed bug	1.2
		Metochus abbreviatus	Large white-spotted seed bug	0.8
	Pentatomidae	*Eurydema rugosa*	Cabbage bug	1.1

(Continue to Table 1.)

Diptera	Agromyzidae	*Liriomyza sativae*	Tomato leaf-minor	1.2
	Bibionidae	*Bibio japonicus*	Love bug	0.8
	Culicidae	*Aedes albopictus*	Asian tiger mosquito	0.5
	Drosophilidae	*Drosophila melanogaster*	Vinegar fly	1.7
	Ephydridae	*Scatella stagnalis*	Greenhouse shore fly	1.7
	Psychodidae	*Clogmia albipunctatus*	Bath room fly	1.7
Hymenoptera	Anthophoridae	*Tetralonia nipponensis*	Long-horned bee	1.6
	Chalcididae	*Brachymeria lasus*	Chalcid wasp	0.9
	Formicidae	*Formica japonica*	Japanese wood ant	0.2
	Sphecidae	*Sphex argentatus*	Digger wasps	4.2
Lepidoptera	Geometridae	*Biston robustus*	Lilac beauty	1.4
	Tineidae	*Tineola bisselliella*	Common clothes moth	1.4
Blattodea	Blattellidae	*Blattella germanica*	German cockroach	1.8
	Blattidae	*Neostylopyga rhombifolia*	Harlequin cockroach	1.2
Thysanoptera	Thripidae	*Frankliniella occidentalis*	Western flower thrips	1.2
Mantodea	Mantidae	*Tenodera aridifolia*	Praying mantis	0.7
Psocoptera	Liposcelidae	*Liposcelis bostrychophilus*	Book louse	0.6
Dermaptera	Anisolabididae	*Dermaptera* sp.	Earwig	0.7
Orthoptera	Tetrigoidea	*Acridium japonicum*	Bolivar	1.1
Isopoda	Armadillidiidae	*Armadillidium vulgare*	Pill bug	1.6
	Rhinotermitidae	*Coptotermes formosanus*	Oriental termite	1.6
Araneae	Araneus	*Araneus ventricosus*	Orb-weaving spider	2.2
	Pardosa	*Pardosa astrigera*	Wolf spider	2.1
	Pisauridae	*Dolomedes sulfureus*	Fishing spider	2.3
	Thomisidae	*Thomisus labefactus*	Crab spider	2.8
	Uloboridae	*Octonoba varians*	Zebra spider	2.3

Acknowledgement

This work was supported in part by JSPS KAKENHI Grant Number 25450375 from the Ministry of Education, Culture, Sports, Science and Technology, Japan

References

[1] Matsuda Y, Ikeda H, Moriura N, Tanaka N, Shimizu K, Oichi W, Nonomura T, Kakutani K, Kusakari S, Higashi K and Toyoda H 2006 A new spore precipitator with polarized dielectric insulators for physical control of tomato powdery mildew *Phytopathology* **96** 967

[2] Matsuda Y, Nonomura T, Kakutani K, Takikawa Y, Kimbara J, Kasaishi Y, Osamura K, Kusakari S and Toyoda H 2011 A newly devised electric field screen for avoidance and capture of cigarette beetles and vinegar flies *Crop Protection* **30** 155

[3] Nonomura T, Matsuda Y, Kakutani K, Kimbara J, Osamura K, Kusakari S and Toyoda H 2012 An electric field strongly deters whiteflies from entering window-open greenhouses in an electrostatic insect exclusion strategy *European Journal of Plant Pathology* **134** 661

[4] Kakutani K, Matsuda Y, Haneda K, Sekoguchi D, Nonomura T, Kimbara J, Osamura K, Kusakari S and Toyoda H 2012 An electric field screen prevents captured insects from escaping by depriving bioelectricity generated through insect movements *Journal of Electrostatics* **70** 207

[5] Matsuda Y, Kakutani K, Nonomura T, Kimbara J, Kusakari S, Osamura K and Toyoda H 2012 An oppositely charged insect exclusion screen with gap-free multiple electric fields *Journal of Applied Physics* **112** 116103

Transient interaction model of electromagnetic field generated by lightning current pulses and human body

T Iváncsy[1], I Kiss[1], L Szűcs[2] and Z Á Tamus[1]

Budapest University of Technology and Economics
[1] Department of Electric Power Engineering
[2] Department of Broadband Infocommunications and Electromagnetic Theory
H-1111 Budapest Egry J. u. 18, Hungary

E-mail: ivancsy.tamas@vet.bme.hu

Abstract. The lightning current generates time-varying magnetic field near the down-conductor and the down-conductors are mounted on the wall of the buildings where residential places might be situated. It is well known that the rapidly changing magnetic fields can generate dangerous eddy currents in the human body.The higher duration and gradient of the magnetic field can cause potentially life threatening cardiac stimulation. The coupling mechanism between the electromagnetic field and the human body is based on a well-known physical phenomena (e.g. Faradays law of induction). However, the calculation of the induced current is very complicated because the shape of the organs is complex and the determination of the material properties of living tissues is difficult, as well. Our previous study revealed that the cardiac stimulation is independent of the rising time of the lightning current and only the peak of the current counts.

In this study, the authors introduce an improved model of the interaction of electromagnetic fields of lighting current near down-conductor and human body. Our previous models are based on the quasi stationer field calculations, the new improved model is a transient model. This is because the magnetic field around the down-conductor and in the human body can be determined more precisely, therefore the dangerous currents in the body can be estimated.

1. Introduction

Persons may suffer lightning injuries in different ways such as: direct strike, contact injury, side flash or splash, ground current, upward streamer or leader. Beside these electrical injuries there is another deleterious effect of lightning strike [1][2]. The lightning current generates time-varying magnetic fields near to the down-conductors. The down-conductors are mounted on the wall of buildings where residential places can be situated. It is well known that rapidly changing magnetic fields can generate dangerous eddy currents in the human body and if the duration and the gradient of the magnetic fields are high enough, the peripheral nerves are excited. Gradient magnetic field having higher intensity can cause cardiac stimulation, which may result in fatal cardiac fibrillation [1]. Our previous investigations revealed that the cardiac stimulation is independent of the rising time of lightning current only the peak of the current counts [3].

In this article the time dependent effect of the lightning current in the down-conductor will be investigated. Upon these calculations the dangerous eddy currents in the human body estimated more precisely. To be able to estimate of the currents generated in the human body by the lightning pulse in the down-conductor it is also necessary to have knowledge about the electric and magnetic properties of the tissues in the body.

For the calculations the median current value of the lightning current was used. It is well known, that the positive lightning has higher energy because of the higher peak currents, and longer rising and decay times [2] [4]. In our case, the polarity of the lightning has no effect on the calculations because the induced field depends highly on the rising steepness of the current, therefore we used the current with the highest probability.

2. Effect of time-varying magnetic fields

The health effects of intense gradient magnetic fields have been investigated since the beginning of the diagnostic application of magnetic resonance (MR) systems. The patients of MR examinations are exposed to time-varying magnetic fields beside the static magnetic and radio-frequency fields. The physiological reactions of intense gradient magnetic fields are peripheral nerve stimulation, muscle movement, discomfort and cardiac stimulation. The interaction of time-varying magnetic field and the human body is based on the induced current according to the Faradays induction law. The induced current depends on the intensity of the gradient field, the circumference of the exposed organ and the conductivity of the tissue, obviously.

Many research have been addressed to reveal the interaction between a patient and time-varying magnetic fields since the beginning of clinical application of magnetic resonance imaging systems. These studies have resulted in many interaction models, which have been summarized and published elsewhere [5].

The threshold level of excitation of nerves depends on the duration of stimulation, i.e. higher-level stimuli are needed for shorter exciting pulses. If a nerve is excited by a time-varying magnetic field this gradient field induces an electric field in it and this induced field can stimulate the nerve. For long gradient magnetic fields ($\tau > 1$ ms) the minimum threshold is expressed by the time derivative of the magnetic flux density ($\mathrm{d}B/\mathrm{d}t$). If the impulse is shorter than 10 µs, the product of the impulse duration and intensity give the threshold for the nerve stimulation [6]. Between these time intervals, the threshold for nerve excitation depends on the duration of the stimuli. This rule of neurophysiology is summarized in a very simple hyperbolic formula, namely in the Lapicque-formula. At excitation of a tissue by a time-varying magnetic field, the threshold can be expressed as a time derivative of magnetic flux density ($\mathrm{d}B/\mathrm{d}t$):

$$\frac{\mathrm{d}B}{\mathrm{d}t} = b \cdot \left(1 + \frac{c}{\tau}\right) \tag{1}$$

where b is the rheobase and c is the chronaxie, τ is the duration of the stimulus. This formula can be used when the exciting duration is higher than tens of µs.

Based on this nerve excitation model and a simple calculation of magnetic field encircling the down-conductor, the adverse health effects can be easily estimated. The previous results of the authors have revealed that positive lightning strikes can generate cardiac muscle stimulation, which can lead to fatal cardiac fibrillation. Moreover, the analysis of different lightning current parameters revealed that the probability of this life threatening interaction does not depend on the rise time of the lightning current [2].

3. Description of the model

The interaction model is a simplified model. The down-conductor is a straight wire. As our previous calculations showed, that the closeness of the ground affects the intensity of the magnetic field. Namely the magnetic field decreases around the ground point. Hence, this phenomenon also decreases the deleterious effects near to the ground [3]. As first the human body is modelled by a cylinder having 60 cm of diameter and 70 cm height. The cubic measures of the selected cylinder are close to the measures of the human torso. The electric parameters of the cylinder have the same parameters as the human muscle. Both the relative permeability (μ_r) and relative permittivity (ε_r) are selected for 1 and the conductivity is set for $\sigma = 0.11$ [S/m]. These values are equal to the measured values of the human muscle tissue [7] .

The organs inside the body have different electric properties. The muscle tissue was chosen as the material for the model because the body mainly consists of different muscles and the conductivity of the muscle tissue is also close to the conductivity of the nerve tissue.

The distance between the down-conductor and the closest point of the body model was set for 1 m in the calculation. The calculation uses a 1.2 µs rise time and 50 µs decay time lightning current pulse with 45 kA peek value. If the distance is higher than 1 m formation of dangerous effects in the human bodyis unlikely [2].

4. Results of the calculation

The down-conductor is not shown on the figures it is positioned outside the image, parallel to the right side. In the steady state model the eddy currents were calculated for 50 kA 50 Hz sinusoidal current. The results are shown in Figure 1. These calculations showed that the eddy currents are higher in parts located near to the down-conductor and decrease moving away from the down-conductor. Modelling the effects of the transient current – in other words calculating

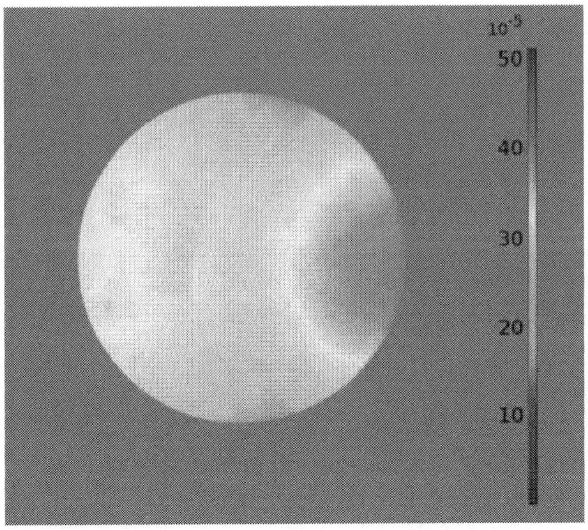

Figure 1. Calculated eddy currents in the model body (top view) for 50 Hz steady current. The measure for the calculated values is A/m^2

the eddy currents generated by a 1.2/50 current pulse with 45 kA peak value – the results are significantly different. The peak current density value is much higher and also the distribution differs a lot. On the top view in Figure 2 it is easy to observe the different current distribution compared to the steady state on Figure 1. In the transient case (Figure 2) the higher current density is located at the perimeters and the distance to the down-conductor has also an effect on the distribution. In the middle part of the model body the current density is much lower. On Figure 3 the calculated results are shown from a perpendicular plane (i.e. side view), where the skin effect caused by the rapidly changing current is clearly visible.

5. Conclusions

One of the dangerous effects of lightning strike is the magnetic field generated by the impulse current flowing in the down-conductor when a lightning strike hits the Franklin rod. There

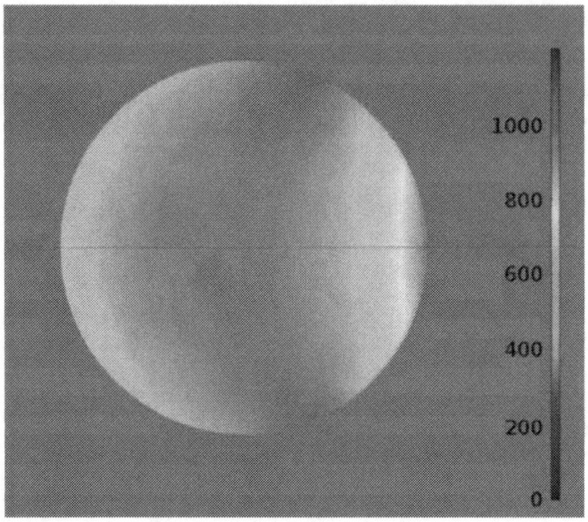

Figure 2. Calculated eddy currents in the model body (top view) for current pulse. The measure for the calculated values is A/m^2

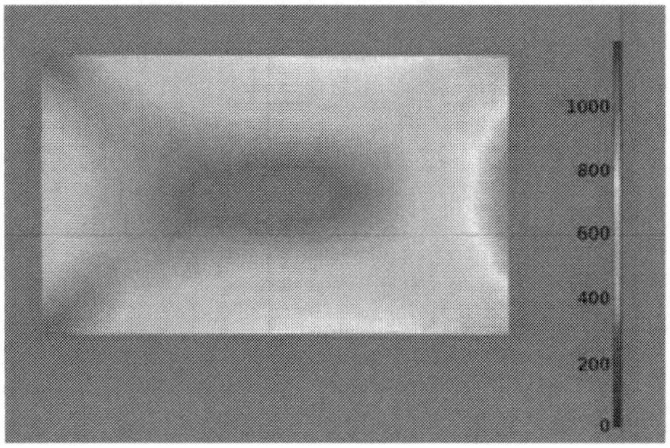

Figure 3. Calculated eddy currents in the model body (side view) for current pulse. The measure for the calculated values is A/m^2

are many factors which should be taken into consideration to determine the danger of the magnetic field around the down-conductor. The calculations showed that the pulse generated eddy currents inside the human body have an uneven distribution. The skin effect is not negligible; the currents on the surface of the body are higher than in the inner parts. Because of this in parts of the body the currents can be higher than the values calculated from the stationer model. Further investigations are needed about the different organs and how the electrical and magnetic parameters of the different tissues influence the eddy current distribution.

References

[1] Tamus Z Á, Novák B, Szabó S, Kiss I and Berta I 2010 *30th International Conference on Lightning Protection* p 1042
[2] Tamus Z Á, Novák B, Szűcs L and Kiss I 2011 *Journal of Physics: Conference Series* **301** 012025
[3] Tamus Z Á, Iváncsy T, Kiss I and Szűcs L 2014 *32nd International Conference on Lightning Protection* p 1495
[4] Walsh K M, Bennett B, Cooper M A, Holle R L, Kithil R and Lopez R E 2000 *Journal of Athletic Training* **35** 471
[5] Schaeffer D J, Bourland J D and Nyenhuis J A 2000 *Journal of Magnetic Resonance Imaging* **12** 20–29
[6] Reilly J P 1989 *Medical & Biological Engineering & Computing* **27** 101–110
[7] Gabriel C, Gabriel S and Corthout E 1996 *Physics in Medicine and Biology* **41** 2231

Frequency dependence and assembly characteristics of silver nanomaterials trapped by dielectrophoresis

R Kataoka[1], H Tokita[1], S Uchida[1*], R Sano[2] and H Nishikawa[2]

[1]Tokyo Metropolitan University, 1-1 Minami Osawa, Hachioji, Tokyo, 192-0397 Japan
[2]Shibaura Institute of Technology 3-7-5 Toyosu, Koto-ku, Tokyo, 135-8547 Japan

[*]s-uchida@tmu.ac.jp

Abstract. For rapid and accurate fabrication of industrial products, the development of effective technique to manipulate and assemble nanomaterials is essential. In the present work, we investigated the dielectrophoretic behavior of silver nanoparticles and nanowires at various driving frequencies and voltage amplitudes, and examined three-dimensional assembly of the nanomaterials using dielectrophoretic devices with microstructures. The collectable range in frequency for nanoparticles and nanowires was from 10 kHz to 1 MHz. In particular, the aggregation shape of nanowires was changed against driving frequency. The bunch of nanowires was bridged between pillars of dielectrophoretic device. When the applied voltage was turned off, the trapped nanowires were entirely released. In pit type device, trapping of nanoparticles was observed in outermost pits of array. The aggregates of nanoparticles were directly fixed to the bottom electrode.

1. Introduction

Nanomaterials have high reactivity and applicability to various objects, since their specific interfacial area is large. The characteristics have been used to improve the performance of coloring materials and cosmetics. Nanomaterials have been also applied to new electronic devices such as high-functioning sensors and specific-shape microelectrodes [1-3]. Furthermore, nanowires and carbon nanotubes were examined for wiring parts of nano-circuits [4]. For rapid and accurate fabrications of above industrial products, the establishment of effective scheme to manipulate and assemble nanomaterials is essential.

Electrical handling by dielectrophoresis (DEP) is a powerful technique of bottom-up process for particulate manipulation. DEP is a phenomenon that the particles under non-uniform electric field migrate along the gradient of electric field. It is possible to control the particle behavior easily and rapidly, since DEP force strongly depends on the electric field gradient and driving frequency [5]. In fact, TiO_2 NWs were collected on the gold plate by DEP, and the formation products were available for sensor electrodes to detect organic vapors [6]. We fabricated three-dimensional DEP (3D-DEP) devices with dielectric pillars by proton beam writing [7] and succeeded in high concentration of bacteria around pillars. However, the suitable conditions of DEP trapping have not been understood well for variously-shaped nanomaterials.

In the present work, we investigated the frequency dependence of the trapping number and collected configuration for silver (Ag) nanoparticles (NPs) and nanowires (NWs). Moreover, we examined the trapping and release operation of Ag NWs around pillar array of DEP device. We also discussed the influence of voltage amplitude and polarity on the assembly state of Ag NPs using a pit type DEP device.

2. Materials and Method

We prepared Ag NPs of 100 nm in diameter, and NWs of 60 nm in diameter and 10 μm in length (Sigma-Aldrich). 20 mg of NPs and 2-20 μl of NWs were suspended each by distilled water. The measurements of conductivities of NPs and NWs were 0.41 mS/m and 0.19 mS/m, respectively.

The DEP device consists of basal plate with microelectrodes, cover plate with microchannel and device holder. Figure 1 shows photographs of the electrode configurations. Interdigitated electrodes were fabricated by deposition of indium tin oxide (ITO) on silica substrate. The electrode width is 110 μm and gap length is 10 μm. With respect to 3D-DEP devices, dielectric pillars were formed in a 200 μm of gap area between electrodes. The diameter, height and pitch of pillar were 3 μm, 10 μm and 24 μm, respectively. The pit type DEP device has ITO plate electrodes and silicon spacer of 500 μm in thickness. Polymethyl methacrylate pits were deposited on the bottom electrode. The diameter, depth and pitch of pit were 4 μm, 10 μm and 10 μm, respectively.

The suspension was delivered into the channel using a peristaltic pump (PSM050DA, Advantec) for interdigitated plate and pillar type DEP devices. Only the suspension in pit type DEP device was filled using micropipettes. The applied voltage was supplied to the electrodes by a function generator (AFG3022, Tektronix). The collected state of nanomaterials was observed with a microscope (LV-100DA, Nikon) with CCD camera (DXM1200C, Nikon).

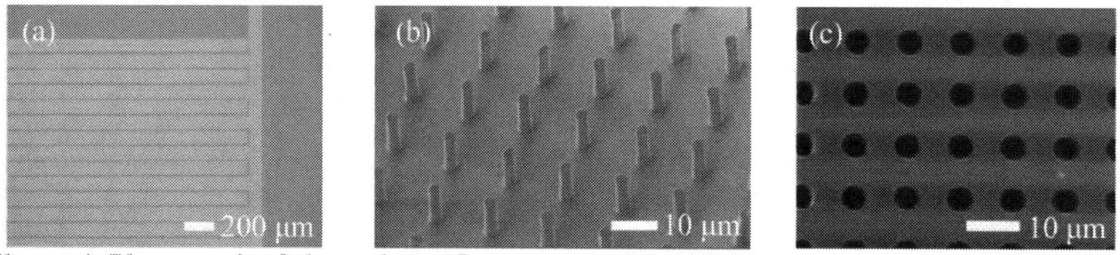

Figure 1. Photograph of electrode configurations. (a) interdigitated plate, (b) pillar and (c) pit.

3. Results and discussion

3.1. Frequency dependence of dielectrophoretic trapping for nanomaterials

We investigated collecting characteristics of Ag NPs at various driving frequencies f from 1 kHz to 10 MHz as is shown in figure 2. Here, the voltage amplitude V_a was 15 V_{pp} and flow rate ϕ was 10 ml/h. Trapping of the particles did not occur at 1 kHz and 10 MHz. In collectable frequency region between 10 kHz and 1 MHz, the NPs were firstly trapped in the gap and then were chained each other. This reason is that electric field was generated locally at trapped particles in upstream part. The thickness of chain gradually increased with frequency.

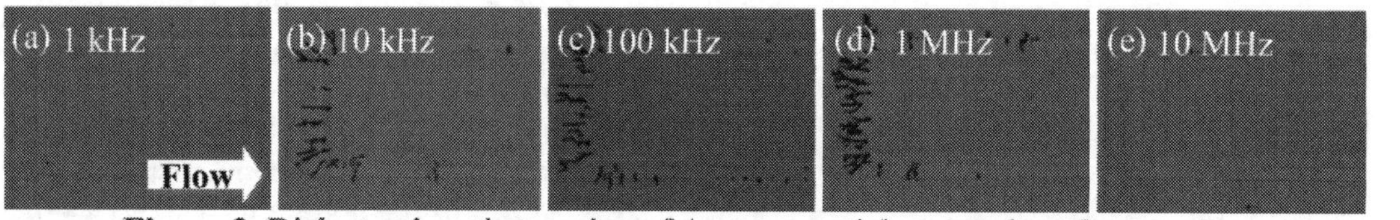

Figure 2. Dielectrophoretic trapping of Ag nanoparticles at various frequencies.

In the case of NWs (figure 3), the collectable frequency region was same as that of NPs. However, the frequency dependency of collected shapes was different from that of NPs. At a 10 kHz of frequency, NWs are collected in a straight line on the gap. In addition, small aggregates were observed in the center of electrode. The wires were arranged as a netlike appearance on overall electrode at 100 kHz. Up to 1 MHz, they were parallelized along the gaps. Gierhart *et al.* showed that the collection amount and shape of NPs depend on the frequency change of direct and mutual DEP forces F_{DEP} [8]. In addition to the above reason, the balance of AC electroosmosis between microelectrodes [9] and external flow would influence the distribution of Ag NWs in the present system. The numerical simulation of particle dynamics, however, should be required for quantitative and morphological assessment.

Figure 3. Dielectrophoretic trapping of Ag nanowires at various frequencies.

3.2. *Numerical analysis of DEP force in 3D-DEP devices*

With an electromagnetic field solver (COMSOL Multiphysics 4.4), we calculated the special distribution of electric field E (figure 4). Here, V_a was normalized to be 1 V. In the pillar type, E at 5 μm height distorts at boundary of pillar and medium. E around nearest pillars from the electrodes was about 1.6 times as large as that around central pillars. F_{DEP} at center and edge were 3.7×10^{-17} N and 2.9×10^{-15} N, respectively. These results suggested that bridging of NWs would occur from pillars near the electrodes. In the case of pit type at 12 μm height from the bottom, outer pits had stronger E than inner pits, although the direction of E was parallel to pits. Since the present pitch of pits was narrow, the influence of nearest pits on distortion of E became large. F_{DEP} of inner and outer pits were 2.2×10^{-16} N and 2.5×10^{-15} N, respectively. The force ratio of inner and outer was one tenth compared with pillar type. The present numerical evaluations should be useful for voltage control of assembly process in 3D-DEP devices.

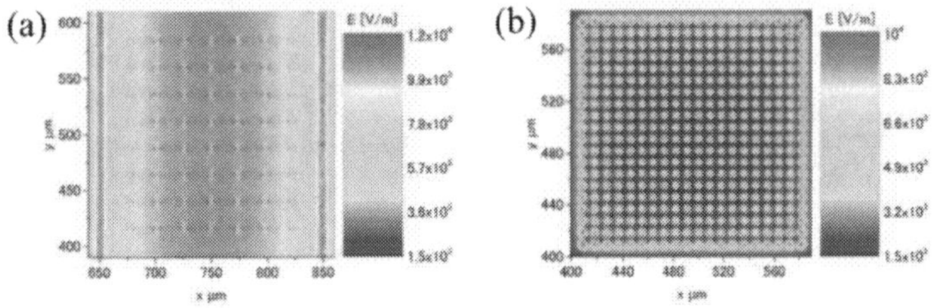

Figure 4. Spatial distribution of electric field. (a) pillar type and (b) pit type.

3.3. *Trapping and release operations of nanowires using pillar type DEP device*

On the basis of above findings, we tried the condensation and release operations of NWs using pillar type DEP device. Figure 5 shows the temporal behavior of Ag NWs between pillars. V_a was 10 V_{pp}, f was 100 kHz and ϕ was 10 ml/h. NWs were collected only around pillars. After 30 min, the wires were bridged between the pillars and weakly-bind chains were formed. When the applied voltage was turned off, the NWs were rapidly released from pillars. In the case of over 10 V_{pp}, the collected wires aggregated each other and the large aggregates broke some pillars.

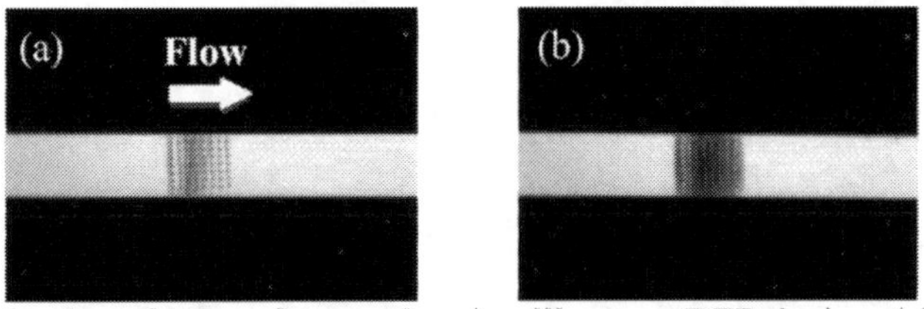

Figure 5. Temporal behavior of trapped nanowires in pillar type DEP device. (a) 5 min and (b) 30 min.

3.4. *Assembly of nanoparticles using pit type DEP device*

To explore effective assembly of nanomaterials, we investigated the voltage dependency of Ag NP trapping in pit type DEP device. Figure 6 shows the collecting states of NPs at 100 kHz. When V_a was

10 V_{pp}, the particles were not collected anywhere. At a voltage of 20 V_{pp}, they were focused in outermost pits at 30 min. However, the bunch of particles grew up to dendritic outside the pit within 1 min, if V_a was raised up to 30 V_{pp}. The suitable threshold of V_a was 20 V under the present condition.

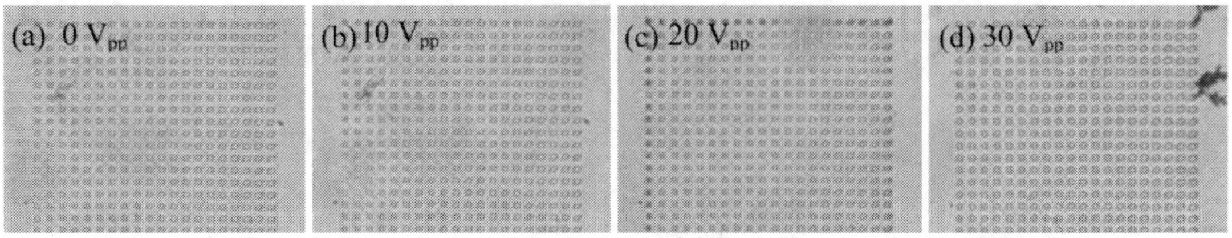

Figure 6. Voltage dependency of Ag nanoparticle trapping in pit type DEP device.

Moreover, we examined superposition of DC offset bias on AC driving voltage to transport NPs equally around every pit. As is shown in figure 7, the collectable points at 1 V bias were randomly observed in inner pits, although the total collection amount was rarely different from that without bias voltage. On the other hand, we obtained thin film of Ag particles on overall of pit array within 1 min by changing the polarity of offset bias. This film was not dispersed even if the applied voltage was turned off. They was then retained its shape even if explored in air.

(a) (b)

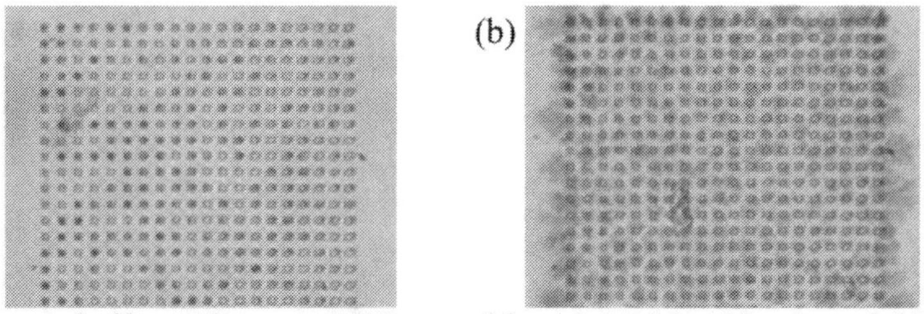

Figure 7. Influence of offset voltage on DEP assembly. (a) 1.0 V at 90 min and (b) -1.5 V at 1 min.

4. Conclusion

We investigated DEP characteristics and effective assembly of Ag nanomaterials using various microelectrodes. The trapping number of NPs depended on applied frequency. In particular, the collecting shape of NWs changed because of their large dipole moment. Using pillar type device, we succeeded in high concentration and rapid release of NWs around pillar. Also, we evaluated the voltage dependency for NP assembly and specified effective value of AC applied and DC bias voltages in pit type device. In the nearest future, we will fabricate functional nanostructures by DEP assembly on the basis of above information of control parameters.

References

[1] Lee H J, Yasukawa T, Suzuki M, Lee S H, Yao T, Taki Y, Tanaka A, Kameyama M, Shiku H, Matsue T 2009 *Sens. Actuators B, Chem.* **136** 320-325
[2] Bhatt K H and Velev O D 2004 *Langmuir* **20** 467-476
[3] Ranjan N, Mertig M, Cuniberti G and Pompe W 2010 *Langmuir* **26** 552-59
[4] Suehiro J 2010 *Biomicrofluidics* **4** 022804
[5] Pohl H A 1978 *Dielectrophoresis* Cambridge University Press
[6] Wang S, Lin Z-X, Wang W-H, Kuo C L, Hwang K C and Hong C-C 2014 *Sens. Actuators B, Chem.* **194** 1-9
[7] Uchida S, Nakao R, Asai C, Jin T, Shiine Y and Nishikawa H 2012 *Intelligent Automation and Soft Computing* **15** 1-13
[8] Gierhart B C, Howitt D G, Chen S J, Smith R L, and Collins S D 2007 *Langmuir* **23** 12450-12456
[9] Ramos A, Morgan H, Green N G, Castellano A 1999 *J. Colloid Interface Sci.* **217** 420-422

Estimation of droplet charge forming out of an electrified ligament in the presence of a uniform electric field

H Osman[1,3], G S P Castle[1], K Adamiak[1], H T Fan[2] and J Simmer[2]
[1] Department of Electrical and Computer Engineering, Western University, London, ON, Canada
[2] General Motors Global, Warren, MI, USA
E-mail address: hosman4@uwo.ca

Abstract. The charge on a liquid droplet is a critical parameter that needs to be determined to accurately predict the behaviour of the droplet in many electrostatic applications, for example, electrostatic painting and ink-jet printing. The charge depends on many factors, such as the liquid conductivity, droplet and ligament radii, ligament length, droplet shape, electric field intensity, space charge, the presence of adjacent ligaments and previously formed droplets. In this paper, a 2D axisymmetric model is presented which can be used to predict the electric charge on a conductive spherical droplet ejected from a single ligament directly supplied with high voltage. It was found that the droplet charging levels for the case of isolated electrified ligaments are as much as 60 times higher than that in the case of ligaments connected to a planar high voltage electrode. It is suggested that practical atomization systems lie somewhere between these two extremes and that a better model was achieved by developing a 3D approximation of a linear array of ligaments connected to an electrode having variable width. The effect on droplet charge and its radius was estimated for several cases of different boundary conditions.

1. Introduction

Much prior research has been completed to investigate the stability and disintegration of a liquid jet into a stream of discrete droplets in a uniform electric field [1]-[7], but relatively little work has been done on exploring the correlation between the droplet charge and droplet radius. Toljic et al. [8] investigated the charge to radius dependency for conductive particles atomized in an electric field between planar electrodes. The results showed that the ligament length has a strong effect on the particle charge to radius dependency. They found that the radius exponent is equal to two when the particle is in direct contact with a planar electrode, and decreases rapidly as the ligament length increases approaching a limit of 1.1. Osman et al. [9] studied numerically the effect of the presence of previously formed droplets, the presence of adjacent ligaments as well as the ligament radius and length on the predicted droplet charging levels for ligaments formed from a planar high voltage electrode. The effect of the ligament radius on the charging level was found to be significant with larger charge resulting when the ligament radius is smaller than the droplet radius. This is important because in practice a narrow necking normally occurs prior to the droplet ejection. They also investigated the dynamic modeling of droplet formation at different inlet velocities. The results showed that the droplet shape prior to detachment is actually a prolate ellipsoid and that the spherical model underestimates the actual charge. Distortion of the droplet shape agrees with previous studies of droplet deformation in an external electric field [10].

[3] To whom any correspondence should be addressed

In this paper, a 2D axisymmetric numerical model was created using COMSOL, a Finite Element commercial software, to calculate the droplet charge formed from the end of a variable length ligament. Two cases have been compared: a ligament connected to a planar high voltage electrode and an isolated ligament supplied with a high voltage. In practice, the ligaments in paint sprayers are formed off the lip of a rotating cup, where the edge thickness is much larger (up to 10 times) than the ligament radius. This situation was simulated by expanding the model into a 3D linear array of ligaments forming from an electrode of variable width. With this, the droplet charge could be determined both for the single ligaments and for cases of an array of adjacent electrified ligaments.

2. Numerical model

The 2D axisymmetric stationary model consisted of a grounded electrode located 25 cm from the high voltage electrode, a cylindrical conductive liquid ligament of variable length and ejected spherical droplets with different radii. Two boundary conditions were considered. First, the ligament was assumed to be electrified through contact with a high voltage planar electrode. Then, the planar electrode was removed and the voltage applied directly to the ligament (Figure 1). The model was then extended to 3D (Figure 2) to enable the investigation of the effect of a linear array of adjacent ligaments on the droplet charge levels by applying the symmetry boundary conditions along the walls of the computational domain. In this latter model, the width of the lower electrode was also varied.

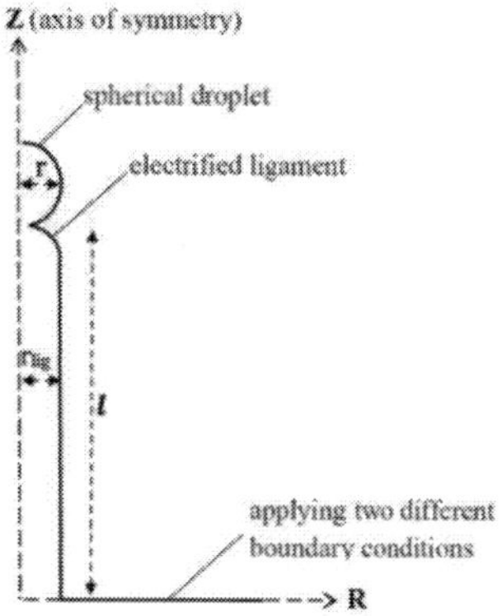

Figure 1. 2D model for a single ligament with two BCs: a) applied high voltage (90 kV) and b) dV/dn=0 for mirror plane.

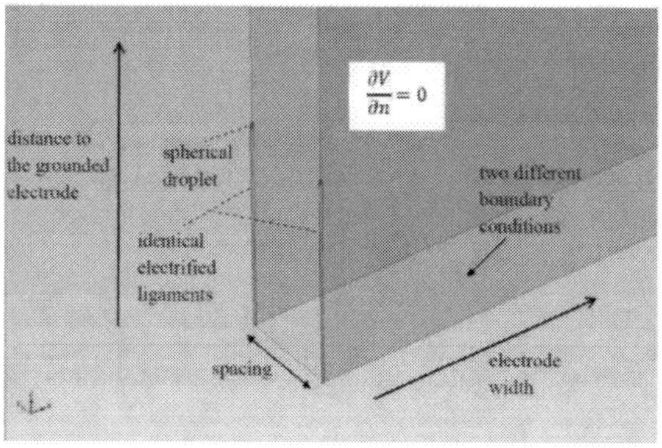

Figure 2. 3D model for an array of electrified ligaments with two BCs: a) applied high voltage (90 kV) and b) dV/dn=0 for mirror plane.

3. Numerical results and discussion

3.1. Droplet charge for a single ligament

Osman et al. [9] previously estimated the electric charge on the surface of a spherical droplet, which is attached to a single ligament connected to a planar high voltage electrode in the 2D axisymmetric model. The simulation results were calculated for different droplet radii (r= 15, 30 and 45 μm) and ligament lengths (from 0 to 4 mm), and it was initially assumed that in all cases the droplet and ligament radii are equal. Their results demonstrated that the charging level increases with the droplet size and increases monotonically, as the ligament length increases. When a single spherical droplet is in direct contact with the surface of the planar electrode, it was confirmed that the charge magnitude coefficient and the radius exponent agree well with Félici's [11] predicted value, confirming the validity of the model.

In this section, the planar high voltage electrode was removed so that this case represents a single ligament-droplet system directly energized with high voltage. The results in this case reveal that the droplet charge also increases with radius, but in the order of 55 to 60 times higher than reported earlier [9] for the case, where the ligament was connected to a planar electrode. Also, the results show an opposite trend, where the charge slightly decreases, as the ligament length increases (Figure 3). These results are somewhat surprising as they show such large differences in the level of charge and the influence of ligament length. However, it appears that the effect of the planar electrode is to partially screen the electric field at the end of the ligament, and this influence decreases as the ligament gets longer. On the other hand, for the case where the ligament stands alone, the field strength is much stronger at the end of the ligament and it decreases slightly with length, as more electric flux is attracted to the surface of the ligament.

3.2. Linear array of electrified ligaments

This section presents a study to examine the effect of a linear array of ligaments on the level of charging over the surface of a spherical droplet of radius equal to 15 μm in absence of the planar electrode. A 3D model described in [9] was modified such that a condition of zero normal partial derivative was applied to the lower electrode and walls of the computational domain, which is equivalent to the symmetry boundary conditions. This assumption creates a model, where a single ligament is accompanied on both sides by an infinite row of identical electrified ligaments. The droplet and ligament radii are assumed to be equal. The charge magnitudes over the droplet surface have been calculated for different values of the spacing between adjacent ligaments and for different ligament lengths (Figure 4).

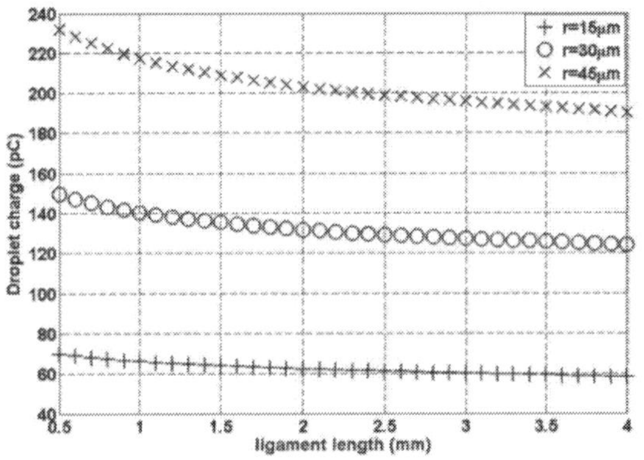

Figure 3. The droplet charge for a single ligament in absence of the planar electrode assuming three different droplet sizes.

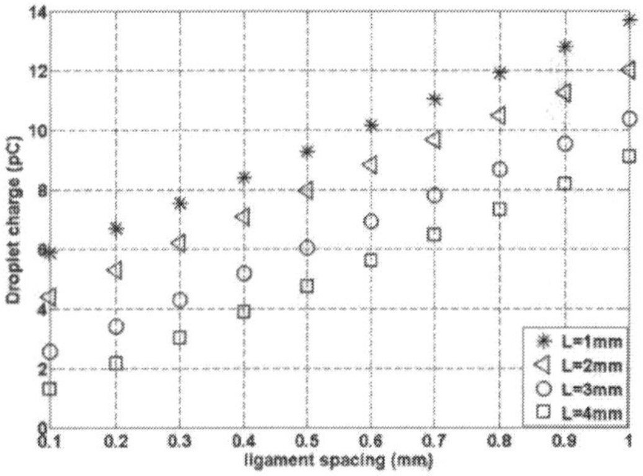

Figure 4. The charge for a 15 μm droplet formed from an array of ligaments of different lengths in absence of the planar electrode.

Similarly to the previous results presented in [9], it was found that the charging level increases, as the spacing between ligaments of fixed length increases, but for the longer ligaments, the droplet charge decreases compared with the shorter ones at a given spacing (Figure 4). A linear array of ligaments of fixed length of 1 mm and spacing of 0.5 mm was then assumed for ten values of droplet radius and the effect of the electrode width on the droplet charging level determined. The electrode width was varied from one extreme condition that of a planar electrode, approximated as a 100 mm strip, to the case of no electrode. These results are shown in Figure 5. For the case of a planar electrode, the results show that the estimated droplet charge increases monotonically with the droplet radius. As the electrode width shrinks to 0.5 mm, the charge magnitude increases by about a factor of 30 and for an even narrower width of 20 μm, where the electrode width is in the same order of the droplet radius, charge is approximately 40 times larger. Finally in the extreme case when the electrode is absent and the ligaments become isolated, the charge magnitude was found to increase up to 50 times the case of planar electrode. These results also enable the value of the dependency of charge on radius to be estimated and show an exponent value of about 1.1 which agrees well with the value reported in [9].

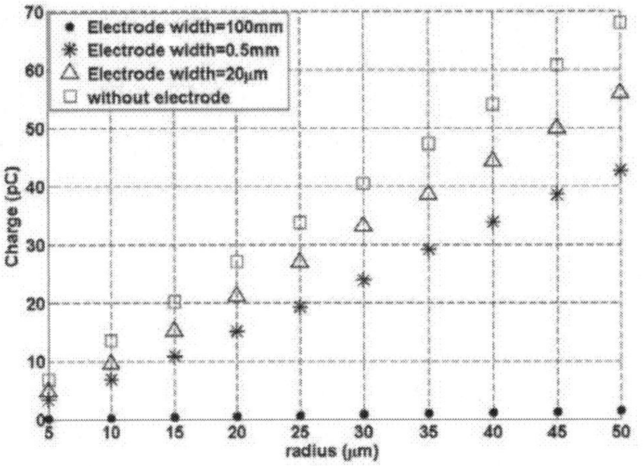

Figure 5. The droplet charge levels for an array of ligaments of 1mm length, spaced 0.5 mm apart assuming variable electrode widths

Conclusions

This study sheds some light on the design parameters that allow the charge level to be predicted for different droplet sizes and ligament lengths. A 2D axisymmetric model was generated for predicting the charge magnitude by assuming different boundary conditions. When the electrode is removed and the ligament itself forms the high voltage electrode, it was found that the droplet charge greatly increases in magnitude by a factor of up to 60 times. Also, a 3D model has been created to examine the effect of the spacing between a linear array of identical ligaments on the droplet charge in the absence of the planar electrode. It was found that the droplet charge increases, as the spacing between adjacent ligaments increases. Also it was demonstrated that the electrode width strongly affects the droplet charge. In particular, as the width of the electrode decreases, the charges on the droplets in the linear array may increase by a factor of up to 50. This large influence of the boundary conditions suggests that the next stage of the modelling, currently underway, is to include the cylindrical geometry of the sprayer as well as the other important parameters particularly the space charge due to previously formed droplets.

Acknowledgment

The authors acknowledge with thanks the joint contributions of the Natural Sciences and Engineering Research Council (NSERC) of Canada for the financial support and CMC Microsystems for providing an access to COMSOL software that facilitated this research.

References

[1] Macky W A 1931 *Proc. R. Soc. Lond.* **133** 565-87
[2] Pfeifer R and Hendricks C D 1967 *Phys. Fluids.* **10** 2149-54
[3] Schweizer J and Hanson D 1971 *J.Colloid Interface Sci.* **35** 417-23
[4] Schneider J M, Lindblad N R, Hendricks Jr C D and Crowley J M 1967 *J. Appl. Phys.* **38** 2599-605
[5] Wong M C Y and Shrimpton J S 2004 *IEEE Trans. on Dielectr. and Electr. Insul.* **11** 362-8
[6] Lopez-Herrera J M and Ganan-Calvo A M 2004 *J. Fluid Mech.* **501** 303-26
[7] Zhakin A I and Belov P A 2013 *Surface Eng. and Appl. Electrochem.* **49** 205-14
[8] Toljic N, Castle G S P and Adamiak K 2010 *J. Electrostatics* **68** 57-63
[9] Osman H, Ghazian O, Adamiak K, Castle G S P, Fan H and Simmer J 2014 *Proceedings of the IEEE/IAS Annual Meeting,* Vancouver, BC, 1-7
[10] Adamiak K and Floryan J M 2011 *IEEE Trans. on Ind. Appl.* **47** 2374-83
[11] Félici N J 1966 *Revue générale de l'électricité.* **75** 1145-60

Carbon nanotubes in electrospun polyethylene oxide nanofibres: *A potential route to conducting nanofibres*

M Nazhipkyzy[1,2], S D Mohan[3,4], F J Davis[3] and G R Mitchell[2,4,]
[1] Institute of Combustion Problems, Al-Farabi Kazah National University Almaty, Kazakhstan
[2] Centre for Advanced Microscopy, University of Reading, Whiteknights, Reading RG6 6AF UK
[3] Department of Chemistry, University of Reading, Whiteknights, RG6 6AD, UK
[4] Centre for Rapid and Sustainable Product Development, Institute Polytechnic of Leiria, 2430-028, Marinhe Grande, Portugal

E-mail: geoffrey.mitchell@ipleiria.pt

Abstract. Polyethylene oxide solution containing multi-walled carbon nanotubes have been electrospun onto a rotating collector to produce highly aligned arrays of electrospun nanofibers ranging in diameters from (200 – 360) nanometres. The addition of a surfactant (Triton X-100) is highly effective in dispersing carbon nanotube within an aqueous solution of polyethylene oxide and the resulting mixture can be electrospun without excessive clumping to produce nanofibers containing high loadings of nanotubes; in this case up to 5% wt thereby providing an effective route to electrically conductive nanofibres.

1. Introduction

Electrospinning is a manufacturing technique that utilises an electric field to prepare micro- to nano-scale polymer fibres from a solution or melt [1]. The process works by application of a high voltage to a polymer solution contained within a spinneret. The droplet at the spinneret tip is extruded as a jet and through the process of jet whipping (referred to as bending instabilities), the jet is stretched into a micro-to-nano scale, solvent evaporates off and a solid polymer fibre is deposited onto a grounded electrode. Electrospun fibres offer some useful applications such as in electronics, biomedical applications, textiles and composite materials [1]. The fibres themselves boast impressive properties, which can be further improved with the addition of nanoparticles, for example carbon nanotubes (CNT's), which have an abundance of properties ranging from thermal, optical, electrical and mechanical [2]. CNT's typically have diameters on the order of several nanometres, but they can be micrometres in length. The tubes exhibit improved electrical and mechanical properties parallel to its axis making alignment desirable [3], which makes the electrospinning process attractive for as a method of alignment due to shear forces and narrowing of the jet which can aid in orientating and confining the tubes parallel to the fibre axis [4]. In addition to alignment, nanoscale particle agglomeration from strong intermolecular forces must be overcome to aid in a uniform dispersion. This can be facilitated through use of surfactants, for example in one such study Rastogi *et al.* showed Triton X-100 providing a good nanotube dispersion [5].

Previous studies incorporating CNT's into electrospun fibres have shown improvements to the electrical conductivity of the fibres. Examples include the addition of nanotubes to polystyrene fibres

[6], polyvinyldiene difluroide [7] and polypropylene [8], all cases showing improvements to fibre conductivity. The key to high electrical conductivity is the network connectivity of the CNT's, with higher loadings naturally leading to an enhanced number of contacts between tubes. There are three possible models of nanotube incorporation outlined in Figure 1. In Type A carbon nanotubes are stabilised by a rigid polymer matrix which solidifies as the fibres are produced; in Type B the orientation is stabilised by cross-linking, in Type C the nanotubes are linked together as a polymer chain and in this scenario, the system resembles a main-chain liquid crystal polymer [9] contained within a polymer matrix. In this work cross-linking was not invoked so the system resembles more closely (a).

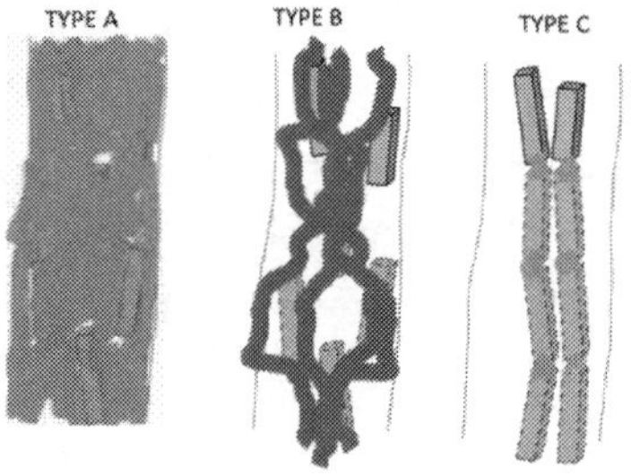

Figure 1: Routes to retaining alignment of the nanotubes in a polymer composite fibre

In this study we propose a methodology for preparing nanofibers containing high concentrations of carbon nanotube loadings using the polymer matrix to retain the alignment of the tubes. For this study we selected polyethylene oxide as the encapsulating polymer as it can be processed in aqueous solution, i.e. without the use of volatile organic compounds and that it readily produces nano scale fibres.

2. Materials and Methods

2.1. Materials
Multi-walled carbon nanotubes (MWCNT) purchased from Sigma-Aldrich with an outer diameter of 6–9nm and a length of ~5μm were used. Polyethyelene oxide (PEO), M_v 900,000gmol^{-1} (Sigma-Aldrich) was selected as the encapsulating material. Solutions were prepared using deionised water as the solvent with the addition of the surfactant, Triton X-100 (TX-100) (Sigma-Aldrich) in selected solutions to aid nanotube dispersion.

2.2. Methods

2.2.1. Solution Preparation
Aqueous (deionised water) solutions/suspensions were prepared by weight with loadings of MWCNTS corresponding to 0, 0.05%, 1%, 3% and 5% wt in the final fibre mat. The mixtures were stirred using a vortex mixer and placed in a sonicator for 4-5 hours to encourage dispersion. 1% wt TX-100 was added to selected solutions prior to stirring and sonication, corresponding to ~16% wt loading of surfactant in the final fibres. Suspensions prepared with surfactant appeared to have a better dispersion of nanotubes based upon a visual inspection. After sonication PEO was added to the suspension at 4.8% wt concentration and allowed to dissolve with further sonication and mixing on the vortex mixer.

2.2.2. Electrospinning
The electrospinning apparatus was arranged in a horizontal configuration. Solutions were loaded into plastic syringes with 22 gauge metallic needle (0.41mm inner diameter) attached and placed into a syringe pump set to 4.07mLhr^{-1}. A Glassman 0 – 30kV high voltage power supply was used to supply the high voltage. Electrospinning was performed using a potential of 15kV at the needle with samples

collected onto a grounded rotating collector, with a surface tangential velocity 7ms^{-1} positioned 15cm from the needle tip. A layer of aluminium foil placed over the surface of the rotating drum was used as the collector.

2.2.3. Electron Microscopy
Scanning electron microscopy (SEM) was performed on sections of the fibre matts produced. The microscopy was performed using a Cambridge Instruments SEM360 in high vacuum mode with an accelerating voltage of 20kV. Fibres were coated with gold prior to examination. Fibres were selected at random from micrographs obtained and measurements made using the software *ImageJ*.

Transmission electron microscopy (TEM) of the fibres was performed using a Philips CM20 at an operating voltages of 200kV with samples prepared on a copper TEM grid, 300 mesh. Samples were prepared for TEM by electrospinning directly onto the grid which was attached to the foil on the rotating collector. TEM on a pure sample of nanotubes was performed on a JEOL 2010 at 200kV. The sample was ground up in isopropyl alcohol with a pestle and mortar then pipetted onto a Lacy carbon TEM grid.

3. Results and Discussion
Preliminary observations of the electrospinning process were conducted by eye. For solutions containing CNT's without the surfactant, dripping occurred from the needle tip throughout the spinning process, where as those solutions with surfactant only the higher CNT loadings, 3% and 5% wt exhibited any observable dripping. The level of dripping still appeared to be to a lesser extent than that of the solutions without surfactant. The dripping was attributed to agglomeration of the nanotubes at the needle tip producing a large scale material that cannot be incorporated into the nanofibre and thus drips off. The presence of the surfactant allowed for a better dispersion of the nanotubes, thereby allowing a more continuous electrospinning process and hence electrospinning of fibres containing higher CNT concentrations. Fibre mats collected for different nanotube loadings were found to vary in appearance with higher loadings having a darker appearance.

Examination of the fibre mats through the use of SEM showed well aligned parallel arrays of fibres for the majority of samples collected, for which two examples are provided in Figure 2. The fibres appear fairly smooth and continuous with the exception of a few beads within the sample. Average fibre diameters for samples varied, increasing with nanotube loadings from ~200nm with no nanotubes (Figure 2a) up to 360nm for 5% CNT loading (Figure 2b).

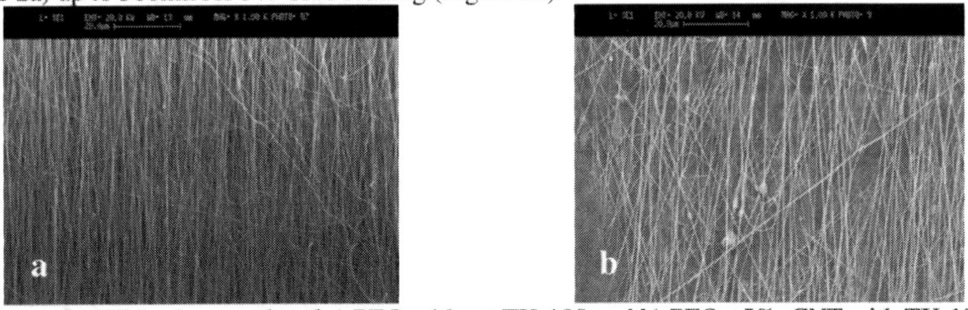

Figure 2: SEM micrographs of a) PEO without TX-100 and b) PEO +5% CNT with TX-100

Figure 3 shows the TEM micrographs for the nanotubes (figure 3a) along with the sample containing 5% wt loading of nanotubes (figure 3b). The mean CNT diameter was measured as 12 nm, slightly larger than the 6–9 nm quoted by Sigma with some diameters measuring near 20 nm. Figure 3b shows that when the nanotubes were straight there is alignment with the fibre, but where they bend there are defects, leading to kinks at the edge of the fibre, similar to observations made by Dror *et al* [4]. The bending can be reduced by the use of stiffer nanotubes such as those prepared through arc discharge, as opposed to the ones used here which appear to be prepared through chemical vapour deposition.

Figure 3: TEM micrographs of a) CNTs b) PEO electrospun fibres containing CNT's

The colouration of the fibre mat and the TEM micrographs show that high loadings of CNT's within nanofibers are possible. There is a visible number of tubes along the fibre axis (Figure 3b) which shows the potential for good connectivity of CNT's in the nanofibers leading to a higher level of conductivity. This was accomplished without need to modify the nanotubes prior to mixing to make them more dispersible in an aqueous media and high levels of loading were achieved, reducing dripping and promoting spinning by the addition of a surfactant.

4. Conclusions

We show that it is possible to electrospin polymer nanofibres containing CNTs from aqueous solution without additional nanotube treatment. The addition of a surfactant provides an effective route to even dispersions of CNT's reducing agglomeration and for allowing a more continuous electrospinning process. With the addition of the surfactant loadings of up to 5% CNT were obtained leading to a potential route for conducting nanofibers with high loadings of nanotubes.

Acknowledgements

MN acknowledges the support of a BOLASHAK scholarship from the Ministry of Education, Kazakhstan for her internship at the University of Reading. The research leading to these results was in part funded by the European Union Seventh Framework Programme (FP7/2007- 2013) under grant agreement n° 263017, Project "NanoCelluComp". Part of the work of GRM was support by the FCT (Portugal). The electron microscopy was performed in the Centre for Advanced Microscopy at the University of Reading and we thank Dr Peter Harris and Matthew Spink for their help.

References

[1] Mitchell GR 'Electrospinning: Principles, Practice and Opportunities' Royal Society of Chemistry 2015

[2] Harris PJF 2009 *Carbon Nanotube Science: Synthesis, Properties and Applications* (Cambridge University Press)

[3] Sun X, Chen T, Yang Z, and Peng H 2013 *Accounts of Chemical Research* **46** 539–49

[4] Dror Y, Salalha W, Khalfin RL, Cohen Y, Yarin AL, and Zussman E 2003 *Langmuir* **19** 7012-20

[5] Rastogia R, Kaushala R, Tripathib SK, Sharmaa AL, Kaura I and Bharadwaja LM 2008 *Journal of Colloid and Interface Science* **328** 421–28

[6] Mazinani S, Ajji A and Dubois C 2009 *Polymer* **50** 3329–42

[7] Wang SH, Wan Y, Sun B, Liu LZ and Xu W 2014 *Nanoscale Research Letters* **9** 522–29

[8] Cao L, Su D, Su Z and Chen X 2014 *Ind. Eng. Chem. Res.* **53** 2308−17

[9] Lenz RW, 1985 *Polymer Journal* **17** 105-15

Mobile robot based electrostatic spray system for controlling pests on cotton plants in Iraq

M Al-Mamury, N Manivannan, H Al-Raweshidy and W Balachandran
Centre for Electronic Systems Research, College of Engineering Design and Physical Science
Brunel University, United Kingdom

Email: Murtadha.Al-Mamury@brunel.ac.uk

Abstract: A mobile robot based electrostatic spray system was developed to combat pest infestation on cotton plants in Iraq. The system consists of a charged spray nozzle, a CCD camera, a mobile robot (vehicle and arm) and Arduino microcontroller. Arduino microcontroller is used to control the spray nozzle and the robot. Matlab is used to process the image from the CCD camera and to generate the appropriate control signals to the robot and the spray nozzle. COMSOL multi-physics FEM software was used to design the induction electrodes to achieve maximum charge transfer onto the fan spray liquid film resulting in achieving the desired charge/mass ratio of the spray. The charged spray nozzle was operated on short duration pulsed spray mode. Image analysis was employed to investigate the spray deposition on improvised insect targets on an artificial plant.

1. Introduction

Due to Iraq's arable land and unique location in Asia continent, and having two grand rivers named Tigris and Euphrates, wide lands can be cultivated. These factors created agricultural diversity in whole of the Iraq. However, a number of pests find suitable places in the plants and damage them badly. Among of these plants, the cotton plant is grown in many parts of Iraq (northern, central and southern) due to appropriate climatic conditions during the summer season. Insect on the cotton leaf known as ***'spodoptera littorals'*** is prominent insects affecting plants production [1]. The purpose of this project is to introduce intelligent robot based electrospray system and to design a new induction charging electrospray nozzle (IESN). The system is designed to achieve at the efficient use of pesticide with minimal drift. The mini-robot will scan the plant to identify the insect location using the CCD camera. Once the target is located the short duration pulsed spray will be activated to spray over the target location. Section 2 explains how a new electrospray nozzle was designed using commercially available hydraulic fan spray nozzle. COMSOL based FEM model was developed and used to design appropriate spray charging electrodes to achieve the desired charge/mass ratio for this application. Section 3 presents the experimental set-up used in the project to validate the performance of the electro spray nozzle. Set-of preliminary experimental results are presented in Section 4. Finally, conclusions drawn are presented in the Section 5.

2. Design of IESN

A new IESN was developed to generate pulsed spray and to be used in conjunction with a micro robot capable of traversing the plant canopy. The charged spray will make efficient use of pesticides in terms of deposition on the target locations cited on the plant and help to reduce the spray drift; thus saving money and time. The design of IESN was implemented in two stages: The first one was to determine spray properties (dimensions of the liquid film spray plume and drop size distribution) of the chosen commercial fan spray nozzle (FSN) and the second one is the design of induction electrodes to impart appropriate charge level onto the spar droplets. Commercial FSN used in this project work was TEEjet -TO80015E, which operates at the pressure between 3 and 4 bars and has spraying capacity of 0.59 l/min. Figure 1 shows the experimental set-up used and the spray film obtained under the operating conditions of 40% relative humidity and 20 ^{0}C. As in the figure 1(b), the spray forms a diverging thin film which has a diverging semi cone of 21.30. This diverging spray helps the spray liquid to spread and hence efficiently deposited onto the plant. The induction charging electrodes were designed to achieve maximum charge transfer from the electrodes to the liquid film so that when the film breaks up to form droplets, each droplet will acquire maximum charge. As shown in the figure 1(b), it is a planer liquid film and hence it requires induction electrode placement on both sides of the film to achieve maximum charge transfer. The electrode dimensions were optimized using COMSOL FEM model. Thin film coating of insulating material was used to cover the outer surfaces of both electrodes to minimize the surface wetting by the fine spray and to avoid electrical break down [2]. Sharp edges were also avoided to minimize electric discharge. The shape of plastic cover was designed using solid work software and made using a 3D printer. 'Image J' software was used to measure the liquid film dimensions. In this case, the dimensions of the thin film are shown in the figure 1(b).

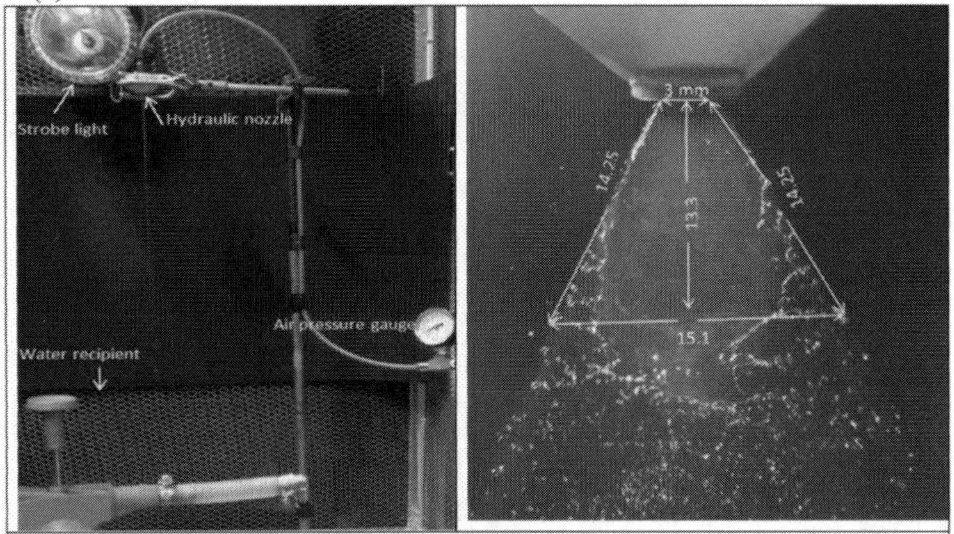

Figure. 1: Experimental characterization (a): experimental arrangement used to characterize the spray and (b) a thin film spray under normal operating condition.

A FEM was developed on COMSOL multi-physics environment and number of simulations was performed for various settings of the electrode parameters. The purpose of this simulation was to determine the parameters of the electro spray nozzle, such as, electrode dimensions, distances between the induction electrodes and nozzle tip, and the value of the high voltage that is required to achieve maximum surface charge density on the liquid film. Figure 2(a) shows a typical COMSOL simulation results which shows electric filed lines between the electrodes and the surface charge density on the surface of the liquid film. The optimum operating condition was found to be 0.5 kV electrodes potential for 10 mm spacing between the electrode and the liquid film. The designed electrodes have the same surface area as the film and of thickness of 1.5 mm. The constructed IESN is shown in figure 2(b).

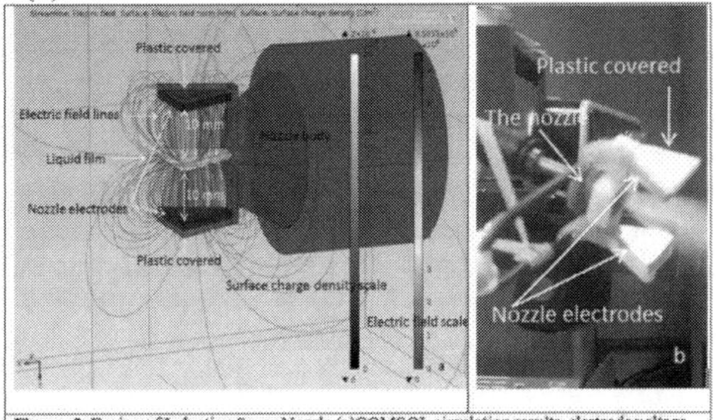

Figure. 2 :Design of Induction Spray Nozzle (a)COMSOL simulation results, electrode voltage 0.5kV, and distance between electrode voltage and thin film spray=10mm (b) Constructed Induction Spray Nozzle.

The surface charge density of the liquid film was calculated in the COMSOL simulation and for the electrodes with separation of 20 mm (as shown in the figure. 2) is 8 μc/m^2.

3. Experimental study

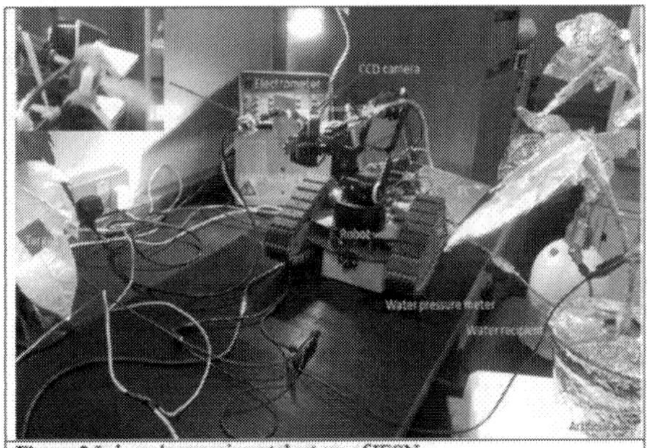

Figure.3:Lab scale experimental set-up of IESN

Experimental set-up used to demonstrate our IESN is shown in figure 3. As shown, a mini-robot with moving arm (AL05D robotic arm and the TTRK-KT tracked electric vehicle from Lynx motion™) is mounted with the newly constructed IESN and the CCD camera, and is connected to a PC. The PC receives images taken by the CCD camera, and then process them to identify and locate the insects (artificial). Finally the PC instructs the mini-robot to move forward or backward and activate IESN to spray the pesticides to the targets (artificial). An advanced imaging processing code was written in MATLAB programming environment to identify the pre-known targets (a various versions of '***spodoptera littorals***') on the artificial plant. Two artificial cotton plants are made using cardboard and their dimensions are similar to that of actual cotton plant. These plant leaves were covered with aluminum foil to make the leaves electrically conducting to mimic the natural plants which are electrically conductive. Artificial cotton plant insects of various shapes and sizes were also made and used in this study [3]. A number of experiments were carried out using ordinary water as the pesticides to spray over the artificial targets.

4. Results

The IESN was set to operate in pulse mode. In the experiments, a single pulse of varying time periods was applied over a target and the volume of water on the ground (wasted pesticides) was measured. The volume of the applied spray was calculated by multiplying the flow rate of the spray nozzle and the pulse duration. Figure 4 shows the results obtained in the preliminary set-of experiments. This figure 4(a) shows normalized spray volume, normalized drift volume and the ratio between drift volume and total spray volume for the pulse durations of 50, 100, 150, 250, 400, 500, 600 and 700 ms. As can be seen in figure 4, the ratio of waste/applied reduces for longer pulse period and further experiments to be performed to find the optimum pulse duration by considering other factors such as leave coverage along with the waste [4]. Experimentally measured the charge-mass ratio of the induction spray nozzle is plotted in figure. 4(b) as a function of spray pulse duration.

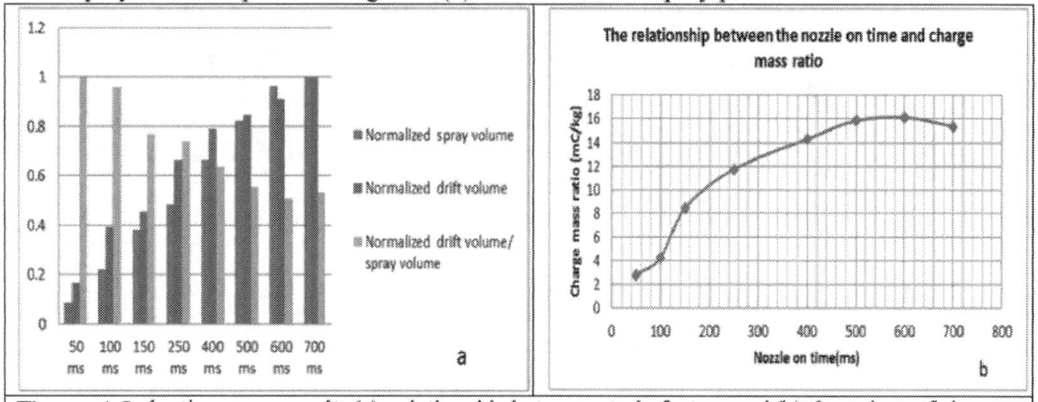

Figure. 4: Induction spray results (a) relationship between study factors and (b) the values of charge mass ratio

5. Conclusions

A mini-robot electrospray system has been designed and preliminary tests have been carried out to evaluate its feasibility. Early experimental results indicate that the system is capable of locating the pests on leaves of a plant canopy and then activate the electrospray for a short duration in an around that location. This minimises the amount of pesticides needed in real operation. Furthermore, this system also helps to minimise ground contamination and spray drift. Further work is being continued to fully evaluate the system under various operating conditions.

Acknowledgements:

M Al-Mamury likes to acknowledge to the ministry of higher education and scientific research of Iraqi government for the scholarship to purse his PhD studies at Brunel University, UK.

References

[1] Ahmad T 1988 Field studies on sex pheromone trapping of cotton leafworm Spodoptera littoralls (Boisd.)(Lep., Noctuidae) *J. Appl. Entomol.* **105** 212-215

[2] Patel M K, Ghanshyam C and Kapur P 2012 Characterization of electrode material for electrostatic spray charging: Theoretical and engineering practices *J. Electrostat.* **75** 55-60

[3] Lojans A and Kakitis A 2012 Evaluation of environment factors influencing robots in greenhouses," in *11th Int. Sci. Conf. Engineering for Rural Development (Jelgava, Latvia)* 151-156

[4] Maski D and Durairaj D 2010 Effects of electrode voltage, liquid flow rate, and liquid properties on spray chargeability of an air-assisted electrostatic-induction spray-charging system *J. Electrostat.* **68** 152-158

Design and performance of a small-scale wind turbine exploiting an electret-based electrostatic conversion

M Perez[1,2,3], S Boisseau[1,2], and J L Reboud[1,3]
[1]Univ. Grenoble Alpes, F-38000 Grenoble, France
[2]CEA, Leti, Minatec Campus, 17 rue des Martyrs, F-38054 Grenoble Cedex 9, France
[3]Univ. Grenoble Alpes, G2ELab, F-38000 Grenoble, France

E-mail: matthias.perez@cea.fr

Abstract. This paper reports on a cm-scale wind turbine coupled to an electret-based electrostatic energy converter for airflows energy harvesting. The device we propose is made of a typical axial turbine to convert the wind energy into a mechanical energy of rotation and of a patterned electret-based electrostatic converter to turn this mechanical energy into electricity. This is actually the first time that the mechanical power extracted by a turbine is turned into electricity with an electret-based electrostatic converter. Several prototypes have been tested on a wind tunnel from 0 to 10 m/s; a power of 200 μW has been extracted on a 4 cm-diameter device at 10 m/s.

1. Introduction

Every day, the world is becoming more instrumented and interconnected; sensors become able to communicate without wire by RF, simplifying their installation (Wireless Sensor Nodes). Yet, there remains the energy supply issue, which involves batteries that have to be recharged. Then, the electronic research is directed towards the miniaturization of components, the reduction of their energy consumption, and ideally energy harvesting for an autonomous and long-life operation. Energy harvesting obviously requires an ambient source of energy to convert into electricity; here, we will focus on kinetic energy harvesting from airflows. Such energy harvesters can be placed on buildings, vehicles, in air conditioning system (HVAC), or in any place subjected to free or forced airflows.

We may find in the state of the art some small-scale electromagnetic wind-turbines [1], but also more original ideas such as flow-induced vibration harvesters, exploiting fluttering effects [2]. The transducer we propose here includes a classic windmill coupled to an electrostatic converter. The turbine is driven by wind energy, and its rotational kinetic energy is directly converted into electric energy through capacitance variations between movable and fixed electrodes. An electret is added to polarize the capacitive structure and to enable a direct mechanical-to-electrical energy conversion.

The next sections are devoted to small-scale electrostatic wind-turbines. Section 2 introduces the theory of electret-based electrostatic conversion which will be applied to the specific case of electrostatic wind-turbines (section 3) to end up with experimental results in section 4.

2. Electret-based conversion: models and equations

An electret is a dielectric material in which electrical charges have been implanted, and can be considered as an "electrical magnet" presenting a permanent electrical polarization [3]. Electrets are generally obtained by Corona discharge [4,5], from thin-film materials such as Teflon®, SiO_2, Si_3N_4,

SiN, Parylene, CYTOP… and can keep their charges during several decades [6]. Electrets are currently used to replace charge-discharge cycles [3] in electrostatic converters. They offer the main benefit of simplifying the power management circuit, thus preventing additional energetic losses. Nevertheless, the surface voltage of electrets is limited for stability reasons [5], and the electrets must be kept in a dry environment to avoid their discharge.

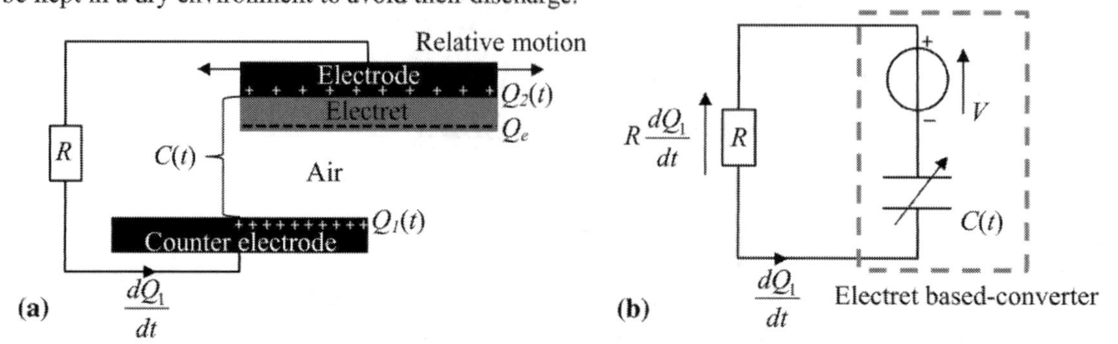

Figure 1. (a) Electret-based converter and (b) equivalent electric model with a simple load R

The basic electret-based converter is represented in Figure 1a, and its equivalent circuit in Figure 1b. We are dealing here with a capacitive structure made of two electrodes separated by an air gap, with one electrode covered by an electret layer. The constant electrical charges Q_e present on the electret instantaneously induces the same amount of charges on the counter-electrode $Q_1(t)$ and the electrode $Q_2(t)$, to satisfy Gauss's law:

$$Q_e + Q_1(t) + Q_2(t) = 0 \qquad (1)$$

Then, the relative motion between the electrode and the counter-electrode induces a reorganization of charges between both electrodes, and thus a current circulation through load R. By implementing Kirchhoff's circuit laws to the electrical circuit presented in Figure 1b, we can thus deduce the following equation (2). This differential equation provides a link between the electret's surface voltage V, the load R, the capacitance variation of the harvester $C(t)$, and the charges on the counter electrode $Q_1(t)$:

$$\frac{dQ_1}{dt} + \frac{Q_1}{RC(t)} = \frac{V}{R} \qquad (2)$$

The average output power P_{elec} harvested in load R during a period of time T, can thus be calculated thanks to a finite element analysis (to determine C(t)) and by calculating the integral of equation (3). More simply, P_{elec} is directly proportional to the capacitance variation C_{max}-C_{min}, the oscillation frequency (f) and the electret's surface voltage squared (V^2):

$$P_{elec} = \frac{1}{T}\int_0^T R\left(\frac{dQ_1}{dt}\right)^2 dt \propto (C_{\max} - C_{\min})V^2 f \qquad (3)$$

3. Electret-based wind-turbine: theory and simulations

Windmills are clearly the most mature technology for "large-scale" applications (kW-MW). The windmill is composed of a rotating part called rotor, and a fixed part called stator. The most common turbines are Horizontal-Axis Wind Turbines (HAWTs) which have the benefit of being highly efficient and the main drawback of having to be pointed in the direction of the wind. In the majority of cases, the electromechanical converter is electromagnetic, and uses electromagnets or permanent magnets as inductors. For "small-scale" applications (µW-mW), the physical approach is very different but windmills still remain a reference, as well as the electromagnetic conversion [1].

The electret-based windmill energy harvester we propose is a simple HAWT (Figure 2). The operating principle is identical as the one mentioned in section 2: the electrets are positioned on the N mobile electrodes of the rotor (red layer in Figure 2a), and the stator includes $2\times N$ electrodes to achieve an optimal energy conversion. Two connection options are possible: (i) the first scheme where one external load is connected between odd and even electrodes of the stator, and (ii) the second scheme where two external loads are connected between the ground (electrodes on the rotor) and respectively odd and even electrodes of the stator.

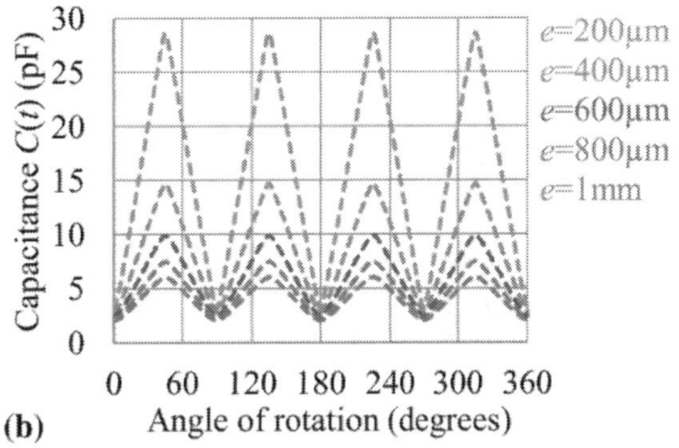

Figure 2. (a) Solidworks view of an electret-based wind-turbine with N=4 mobile electrodes. (b) Evolution of the capacitance for various air-gaps (COMSOL Multiphysics® simulations)

Contrary to the frequency and the torque that depend on the design of the turbine and wind speed, capacitances variations only depend on the spacing between the stator and the rotor and the number of mobile electrodes N. COMSOL Multiphysics® simulations have enabled us to estimate the capacitance variations for a converter of 40 mm of diameter, with various air gaps. We can note that the capacitance variation is almost triangular (Figure 2b), which makes sense given the triangular evolution of surfaces facing each other. Figure 2b also reveals the impact of the air-gap on the capacitance variation, and thus directly on the output power (equation (3)). We will see in the next section that the air-gap is the most critical parameter and actually very difficult to minimize due to this fabrication process.

4. Electret-based wind-turbine: experimental results

The mechanical part of the harvester has been fabricated by a rapid prototyping machine (Figure 3a). The turbine has a diameter of D=40 mm, and a total depth of H=10 mm (Figure 3a); then the total volume of the prototype is about 12.5 cm^3. All the electrodes are covered by a 200 µm-thick copper film, and then by a 127 µm-thick Teflon PTFE layer for mobile electrodes only. The tests were conducted on a wind tunnel. We have decided to fix the number of mobile electrodes at N=4, and the same number of blades to mechanically hold the electrodes (green lines in Figure 3a).

As summarized by equation (3), the power extracted depends on three parameters. First, the air gap thickness which is directly linked to the capacitance variation C_{max}-C_{min}, and limited to 500 µm due to the chosen fabrication process. Secondly, the electret's surface voltage V which should stay under 1400 V for a 127 µm-thick Teflon PTFE electret (Stability issues [7]). Thirdly, as the number of blades has been set to N=4, the operating frequency only depends on the geometry of the blades. Several blade angles α have been tested (Figure 4a) and it appears that the angle α =10° configuration enables the highest rotational speed between 0 and 10 m/s. We can notice on Figure 4a a linear relationship between the flow speed and the rotational speed. In Figure 4b, the evolution between the output power and the electret surface voltage is quadratic, as predicted by equation (3).

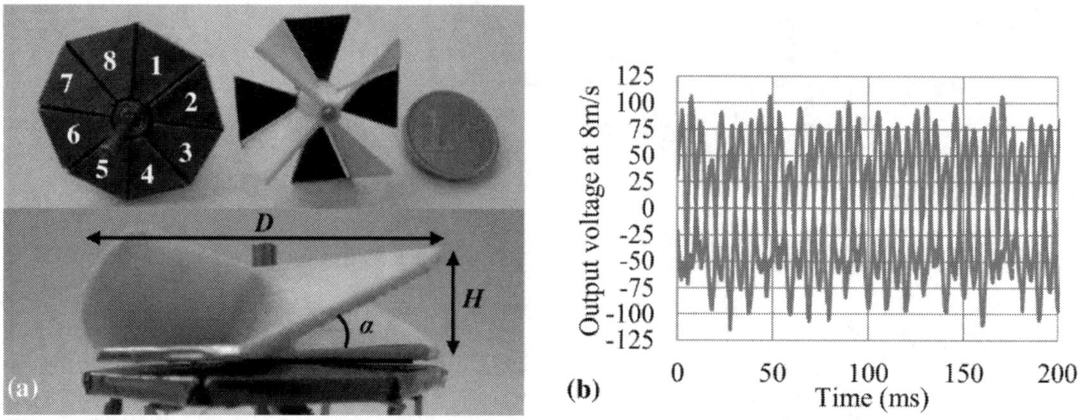

Figure 3. (a) Example of a demonstrator fabricated by a rapid prototyping machine and (b) output voltage at 8m/s, on two loads of R=110MΩ, and V=-1000V

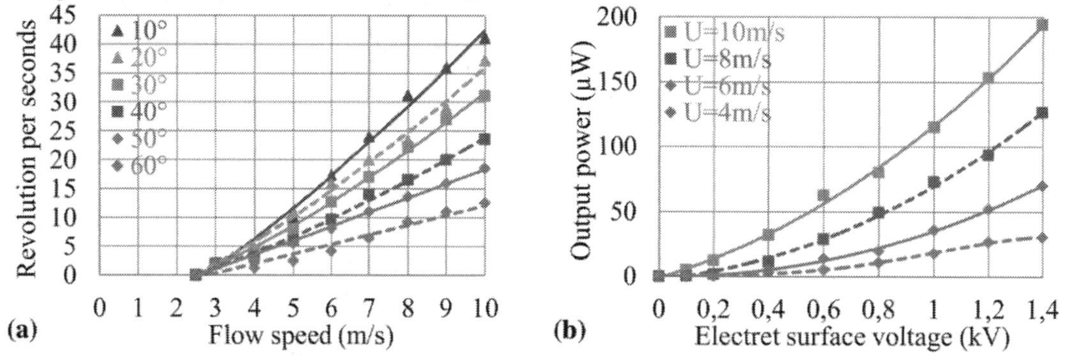

Figure 4. (a) Rotational speed as a function of the flow speed at different blade angles α (V=-1000V), and (b) Output power vs electret's surface voltage and flow speeds (α=10°, R=110MΩ)

5. Conclusions

We have presented the first electrostatic windmill exploiting an electret-based converter. The prototypes realized have enabled to extract a maximum output power of P_{elec}=200 μW at 10 m/s. At these sizes, the results are not yet equivalent to what was accomplished with electromagnetic converters, mainly because of the thick air gap which strongly reduces the capacitance and its variation. However, this harvester provides a high voltage, a reasonable amount of power compatible with autonomous sensors application, and a low manufacturing cost. Research is currently underway to improve the harvester's aerodynamics by replacing the current stator by lateral electrodes, and to increase the output powers by reducing the air gap.

References

[1] Bansal A et al. 2009 Cm-scale air turbine and generator for energy harvesting from low-speed flow *Transducers*, pp. 529-532
[2] Perez M et al. 2015 An electret-based aeroelastic flutter energy harvester *Smart Materials and Structures*
[3] Boisseau S et al. 2012 Electrostatic Conversion for Vibration Energy Harvesting *Small-Scale Energy Harvesting, ed. M. Lallart*, Intech, doi:10.5772/51360
[4] Boisseau S et al. 2013 Stable DRIE-patterned SiO2/Si3N4 electrets *Transducers*, pp. 1942-45
[5] Sessler G et al. 1998 Electrets 3rd edition, Laplacian Press
[6] Kressmann R et al. 1996 Space-charge electrets IEEE Trans. Diel. Electr. Insul. **3**, 5, pp. 607-23
[7] Boisseau S et al. 2011 Cantilever-based electret energy harvesters, *Smart Mater. Struct.* **20** 105013

Deep dielectric charging and breakdown of lunar polar regolith

A P Jordan[1], T J Stubbs[2], J K Wilson[1], N A Schwadron[1], and H E Spence[1]

[1]EOS Space Science Center, University of New Hampshire, Durham, NH, United States
[2]NASA Goddard Space Flight Center, Greenbelt, MD, United States

Email: a.p.jordan@unh.edu

Abstract. Galactic cosmic rays (GCRs) and solar energetic particles (SEPs) penetrate the regolith (layer of soil and dust) covering the Moon's surface and cause deep dielectric charging. To gain insight into this process, we have developed a data-driven, deep dielectric charging model using data from the Cosmic Ray Telescope for the Effects of Radiation (CRaTER), which is onboard the Lunar Reconnaissance Orbiter (LRO), and the Electron, Proton, and Alpha Monitor (EPAM) on the Advanced Composition Explorer (ACE). The model results indicate that GCRs produce a persistent electric field (~700 V m^{-1}) within the top tens of centimeters of regolith, while large SEP events could potentially generate episodic subsurface electric fields ($\geq 10^6$ V m^{-1}) capable of causing dielectric breakdown within the top millimeter of regolith. We also propose that this "breakdown weathering" may have significantly affected the regolith in the Moon's permanently shadowed regions (PSRs).

1. Introduction

Deep dielectric charging is well-known as an important process to mitigate in spacecraft engineering, but its effects have rarely been studied for the many airless bodies that exist throughout the solar system. Because these bodies lack protective atmospheres, their surfaces are directly exposed to the space environment. This allows dielectric surface materials on airless bodies to be charged by penetrating energetic charged particles, which creates subsurface electric fields due to the materials' insulating properties. In some radiation environments, these electric fields could be capable of causing dielectric breakdown, or sparking [1]. The Moon's regolith is a layer of dielectric soil and dust covering the surface, and it is expected to have an extremely low conductivity at the cold temperatures that occur near the poles, especially within permanently shadowed regions (PSRs) that never receive direct sunlight. Therefore, it is anticipated that energetic charged particles from the space environment can significantly charge the subsurface regolith in lunar PSRs. In this paper, we briefly review our work done to address this gap in understanding deep dielectric charging in the lunar regolith [2, 3], and we refer readers to both those papers for further details.

2. Energetic charged particles

Charging and breakdown in the regolith depend, in part, on the flux and fluence (time-integrated flux) of the incident energetic charged particles. At the Moon, there are two populations to consider. The first is galactic cosmic rays (GCRs), which are typically accelerated in supernovae shocks. For this work, we consider only the proton component (~87% of all GCRs), which has a peak flux at ~200 MeV and a total flux of ~2–4 particles cm^{-2} s^{-1} [4]. GCRs penetrate and charge the regolith down to tens of centimeters [5]. For GCR data, we use the Cosmic Ray Telescope for the Effects of Radiation (CRaTER), which is onboard the Lunar Reconnaissance Orbiter (LRO) [6].

The other population is solar energetic particles (SEPs), which comprise ions (~50 keV–10 GeV) and electrons (~1 keV–10 MeV) in roughly equal quantities; their peak fluxes are at much lower energies than that of the GCRs [7]. SEPs are accelerated up to even relativistic energies in solar flares and the shocks of coronal mass ejections. They occur sporadically and their fluxes can exceed GCR fluxes by many orders of magnitude. For SEP data, we use LRO/CRaTER and the Electron, Proton, and Alpha Monitor (EPAM) on the Advanced Composition Explorer (ACE) [8]. SEPs detected by these instruments penetrate about 1 mm into the regolith, with the electrons typically going slightly deeper [2].

3. Electrical properties of the lunar regolith

Charging and breakdown in the regolith also depend on its electrical properties. We assume regolith to be ohmic, that is, having a discharging timescale $\tau = \varepsilon/\sigma$, where ε is its permittivity and σ is its conductivity. The regolith typically has a permittivity of $\varepsilon \approx 2\varepsilon_0$ [9]. Its electrical conductivity depends on temperature: $\sigma = \sigma_0 e^{\alpha T}$, where $\sigma_0 = 6\times10^{-18}$ S m^{-1}, $\alpha = 0.0237$ K^{-1}, and T is the temperature in kelvin [10]. At the temperatures found in PSRs (~50 K) [11], the conductivity is predicted to be ~10^{-17} S m^{-1}, so the regolith's discharging timescale is expected to be ~20 days.

The regolith's intrinsic properties make it conducive to dielectric breakdown for two main reasons. First, its grains tend to be jagged in shape, and such cusps can increase local electric fields by 1–2 orders of magnitude with respect to the overall electric field [12]. Second, its grains frequently contain inclusions, usually solid but sometimes gaseous [13, 14]. These inclusions also can increase local electric fields, thus decreasing the regolith's dielectric strength [e.g., 15].

4. Modeling deep dielectric charging of regolith

We developed a one-dimensional, time-dependent, data-driven model to estimate the electric fields in the regolith due to GCRs and SEPs (for more details, see [2]). The model's inputs and outputs are shown in Figure 1 for four cases. J_{EP}^+ and J_{EP}^- are the incident current densities from positively and negatively charged energetic particles, respectively, and can be provided by spacecraft measurements (e.g., LRO/CRaTER and ACE/EPAM). κ^+ and κ^- are the areal charge densities of the two charged layers. J_{gap} is the current density between the two charge layers, and $J_{interior}$ is the current density below the deeper, negative charge layer.

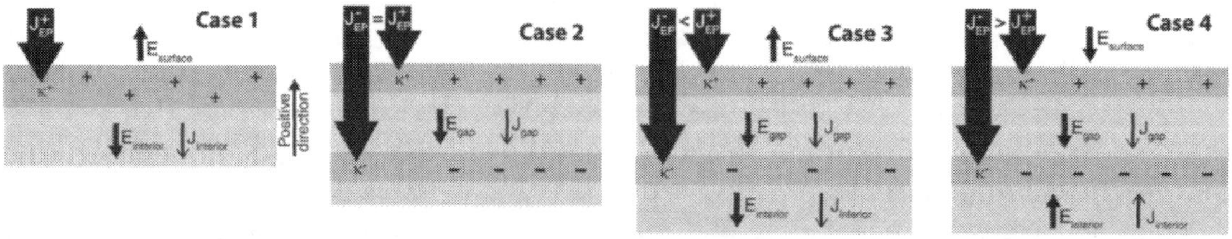

Figure 1. Cartoon showing the charging model's inputs and outputs for four possible cases: (1) only GCRs, (2) equal fluxes of SEP protons and electrons, (3) SEP protons with a greater flux than the electrons, and (4) SEP protons with a smaller flux than the electrons. (Adapted from Figure 2 of [2].)

After demonstrating the model's characteristics with simple test scenarios, we then used spacecraft measurements. We input GCR data from LRO/CRaTER from June 2009 to March 2011 and found that the GCRs created a persistent electric field peaking at 700 V m^{-1} in the regolith within PSRs during solar minimum. We then input ACE/EPAM data from three very large SEP events: the Bastille Day storm in July 2000, the storm in November 2001, and the Halloween storms of 2003. The maximum subsurface electric fields were estimated to be on

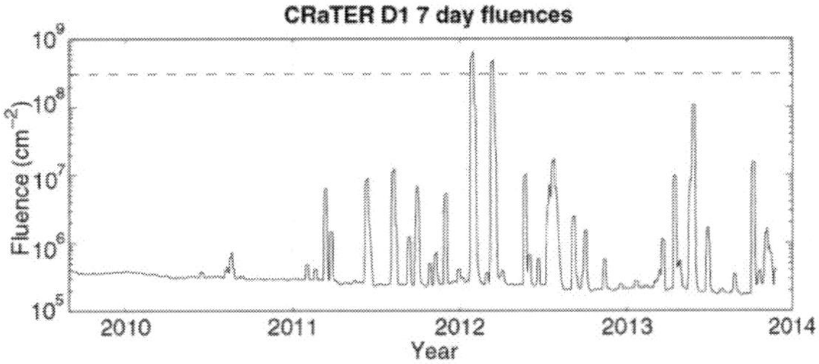

Figure 2. CRaTER 7 day fluences measured by a single detector (D1). Spikes above the GCR background are SEP events. The dashed line indicates the breakdown threshold, which was exceeded by two events. (Adapted from Figure 12 of [3].)

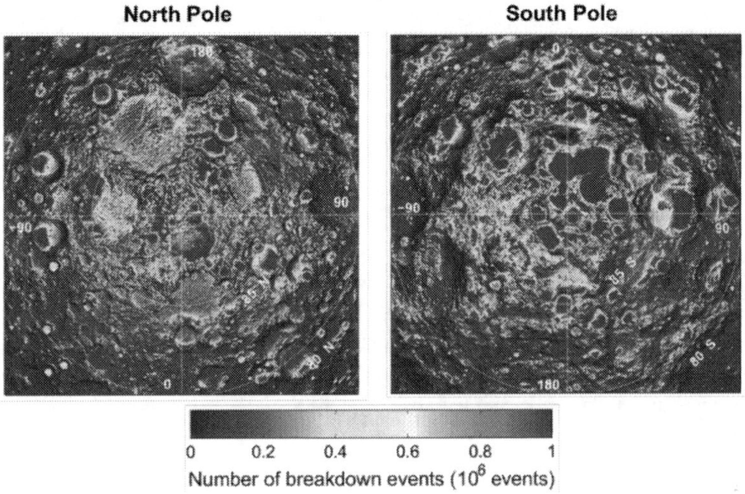

Figure 3. Estimated number of breakdown-inducing SEP events in meteoritically mixed regolith near both lunar poles. (Adapted from Figure 8 of [3].)

the order of 10^6–10^7 V m^{-1} — possibly strong enough to induce dielectric breakdown. The fluence required to cause breakdown in most solids is ~10^{10}–10^{11} particles cm^{-2}, if deposited well within the discharging timescale [e.g., 16]. This criterion was met in all three events.

We also use LRO/CRaTER data as a proxy for the fluences of lower energy, and thus more numerous, SEPs [3]. We find that, since LRO arrived at the Moon, two SEP events (in January 2012 and in March 2012) have occurred that may have caused breakdown within PSRs (Figure 2). Therefore, this indicates that breakdown could be an ongoing space weathering process in the Moon's polar regions.

5. Breakdown weathering of lunar regolith

We assessed the likely prevalence of dielectric breakdown at the Moon by estimating the frequency of breakdown-inducing SEP events. This was achieved by combining maps of the annual average, near-surface temperature at the Moon's polar regions [11] with the Jet Propulsion Laboratory (JPL) fluence model, which estimates the frequency of SEP events with a given fluence [17]. We convert the temperature maps to regolith discharging timescales using the relation described in Section 3. We then adapted the proton fluence model to find the frequency at which breakdown-causing SEP events occur (about one event per year within PSRs). Finally, we incorporate the effect of meteoroid impacts, which thoroughly mix or "garden" the regolith, thus bringing deeper material to the surface [18]; all gardened regolith has been exposed to SEPs for roughly 1 Myr [19]. Therefore, the gardened regolith within PSRs has likely experienced ~10^6 breakdown-inducing SEP events (Figure 3).

This repeated breakdown may be an important form of space weathering. It can fragment regolith grains, particularly along mineralogical boundaries [15], which may explain the unexpectedly porous regolith possibly detected by LRO's Lyman Alpha Mapping Project (LAMP) [20]. We are currently working to better understand how breakdown weathering may affect the regolith, and determine whether breakdown events could be observed with existing lunar spacecraft or with ground-based instruments.

6. Conclusion

We have developed a data-driven model whose results indicate that both GCRs and SEPs deep dielectrically charge the lunar regolith within PSRs. GCRs are predicted to produce a persistent electric field (~700 V m^{-1}) within the top tens of centimeters of regolith, while large SEP events may create episodic subsurface electric fields ($\geq 10^6$ V m^{-1}) — capable of causing dielectric breakdown within the top millimeter of regolith. We also show that meteoritically mixed regolith within PSRs has likely experienced ~10^6 breakdown-inducing SEP events. This breakdown weathering may have significantly affected the physical and chemical properties of the regolith in the Moon's PSRs.

Acknowledgements

This work was supported by NASA grants NNG11PA03C, NNX10AB17A, and NNX14AG13A. The authors thank the ACE/EPAM team and Principal Investigator Dr. Robert Gold of JHU/APL for providing the ACE data via CDAWeb and the LRO/Diviner team and Principal Investigator David Paige for providing the Diviner Level 4 Polar Resource Products at the NASA Planetary Data System.

References

[1] Campins H and Krider E P 1989 *Science* **245** 622–4 doi:10.1126/science.245.4918.622
[2] Jordan A P *et al* 2014 *J. Geophys. Res. Planets* **119** 1806–21 doi:10.1002/2014JE004648
[3] Jordan A P *et al* 2015 *J. Geophys. Res. Planets* **120** 210–25 doi: 10.1002/2014JE004710
[4] Smart D F and Shea M A 1985 *Handbook of Geophysics and the Space Environment* ed. Jursa A S (U.S. Air Force Geophysics Laboratories) pp 6-1–6-29
[5] Walker R M 1980 *The Ancient Sun: Fossil Record in the Earth, Moon and Meteorites* ed Pepin R O, Eddy J A, and Merrill R B (New York: Pergamon) pp 11–28
[6] Spence H E *et al* 2010 *Space Sci. Rev.* **150** 243–84 doi:10.1007/s11214-009-9584-8
[7] McGuire R E and von Rosenvinge T T 1984 *Adv. Space Res.* **4** 117–25, doi:10.1016/0273-1177(84)90301-6
[8] Gold R E *et al* 1998 *Space Sci. Rev.* **86** 541–62 doi:10.1023/A:1005088115759
[9] Olhoeft G R and Strangway D W 1975 *Earth Planet. Sci. Lett.* **24** 394–404 doi:10.1016/0012-21X(75)90146-6
[10] Olhoeft G R, Frisillo A L, and Strangway D W 1974 *J. Geophys. Res.* **79** 1599–1604 doi:10.1029/JB079i011p01599
[11] Paige D A *et al* 2010 *Science* **330** 479–82 doi:10.1126/science.1187726
[12] Bahder G *et al* 1982 *IEEE T. Power Ap. and Syst.* **6** 1379–90
[13] Roedder E and Weiblen P W 1970 *Geochim. Cosmochim. Ac. Suppl.* **1** 801–37
[14] Funkhouser J *et al* 1971 in *Proc. Lunar Sci. Conf. 2nd* **2** 1381–96
[15] Lisitsyn I V *et al* 1998 *J. Appl. Phys.* **84**(11) 6262–7
[16] Green N W and Dennison J R 2008 *IEEE T. Plasma Sci.* **36** 2482–90 doi:10.1109/TPS.2008. 2003442
[17] Feynman J *et al* 1993 *J. Geophys. Res.* **98** 13281–94 doi:10.1029/92JA02670
[18] Arnold J R 1975 in *Proc. Lunar Sci. Conf. 6th* pp 2375–95
[19] Jordan A P *et al* 2013 *J. Geophys. Res. Planets* **118** 1257–64 doi:10.1002/jgre.20095
[20] Gladstone G R *et al* 2012 *J. Geophys. Res. Planet* **117** E00H04 doi:10.1029/2011JE003913

The effect of electrostatic charges on the removal of radioactive aerosols in the atmosphere by raindrops

M Sow and P Lemaitre

Institut de Radioprotection et de Sûreté Nucléaire (IRSN), PSN-RES, SCA, LECEV

Gif-sur-Yvette, 91192, France

mamadou.sow@irsn.fr

Abstract. In this article, we report results of self-charged water drop generated by hypodermic needle over charge values comparable to those reported in the literature during stormy rainfall. We also controllably charged aerosols particles by corona discharge and evaluate how it affects their collection efficiency. Electric charges on drops and aerosols are precisely monitored by high resolution electrometers. Our preliminary results tend to accredit the impact of electric charges in collection efficiency.

1. Introduction

Measurements of soil contamination by cesium particles in the first days following the Fukushima accident have shown that higher doses were located downwind of the radioactive plumes after snow and rainfall events. In France, given the location of the nuclear plants, the probabilities of occurrence of rainfall coupled to a nuclear accident are real. To establish procedures that should protect people against exposure to radionuclides, it is essential to understand the interactions between aerosols and precipitation that determine their removal from the atmosphere. The capture of aerosol particles by raindrops (washout) has been studied exhaustively in laboratory under simulated storm conditions, and by modeling. However, to date there are discrepancies on the estimation of the aerosol capture efficiency between experimental studies and modeling results that tend to underestimate it. We believe that this discrepancy is partly related to the non-inclusion of electrostatic interactions in the calculations. More importantly, it is well recognized that raindrops as well as aerosol particles naturally and systematically carry electric charges through various mechanisms in the atmosphere [1], which could leads to strong electrostatic interactions. For electrically charged drops, Barlow and Latham [2] in their laboratory experiments on sub-micron aerosols exhibit significant increase in the washout efficiency. More recently, Tinsley et al. [3] model calculations in the same size range pointed out that image charge effect due to aerosol particles charging can significantly enhance their collision efficiency with raindrops compared to uncharged particles, even if the particles and drops carry electric charges of same polarity. Indeed, radioactive aerosols positively self-charge because of emission of electrons during decay [4] and non-radioactive aerosol get electric charges by adsorption of atmospheric ions resulting from cosmic radiations [5].

We are currently investigating the influence of electric charges on both water drops simulating rainfalls and aerosol particles of micron size range on the washout efficiency in BERGAME facility –

French acronym for Facility to study aerosol washout and measure collection efficiency – developed at IRSN (figure 1).

This article focuses on our first results on the evolution of washout efficiency when aerosol particles are controllably charged by a bipolar corona charger. The initial charge state of the drops is monitored with a Faraday pail but it is not intentionally charged. Other experiments on the influence of drops charge on washout efficiency for neutralized aerosols are presently explored but not reported here.

2. Experimental arrangement

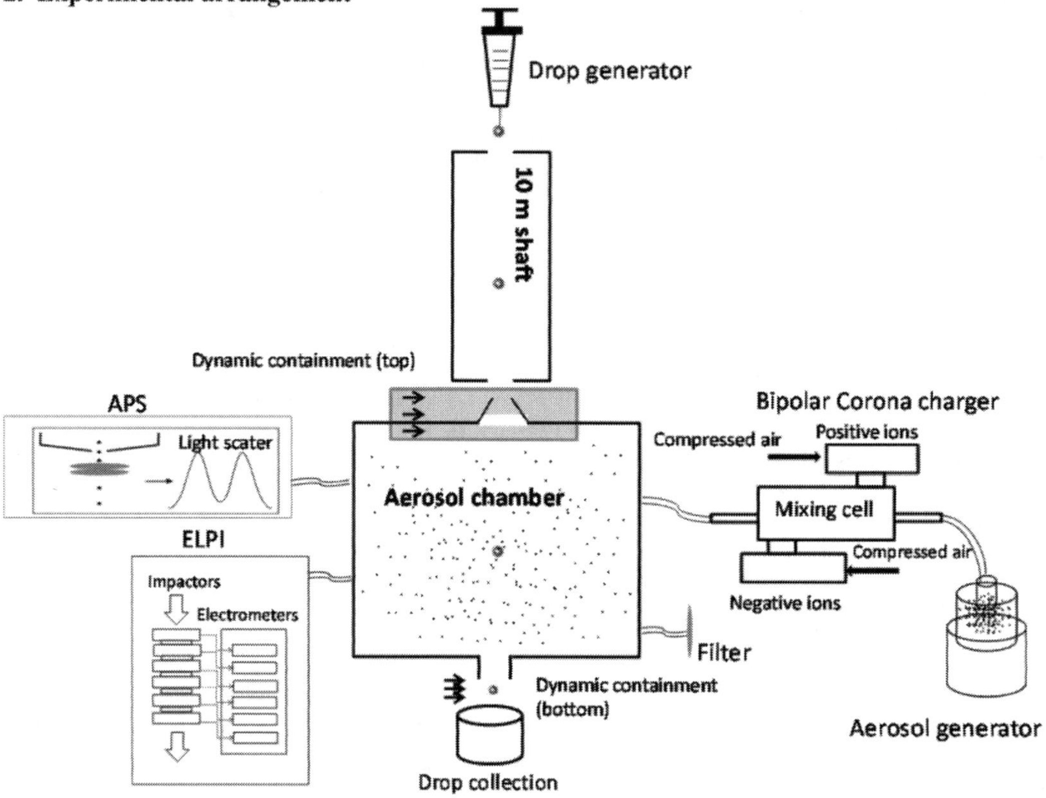

Figure 1 BERGAME experimental set up

BERGAME has been designed specifically to assess the washout efficiency of aerosol particles by raindrops. It consists of a shaft of ten meters high with a square cross section of 0.45 × 0.45 m² through which millimeter drops free fall toward their terminal velocity before entering in a seeded aerosol atmosphere, and collected beneath.

Drops are generated from a hypodermic needle filled with water in a constant level that shapes them uniformly. Drop diameter and velocity are measured inside the aerosol chamber by shadow imaging and image velocimetry (PIV) respectively. The aerosol chamber is a stainless steel rectangular cube of 1 m height with a cross section of 0.8 × 0.8 m². It is equipped with two apertures on the top and the bottom that allows drops to enter and exit the chamber; dynamic containment systems prevent aerosol particles from reaching the shaft.

The aerosols used in this study are soda fluorescein particles that are easily detectable by fluorescence spectroscopy. Aerosols are generated with ultrasonic nebulizers (Sinaptec GA 500) that produce particles of micron size nearly mono-dispersed. Particle size in the aerosol chamber is measured with both APS (Aerodynamic Particle Sizer) and ELPI (Electrical Low Pressure Impactor). To evaluate the mass density of aerosol particles inside the chamber, a known volume is pumped through a particle air filter. The mass of soda fluorescein collected on the filter and by the drops are together measured with fluorescence spectroscopy to determine respectively the mass concentration of soda fluorescein particles in suspension inside the aerosol chamber, and the mass concentration of soda fluorescein in the drops. The washout efficiency (E) is simply defined as the ratio of the mass of aerosol particles (AP) collected by the drop during its fall to the mass of aerosol particles in the volume swept out by a drop of same equivalent spherical diameter ($E = \frac{m_{AP,collected}}{m_{AP,swept}}$).

The aerosol charges are measured with an impactor ELPI. This instrument consists of 13-stage cascade low pressure impactor for a size range from 30 nm to 10 µm. Electrometers are used to measure the current signals on the 12 impactor stages (the first stage is a pre-selector). They can read current signals in the range of 10 000 to 400 000 fA. In usual operation, aerosols are preloaded by a corona charger with a known charge efficiency function from which a computer program deducts the number of particles impacting each stage of the impactor. To assess the aerosol charges, the ELPI is operated with the charger off, so the measured current is only due to aerosol charges on each stage.

Our aerosol particles are subject to contact electrification and are usually highly charged when they exit from the generator. To pilot the charging, particles are allowed to pass through a neutralizer (Topas - Electrostatic Aerosol Neutralizer Series EAN) which is equipped with two corona chargers that generate positive and negative ions in large amount that can either neutralize them or impose a controlled charge level and polarity.

3. Results on aerosol washout efficiency

We calculated washout efficiency for non-charged drop sizes of 2.5 mm, of 3 different aerosol mean size diameters and varied the number of elementary charges on the particles (figure 2). For nearly neutral particles, the washout efficiencies increase with diameter (2.10-3, 4.10-3, 38.10-3 for 1.4 µm, 2.34 µm, 3.5 µm respectively). These sizes are in the range where the main mechanisms responsible of the washout are impaction and interception of the particles by the drop due to gravity. So it is not surprising that washout efficiency increases as the size becomes larger. If the charge level is raised in small steps, the washout efficiency increase very quickly and starts decaying for less than 10 elementary charges and stabilizes after around 20 elementary charges. We can also observe that the influence of the aerosol charge level seems to diminish as the particle size enlarges. This can easily be interpreted by smaller particles having higher charge density compared to the bigger particles for the same number of charges. This trend agrees partly with Tinsley et al. [3] predictions. Indeed these authors showed a rapid increase of the washout efficiency when the drop and the aerosols particles have electric charges of comparable magnitudes due to short range image force interactions, this, even if the drop carries a net charge that is of same sign as the particles. If the charges on the particles are too high, the Coulombic long range repulsion becomes strong enough to prevent particles from being caught by the drop. However, the washout efficiency in these calculations keeps decreasing with increasing charges on the particles until a zero value, while our results show an asymptotic behavior.

These preliminary results seem to confirm that electric charges need to be considered when investigating aerosol atmospheric washout by rainfall. For aerosol particles of around 1 µm in diameter, the washout efficiency has almost tripled when they carry few elementary charges.

Further experiments are currently performed for more aerosols sizes, including the sub-micron range to consolidate these observations.

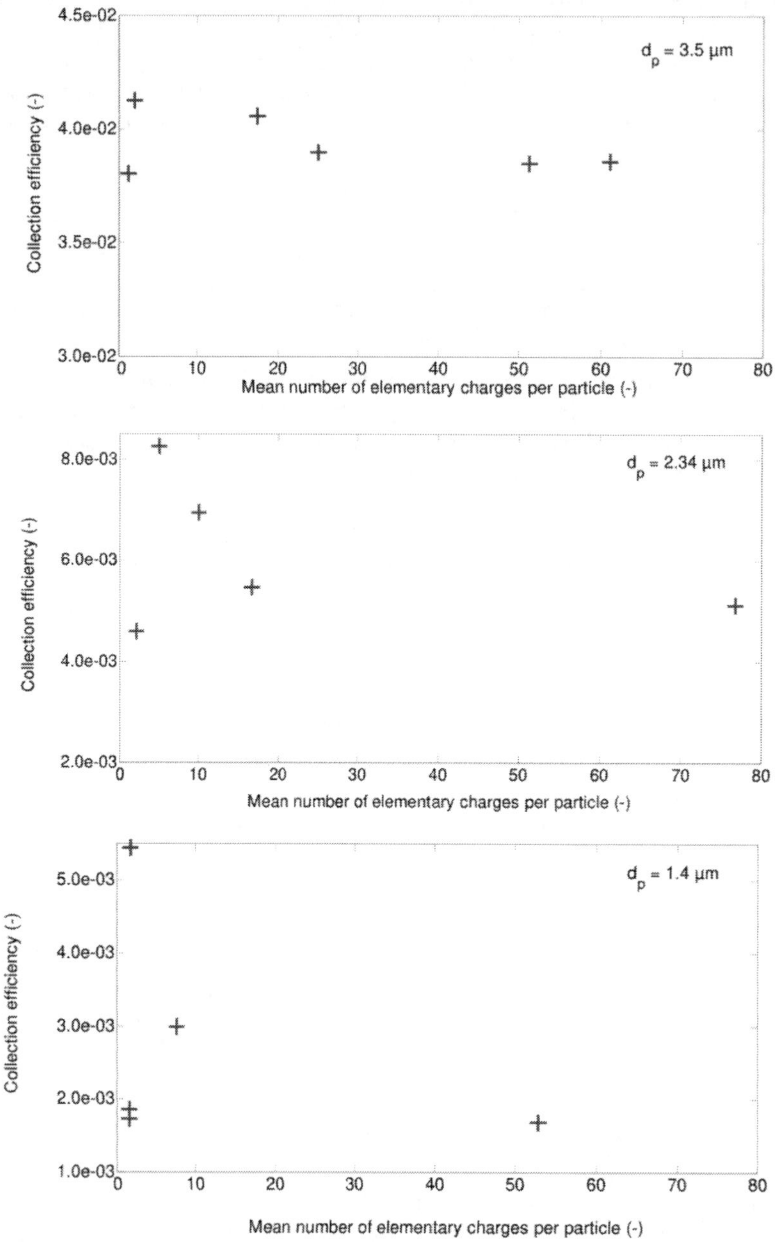

Figure 2 Measured washout efficiency vs number of elementary charges on the aerosols particles for 3 mean size diameters, for a drop diameter of 2.5 mm. For each datapoint 200 drops were collected and analyzed.

References

[1] Williams E R 2009 *Atmos. Res.* **91**(2) 259-271
[2] Barlow A K and Latham J 1983 *Q. J. R. Met. Soc.* **109**(462) 763-770
[3] Tinsley B A, Rohrbaugh R P, Hei M. and Beard K V 2000 *J. Atmos. Sci.* **57**(13) 2118-2134
[4] Gensdarmes F, Boulaud D and Renoux A 2001*J. Aerosol Sci.* **32**(12) 1437-1458
[5] Fuchs N A 1963 *Geofis. pura appli.* **56**(1) 185-193

Physico-chemical qualification of a universal portable sampler for aerosols and water-soluble gases

Jean-Maxime Roux[1], Roland Sarda-Estève[2]

[1] DTBS/CEA-Leti, Minatec Campus, 17 rue des Martyrs, 38054 Grenoble - France

[2] LSCE, CEA-Orme des Merisiers, 91191 Gif sur Yvette - France

Abstract. Developing a universal portable air sampler based on electrostatic precipitation. The challenge is to collect micro and nanoparticles, microorganisms as well as toxic molecules with a portable device. Electrostatic precipitation is an efficient and gentle method to collect airborne microorganisms and preserve their cultivability. But the collection of toxic gases required is not possible in such a device. The collection of such gases requires a liquid into which they have to be solubilized. Two concepts are being evaluated. The first one is based on electrospray. The goal is to investigate the collection efficiency of water-soluble gases. The second concept is based on the semi-humid electrostatic precipitator. Their high collection efficiencies for particles were already demonstrated. In the present study they are both tested with water-soluble gases. Concentrations are measured in the liquid solution by Ion Chromatography and in the gas phase by Proton Transfer Reaction Mass Spectrometry.

1. Introduction

As part of the overall security program for the fight against the spread of toxic agents (CBRN-E) the Center of Atomic Energy (CEA) is developing a universal portable air sampler based on electrostatic precipitation. The challenge is to collect micro and nanoparticles, microorganisms as well as toxic molecules with a portable device.

Electrostatic precipitation is an efficient and gentle method to collect airborne microorganisms and preserve their cultivability. But the collection of toxic gases required is not possible in such a device. The collection of such gases requires a liquid into which they have to be solubilized. Tepper et al. [1-2] showed that it is possible to collect particles and soluble gases with an electrospray based device. Later Elio et al. [3] used an electrospray to make a wind generator that is able to collect airborne particles while the collection electrode is continually rinsed. Another way of adding water into an electrostatic precipitator is to inject water vapor. Galbrun et al. [4] demonstrated the collection efficiency of such a device for nanoparticles. Following their work a fairly simple universal portable sampler for aerosols and water-soluble gases may be developed.

The two concepts shown on figure 1 are being evaluated. The first one is the configuration previously studied by Elio et al. [3] for the collection of particles. The goal is to investigate the collection efficiency of water-soluble gases. The second concept is based on the semi-humid electrostatic precipitator of Galbrun et al. [4]. Their high collection efficiencies for particles were already demonstrated. In the present study they are both tested for the collection of water-soluble gases. Concentrations are measured by ion chromatography in the liquid solution obtained by rinsing the collection electrode. Concentrations of volatile organic compounds (VOCs) are also measured in the gas outlet of the experimental device by Proton Transfer Reaction Mass Spectrometry.

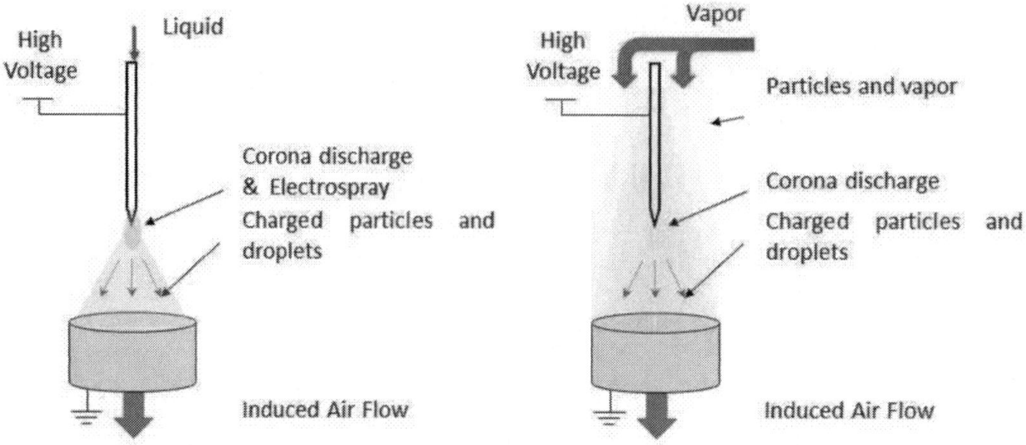

Figure 1. Schematic description of the two evaluated configurations: left, electrospray based design; right, electrostatic precipitation assisted by the presence of vapor.

2. Materials and methods

The two concepts presented above on Figure 1 are first tested with water-soluble gases injected in the device air inlet as shown on Figure 2. Two very different gases were chosen: hydrochloric acid (HCl), a water-soluble gas with a great affinity for water; methanesulfonic acid (MSA), a much more complex molecule with a reduced affinity for water compared to HCl. A volume of about 30 ml of both gases was slowly injected with a syringe into the air inlet. The gas was aspired by the device which air flow relies on the ionic wind between the discharging electrode and the collecting electrode [3]. Afterwards the collecting electrode was rinsed with pure water and the resulting solution was analyzed by ion chromatography (Dionex ICS 2000).

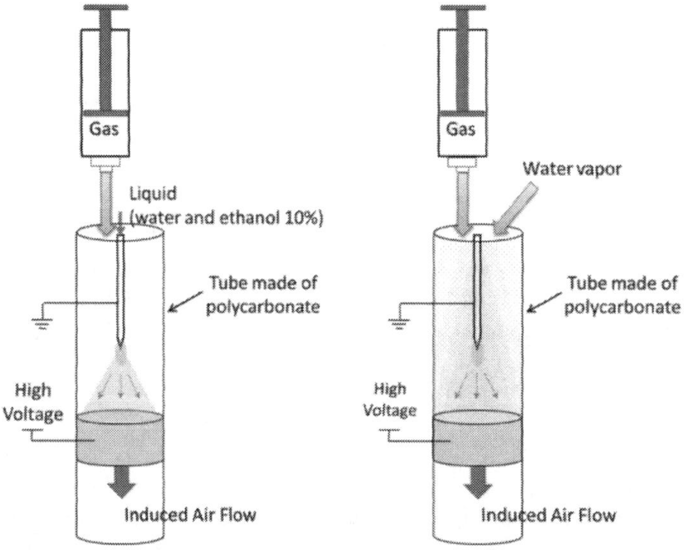

Figure 2. Schematic description of the experiments made to investigate the gas collection capacity: left, electrospray based design; right, electrostatic precipitation assisted by the presence of vapor.

When the device was used in the electrospray configuration, the liquid injected into the capillary was a mixture of water and ethanol (10%) [1] at a flow rate of about 10 µl/min. When the device was used in the vapor based configuration, water vapor was injected at 250 µl (liquid)/min.

Gases in the outlet were also analyzed by Proton Transfer Reaction Mass Spectrometry.

3. Results and discussion

Ion chromatography shows that a very high amount of chlore is present in the rinsing solutions when vapors of hydrochloric acid were injected. Also a high amount of methanesulfonic acid is present in the rinsing solutions when vapors of methanesulfonic acid were injected. These results, obtained in the two studied configurations, are summarized and shown on Figure 3. So adding an aqueous solvent by the mean of an electrospray or by the mean of an evaporator is therefore efficient to collect water-soluble gases.

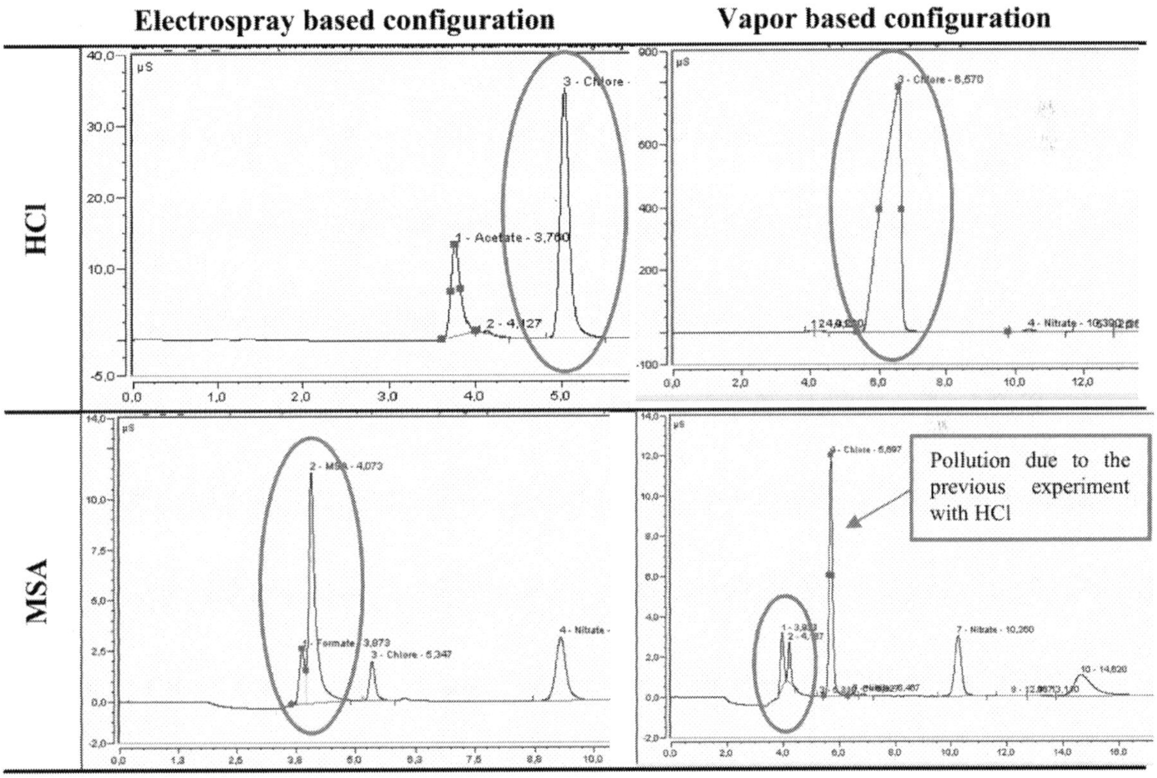

Figure 3. Concentrations of ions in the rinsing solution for each configuration and for gases tested in the experiments.

Next, the amount of volatile organic compounds was measured by Proton Transfer Reaction Mass Spectrometry in the gas outlet, when ambient air only was present around the experimental device. It revealed that methanol was produced in the electrospray based configuration. Also it revealed that acetone was produced when water vapour is injected in the electrostatic precipitator. These gases were not detected prior to the injection of ethanol (10%) or water vapour.

Ethanol was used by Tepper et al. [1]. It reduces the surface tension of aqueous solutions so that smaller droplets can be produced by electrospray based devices. Here the presence of methanol may be understood as a breakdown of the ethanol in the corona discharge. Since methanol is a toxic compound a new way of reducing the surface tension of water has to be found.

The presence of acetone when water vapour is injected in the electrostatic sampler is surprising. Traces of it can be found in ambient air but here it is due to an unknown chemical reaction in the semi-humid electrostatic sampler. It is also a toxic compound and its production must be avoided. This point is critical and further researches must be led to find a way to suppress the presence of this compound.

4. Conclusion

The present study shows that adding an aqueous solvent by the mean of an electrospray or by the mean of an evaporator is efficient to collect water-soluble gases. But, for both configurations, toxic volatile organic compounds were found in the gas outlet. It is therefore required to find a way to avoid the production of these compounds to be able to obtain an electrostatic sampler that is able to collect gases.

Acknowledgments

This work was supported by the CBRN-E R&D research program from CEA.

References

[1] Tepper G, Fenn J, Kessick R, Pestov D, Anderson J 2006 A New Air Sampler based on Electrically Charged Liquid Nanodroplets *Sixth IEEE Conference on Nanotechnology* 781 - 782

[2] Tepper G, Kessick R 2008 A study of ionization and collection efficiencies in electrospray-based electrostatic precipitators, *Journal of Aerosol Science* **39** 609-617

[3] Quinton E, Achard J L, Roux J M 2013 Ionic wind generator derived from a liquid filled capillary pin. Application to particle capture, *Journal of Electrostatics* **71** 963-969

[4] Galbrun E, Fouillet Y, Glière A, Achard J L, Rouillon F, Saint-Bonnet P, Peponnet C 2006 Development of a New Type of Aerosol Collector with yields Independent of Nanoparticle Size *Proceeding of the international Conference on Tracers and Tracing Methods* Autrans (France)

Modulation of urban atmospheric electric field measurements with the wind direction in Lisbon (Portugal)

H G Silva, [1,2] **J C Matthews,**[1] **R Conceição,**[2] **M D Wright,**[1] **S N Pereira,** [2] **A H Reis,**[2] **D E Shallcross**[2]

[1]Atmospheric Chemistry Research Group, University of Bristol, Cantock's Close, Bristol, BS8 1TS, UK
[2]Departamento de Física, ECT, Instituto de Ciências da Terra, IIFA, Universidade de Évora, Rua Romão Ramalho 59, 7002-554 Évora, Portugal

Email: hugo.silva@bristol.ac.uk

Abstract. Atmospheric electric field measurements (potential gradient, PG) were retrieved in the urban environment of the city of Lisbon (Portugal). The measurements were performed with a Benndorf electrograph at the Portela Meteorological station in the suburbs of the city (NE from the centre). The period of 1980 to 1990 is considered here. According to wind direction, different content and types of ions and aerosols arrive at the measurement site causing significant variations to the PG. To the south there are significant pollution sources while to the north such sources are scarcer . The Iberian Peninsula is found east of the station and the Atlantic Ocean covers the western sector, Wind directions are divided in four sectors: *i*) NW: $270^o \leq \theta \leq 360^o$; *ii*) NE: $0 \leq \theta \leq 90^o$; *iii*) SE: $90 \leq \theta \leq 180^o$; *iv*) SW: $180^o \leq \theta \leq 270^o$. Analysis of weekly cycle, caused by anthropogenic pollution related with urban activity, was undertaken for each wind sector. NW sector has been shown to be less affected by this cycle, which is attributed to the effect of marine air. The daily variation of NE sector for weekends reveals a similar behaviour to the Carnegie curve, which corresponds to a clean air daily variation of PG, following universal time, independent of measurement site.

1. Introduction

Surface atmospheric electric field (potential gradient, PG) measurements in urban environments are affected by anthropogenic action [1]. The main agents of these influences are pollutant aerosols from traffic, heating and industrial activity. According to the geographical location of pollutant sources around a measurement point, different winds bring different ion and aerosol contents; for that reason a significant modulation of PG is expected according to wind direction. Such modulation has been observed in different situations: high-voltage power-lines [2], traffic routes [3] and coastal regions [4].

The modulation caused by the wind direction can be of great importance as it varies the measurement conditions, for example, in an urban environment air with different pollutant content can reach the measurement site. The comprehension of such modulation can be achieved in long time series of PG, as in the case of Portela-Lisbon historical records [5]. This is because it enables the selection of a long time series measured with a characteristic pollution type, if the proper wind sectors are selected. Obviously, such possibilities are of huge value, for instance, in studies related to the Global Electric Circuit where pollution drastically perturbs any results and clean air is desirable.

For that reason, in the present work the historical PG time series of Portela-Lisbon is divided in four different wind sectors and the hypothesis of wind direction modulation is tested.

2. Data and methodology

PG was recorded with a Benndorf electrograph coupled to a radioactive probe installed at 1 m height at Portela meteorological station (suburbs of Lisbon, Portugal); more details can be found in [1,5]. Due to the nuclear fallout of the 1960's and 1970's, only the period from 1980 to 1990 is considered. Moreover, only non-negative values of PG are selected as negative ones often correspond to precipitation and cannot be considered. Otherwise, high positive PG values can occur as a consequence of pollution, as in case of fire. This phenomenon occurred during the historical Chiado's fire in Lisbon downtown [6]. Lisbon is situated on the north margin of the Tagus River and Portela station (close to Lisbon airport) is located to its northeast. Traffic and urban heating are the main causes of pollution in the city. In the 1980's the most significant pollution sources could be found in the south margin of the river (Setubal region). At the east of Lisbon the Tagus river basin and the Iberian Peninsula can be found, while the Atlantic Ocean is found to the west . Thus it is expected that winds from different directions should modulate PG as they are expected to contain different air contents. At Portela south winds bring pollution from the city mixed with the pollution from industry, west winds transport some pollution from the city mixed with marine aerosols, north winds carry less pollution mainly from the city activity and the east winds carry also continental aerosols.

To evaluate the effect of the anthropogenic pollution in each sector the procedure developed in [1] is used. It is based on the signature left by the weekly cycle of urban activity in air pollution and hence in PG. In [1] the PG was separated into workdays, Mondays to Fridays, $PG_{workdays}$ (WD) and weekends, Saturdays and Sundays, $PG_{weekends}$ (WK). It is expected that PG has higher values for workdays if compared with PG for the weekends, because workdays have more pollution. Thus, to evaluate the difference between workdays and weekends, the relative difference is used:

$$\Delta PG\,(\%) = (PG_{\text{workdays}}/PG_{\text{weekends}}-1)\times 100. \quad (1)$$

The comparison of the daily cycle for workdays and weekends is also very informative and such method is used here. Finally, the Lomb-Scargle Spectrum (LSS) technique is used to identify the presence and importance of the weekly cycle in each wind sector. This technique was used in [1] and described in detail there. The program used is an LSS implementation developed for MATLAB® [7].

3. Results and discussion

3.1. Boxplot analysis

Figure 1a shows the boxplots of PG values for each wind direction separated by workdays (WD) and weekends (WK), for clarity outliers are not represented. It is evident that: 1) NW sector shows little variation between workdays and weekends having a small relative difference of the medians for workdays, 70 V/m, and for weekends, 60 V/m, ΔPG ~17%, consistent with low levels of anthropogenic pollution; 2) NE sector presents more variability in the workdays if compared to the weekends. The median for workdays is 87 V/m and for weekends is 60 V/m, corresponding to ΔPG ~45%, which is a high value and is mainly due to the low median value for the weekends; 3) SE sector shows the higher variability of PG for workdays of all sectors, again the median for the weekends, 80 V/m, is lower than that of the workdays, 104 V/m, having ΔPG ~30%. This behavior shows evidence of a significant influence of anthropogenic pollution; 4) SW sector has a median PG value for the workdays of 98 V/m, and for the weekends, 80 V/m, i.e., ΔPG ~23%. These results show that all the sectors are affected by the weekly cycle; nevertheless, the NW tends to be the least affected. This feature can be interpreted as a consequence of lower pollution levels and the presence of marine air due to the western winds from the Atlantic Ocean. In fact, marine air is known to increase atmospheric conductivity [4] with the consequently reduction PG.

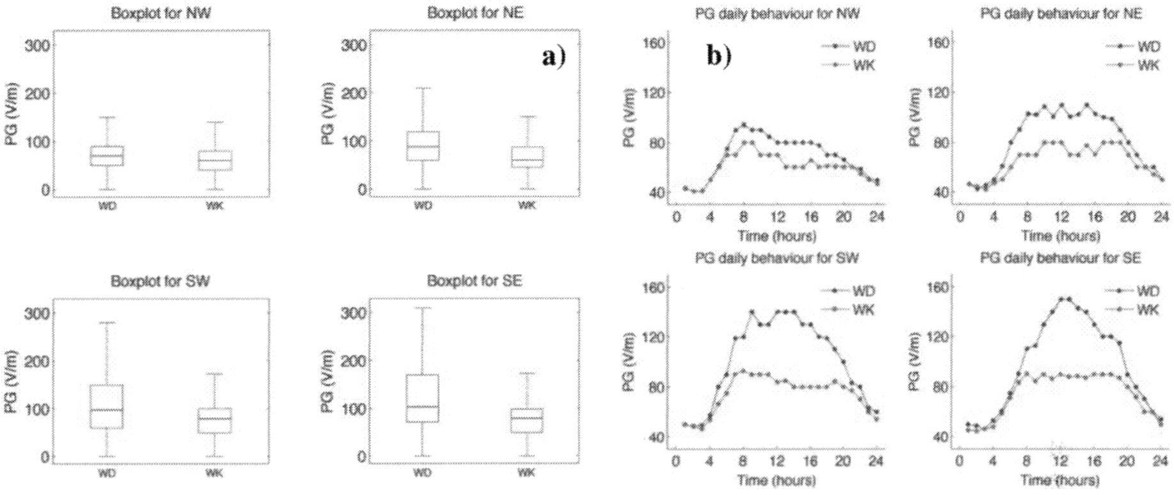

Figure 1 a) Boxplots of the PG for the different wind sectors divided in workdays (WD) and weekends (WK); b) Daily behaviour of the median PG values for 1980 to 1990 separated in workdays (WD) and weekends (WK).

3.2. Daily behaviour

Figure 1b presents the daily behaviour of the median PG values for the studied period and considering workdays and weekends in separate. On the one hand the northern sectors, NW and NE, show little difference between the daily behaviour of the workdays and the weekends. Actually, the NW sector shows the lowest values of the sectors. On the other hand the southern sectors show the most significant difference from the workdays to weekends, in particular, the SE and SW sectors show a prominent disparity in the period 8.00 a.m. – 8.00 p.m., the period of greatest anthropogenic activity. This aspect may be interpreted as the signature of pollution coming from industries south of Tagus River, which are most active in that period. Lastly, the daily variation of NE sector for weekends reveals a similar behavior to the Carnegie curve, the background 24 cycle of atmospheric electric field caused by variations in global thunderstorm activity [8].

3.3. Lomb-Scargle Spectra

The LSS for each of the four wind sectors are depicted in figure 2, where a clear modulation with wind direction can be observed. The NW and NE have higher slopes with n-exponents around, 0.61 and 0.50, correspondingly. The SW and SE have lower n-exponent values, 0.39 and 0.33, respectively. These lower values are compatible with measurements taken under more polluted air as pollution tends to induce spectral dispersion which flattens the spectra [1,9]. Moreover, the presence of the weekly cycle is observed in all wind sectors, less pronounced in the case of NW sector. This observation is compatible with low ΔPG found for this sector and could be another indication of the effect of marine air as it tends to increase the conductivity, in contrast to the effect of pollution which has the opposite effect. The NE sector shows a marked weekly cycle even though the pollution levels should be lower as compared with the sectors from the south. This can be an indication that pollution from the city, mainly traffic, should be the key pollution source in this sector. Otherwise, pollution from the south is a result of industrial activity and, in principle, less affected by the weekly cycle as many industries continue in activity at the weekends. Finally, all sectors, except the SE, clearly show, besides the daily cycle, a half-day cycle.

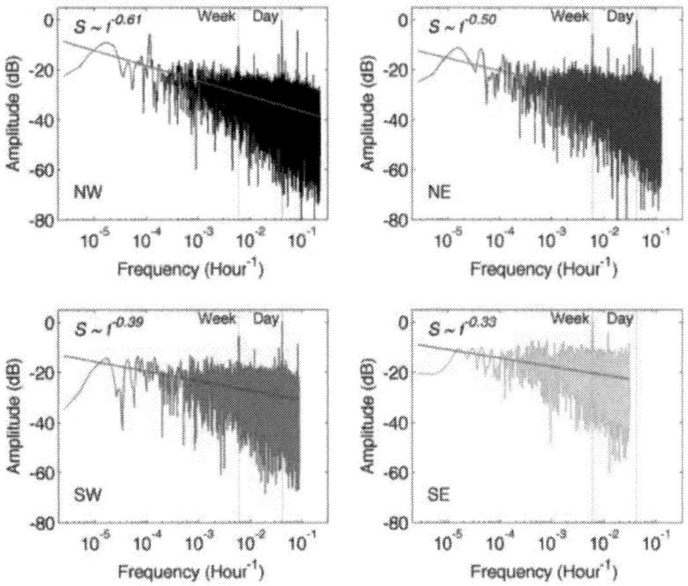

Figure 2. Lomb-Scargle Spectra corresponding to the four wind sectors. The following parameters were used *hifac*=1 (that defines the frequency limit as hifac times the average Nyquist frequency), *ofac*=4 (oversampling factor). This exponent, (n-exponent) is defined from the asymptotic spectral amplitude, S, with the frequency, f, usually written as $S \sim f^{-n}$.

4. Conclusion

A clear modulation of atmospheric electric field measurements with wind sector/direction is observed in the mean daily cycles. NW was found to be the sector less influenced, which has been attributed to the effect of more conductive marine air. The daily variation of NE sector for weekends reveals a similar behavior to the Carnegie curve. Finally, the effect of pollution of industries located to the south of Lisbon on PG is found to be strong in the period 8.00 a.m. – 8.pm.

Acknowledgments

HGS and SNP acknowledge the support of Science and Technology Foundation (FCT) for the Post-Doctoral fellowships, SFRH/BPD/63880/2009 and SFRH/BPD/81132/2011. The authors acknowledge the support from the FCT/FEDER-COMPETE project EAC (PTDC/GEO-FIQ/4178/2012) FCOMP-01-0124-FEDER-029197. The authors are grateful to Claudia Serrano, Samuel Bárias, and Doctor Mário Figueira in recording and preparing the data.

References

[1] Silva H G, Conceição R, Melgão M, Nicoll K, Mendes P B, Tlemçani M, Reis A H, Harrison R G 2014 *Env. Res. Let.* **9** 114025
[2] Matthews J C, Ward J P, Keitch P A, Henshaw D L 2010 *Atmos. Environ.* **44** 5093-100
[3] Israelsson S, Lelwala R 1999 *Atmos. Res.* **51** 301–307
[4] Wilding R J, Harrison RG 2005 *Atmos. Environ.* **39** 5876–5883
[5] Silva H G, Conceição R, Wright M D, Matthews J C, Pereira S N, Shallcross D E 2015 *Jour. Aero. Sci.* **85** 42
[6] Conceição R, Melgão M, Silva H G, Nicoll K, Harrison R G, Reis A H 2015 *Air Qua. Atmos. Health* **8**
[7] http://www.mathworks.com/matlabcentral/fileexchange/993-lombscargle-m
[8] Harrison R G 2013 *Surv. Geophys.* **34** 209–232
[9] Tchepel O, Borrego C 2010 *Jour. Environ. Monit.* **12** 544–55

Spectral response of atmospheric electric field measurements near AC high voltage power lines

H G Silva,[1,2] J C Matthews,[1] M D Wright,[1] D E Shallcross[1]

[1]Atmospheric Chemistry Research Group, University of Bristol, Cantock's Close, Bristol, BS8 1TS, UK
[2]Departamento de Física, ECT, Instituto de Ciências da Terra, IIFA, Universidade de Évora, Rua Romão Ramalho 59, 7002-554 Évora, Portugal

Email: hugo.silva@bristol.ac.uk

Abstract. To understand the influence of corona ion emission on the atmospheric electrical field, measurements were made near to two AC high voltage power lines. A JCI 131 field-mill recorded the atmospheric electric field over one year. Meteorological measurements were also taken. The data series is divided in four zones (dependent on wind direction): whole zones, Z_0; zone 1, Z_1; zone 2, Z_2; zone 3, Z_3. Z_3 is the least affected by corona ion emission and for that reason it is used as a reference against Z_1 and Z_2, which are strongly influenced by this phenomena. Analysis was undertaken for all weather days and dry days only. The Lomb-Scargle strategy developed for unevenly spaced time-series is used to calculate the spectral response of the aforementioned zones. Only frequencies above 1 minute are considered.

1. Introduction

The atmosphere is weakly conducting due to the natural presence of atmospheric air-ions, which are produced by cosmic ray ionisation and ground based radioactive decay (such as Radon) near the ground. Thunderstorm activity moves charge from the ground to the ionosphere, a conducting layer of the atmosphere, which dissipates back to ground through the atmosphere completing the Global Electric Circuit [1]. The difference in potential between the ionosphere and the ground (V_I) cause a Potential Gradient[1] (PG) that can be measured at the Earth's surface using electric field mills [2].

The Earth's PG can be modified by changes in air conductivity: reduction of air ions caused by increased aerosol concentration [3]; charged clouds overhead [4] and local sources of space charge, such as high voltage power lines (HVPL) [5]. HVPL can produce so-called 'corona ions' when the electric field surrounding the cable is large enough to cause acceleration of ions leading to corona avalanching. On a 50-Hz cable, these ions largely oscillate around the cable, but some can escape in a cross wind, both positive and negative ions can be released into the environment [6].

To measure the effect of HVPL on the electrical environment over a full year, a fixed site monitoring station (FSMS) was set up in South Gloucestershire, UK, to monitor changes in PG. PG was measured at 1 Hz using a JCI 131 electric field meter at 2 m height, and local weather conditions were monitored and recorded as 10 minute averages [5].

[1] PG is defined as dV_I/dz and the vertical component of atmospheric electric field is related with this variable through $E_z = -PG$.

Discontinuities on the line such as dirt or raindrops can enhance the field on the surface of the line and increase the amount of ions produced; hence weather conditions can significantly affect the amount of ions released by the line, with higher and more negative fields being seen in wet conditions [7]. If a predominance of one polarity of ions is observed then the ions carried overhead cause significant regions of space charge that can be observed as fluctuations of electric field at ground level. As the cloud of space charge is affected by local meteorology, the time of day also has an effect on the electric field, with fluctuating fields being more seen during the day [8].

2. Data analysis

Previous analysis of data from the FSMS have looked at 10 minute averages and SD of the PG, and information within the finer structure at 1 minute has not been considered yet. This analysis looks at the data average to 1 minute to investigate the finer structure and spectral content.

As with previous research [5, 7, 8], the analysis is undertaken on four different data sets dependent on wind direction: all data, i.e. whole zones (Z_0); zone 1 (Z_1) from SSW to WNW, 175 m downwind of 400 kV HVPL; zone 2 (Z_2) from ESE to S, 750 m downwind of 275 kV HVPL; zone 3 (Z_3) from NW to E, upwind of both HVPL. The last zone, Z_3, is the least affected by space charge produced by HVPL (as it is upwind the power-line) and is used as a reference for Z_1 and Z_2. A full discussion of the sources of aerosol and predominance of a particular weather by wind direction is given elsewhere [5] but it is important to note that stronger winds and more clouds come from the south west (influencing Z_1) than other zones and that there is a source of aerosol in the M48 motorway, also to the south west. Winds in Z_2 are the least present and, for that reason, less data is available for this zone and this will have consequences in the results shown below. Data from the FSMS were averaged into 1-minute samples and wind directions attributed accordingly (assuming that the wind add the same direction the 10 minutes before). Data were further divided into all weather and dry days. By dry days we mean days in which no rain was measured in our weather station. This is distinct from the usual 'fair weather days' encountered in studies regarding atmospheric electricity, as charged clouds may still be present.

The daily average was found for each zone to see the daily cycle. As the PG was been divided into 3 zones according to wind direction, and as some data was removed due to equipment failure, each of the sub-set of PG is a non-continuous time series, and therefore frequency analysis using typical Fourier analysis is not possible. The Lomb-Scargle Spectra (LSS) technique was developed for interrupted data sets in astrophysics [9, 10], and has been used in atmospheric electricity [3]. The LSS is similar to the Fourier transform, but it estimates the frequency spectrum based on a least squares fit of the sinusoids; the LSS spectrum converges to Fourier transform spectrum in the limit of evenly spaced observations. We use MATLAB to implement the LSS [11]. The following parameters were used *hifac*=1 (that defines the frequency limit as *hifac* times the average Nyquist frequency), *ofac*=4 (oversampling factor). It is worthwhile to mention that the resulting spectra depends on the amount of data used, for a times-series with many discontinuities the LSS will only have access to lower frequencies as it loses the high frequency resolution. This is the case of Z_2 time-series and for that reason less attention will be given to this zone.

3. Results and discussion

Figure 1a shows the daily average for zones Z_1, Z_2 and Z_3 at a 1-minute resolution. These data have previously been presented as 1 hour averages where the negative values of PG overnight in Z_1 were attributed to the typical increase in humidity increasing negative corona production [6]. However, the previous analyses had a much lower temporal resolution. Z_2 shows a very noisy behaviour oscillating from negative to positive, though the noise may be an artefact of lower sample numbers from this wind sector. Z_3, despite the noise, shows a 24 hour cycle similar to the Carnegie curve, the global background atmospheric electric field [1]. This is consistent with measurements in a typical rural background electric field profile. Figure 1b shows that the spectra for all zones. Z_1 and Z_3 have a peak at the daily cycle consistent with the daily behaviour shown in Figure 1a. There is not a clear signal for the weekly cycle in any of the zones, which is indicative of urban pollution. Nevertheless, Z_3 evidences a small half-day cycle characteristic of pollution from traffic [12] and could be a result of nearby roads. As usual in spectral analysis the *n*-exponent is determined for each zone (above the threshold frequency of 1 day^{-1}). This exponent is defined from the asymptotic spectral amplitude, *S*, with the frequency, *f*, usually written as $S \sim f^{-n}$ [3]. The values found are: ~1.54, ~1.05, ~0.48 and ~1.01 for Z_0, Z_1, Z_2, and Z_3, respectively. These are relatively similar values, as compared with ones for spectra calculated from hourly values [3]. It is observed that the values found for Z_1 (more affected by the HVPL) are similar to Z_3 (least affected by the HVPL). This is an indication that rainy weather masks the impact that HVPL have in the PG spectral response reducing its response to higher frequencies. This becomes clearer when considering the results from dry days below. The spectrum of Z_2 is much smaller than the other two, not because of a physical reason, but because this zone has less observations and the LSS algorithm cannot access lower frequencies, as mentioned above.

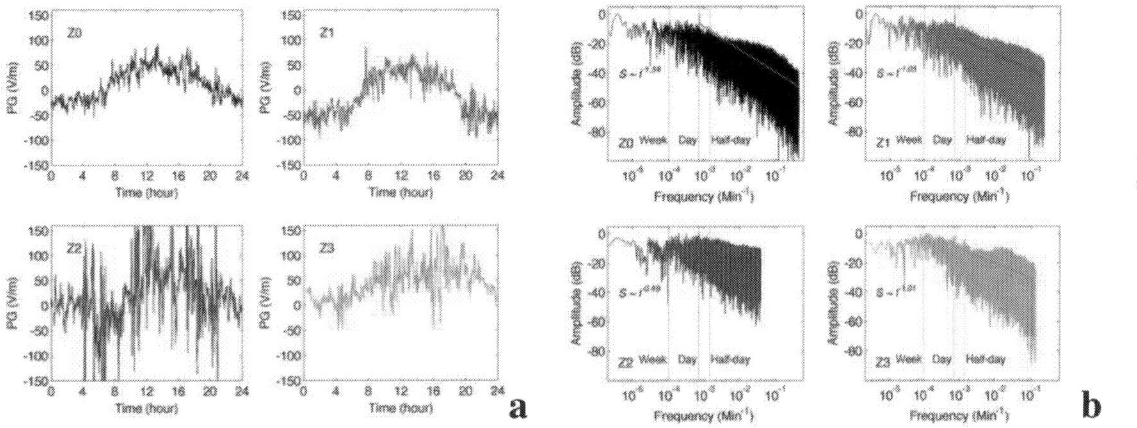

Figure 1. a) Daily behaviour for 1-minute average data in all weather conditions; b) Lomb-Scargle Spectra for all weather conditions. The grey lines represent the fit to the asymptotic spectral amplitude, *S*.

Figure 2a shows the daily average time series of the three zones but for dry days. The results are similar to those for all weather but exhibit less noise. This indicates that the increase in noise in Figure 1a can be attributed to the increase in PG fluctuations caused on rainy days [7]. Z_1 seems to present a clear feature every approximately half-hour. This periodicity appears in the LSS and can be tentatively attributed to the effect of HVPL. Z_2 shows oscillating/noisy negative-positive values from 00:00 until 08:00 returning to a normal behaviour after that. Z_3 shows a daily behaviour similar to the Carnegie curve [1] confirming that this is the zone least affected by the HVPL. However, a peak is observed around 22:00 in this zone that coincides with strong oscillations in Z_1, this may be due to extreme weather outliers. Figure 2b shows the LSS for dry days. The *n*-exponent values, in this case, are: ~0.78, ~0.81, ~0.87 and ~1.09 for Z_0, Z_1, Z_2, and Z_3, respectively. Zones 1 and 2 have similar values and could be an indication of HVPL effect, despite Z_2 having less data. This means that the HVPL would increase the PG response to higher frequencies with a slight tendency to flatten the spectra. This is a consequence of the dispersion of the spectral energy to higher frequencies and could be another result of HVPL activity. It should be noted that the *n*-exponent for Z_1 is reduced to values around ~0.8 while Z_3 shows more or less the same value compared with the spectra in Figure 1b. This could be an indication that rainy days mask the spectral dispersion as mentioned above. Z_1 exhibits an increase in amplitude at a half-hour periodicity that is not present in Z_3 and is suppressed in the all-weather Figure 1b. This could correspond to the features observed in Figure 2a.

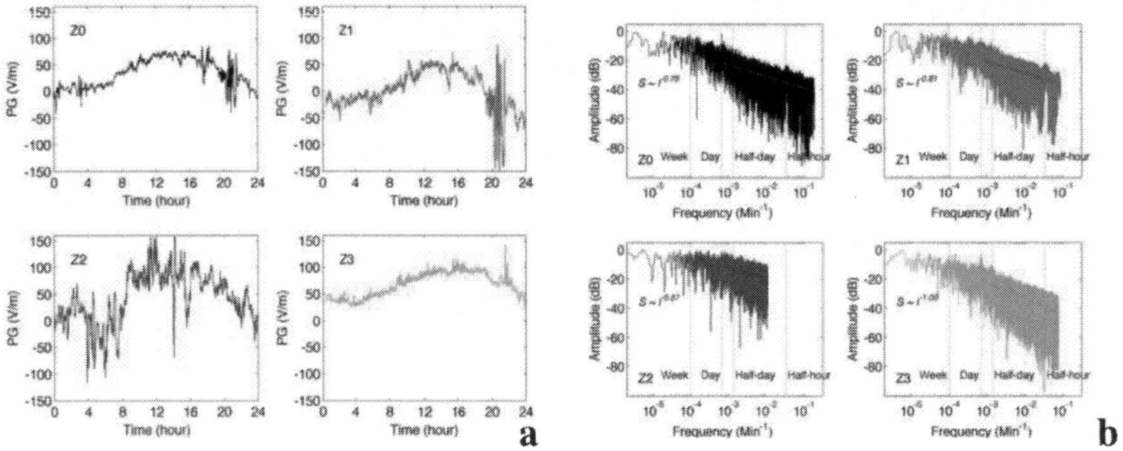

Figure 2. **a)** Daily behaviour for 1-minute average data for dry days only. b) Lomb-Scargle Spectra for dry days only. The grey lines represent the fit to the asymptotic spectral amplitude, *S*.

4. Conclusions

The effect of HVPL is studied with 1-minute resolution of PG measured in a FSMS. Several features are seen. Comparison between all weather and dry days only is made. All zones evidence a daily cycle for both conditions. It is seen that rainy days tend to increase noise in all zones and reduce the high frequency response of Z_1. There is evidence that rain masks the effect of HVPL has in the LSS of Z_1, while for dry days the presence of the HVPL tend to flatten the spectra promoting the dispersion of spectral energy for higher frequencies. The spectra for Z_3 is almost unaffected by weather conditions and present a half-day periodicity tentatively attributed to traffic pollution. Finally, a clear perturbation is observed in the daily behaviours of Z_1 and Z_3 for dry days only at around 22:00, which may be due to extreme weather outliers.

Acknowledgments

HGS acknowledges the support of Science and Technology Foundation (FCT) for the Post-Doctoral fellowship, SFRH/BPD/63880/2009 and the project EAC (PTDC/GEO-FIQ/4178/2012) FCOMP-01-0124-FEDER-029197. FSMS data were collected as part of a project funded by Children with Cancer UK, registered charity number 298405.

References

[1] Harrison R G 2013 *Surv. Geophys.* **34** 209–232
[2] Chubb J N and Butterworth G J 1979 Electrostatics 1979 *Inst. Phys. Conf. Ser.* **48** 85
[3] Silva H G, Conceição R, Melgão M, Nicoll K, Mendes P B, Tlemçani M, Reis A H and Harrison R G 2014 *Env. Res. Let.* **9** 114025
[4] Harrison R G 2011 *J. Phys: Conf. Ser.* **301** 012001
[5] Matthews J C, Ward J P, Keitch P A and Henshaw D L 2010 *Atmos. Environ.* **44** 5093
[6] Wright M D, Buckley A J, Matthews J C, Shallcross D E and Henshaw D L 2014 *Atmos. Environ.* **95** 296-304
[7] Matthews J C 2012 *J. Atmos. Solar-Terr. Phys.* **77**, 235-240
[8] Matthews J C 2012 *Atmos. Res.* **113** 68-79
[9] Lomb N R 1976 *Astrophys. Space Sci.* **39** 447-462
[10] Scargle J D 1982 *Astrophys. J.* **263** 835
[11] http://www.mathworks.com/matlabcentral/fileexchange/993-lombscargle-m
[12] Tchepel O and Borrego C 2010 *J. Environ. Monit.* **12** 544–55

Preliminary results on soil-emitted gamma radiation and its relation with the local atmospheric electric field at Amieira (Portugal)

F Lopes,[1,2] H G Silva, [2,3] S Bárias,[2] S M Barbosa[1,4]

[1] Departamento de Engenharia Geográfica. Geofísica e Energia, Instituto Dom Luiz, Faculdade de Ciências da Universidade de Lisboa, Rua Ernesto de Vasconselos, Campo Grande, Edf. C8, 1749-016 Lisboa, Portugal
[2] Departamento de Física, ECT, Instituto de Ciências da Terra, Universidade de Évora, Rua Romão Ramalho 59, 7002-554 Évora, Portugal
[3] Atmospheric Chemistry Research Group, University of Bristol, Cantock's Close, Bristol, BS8 1TS, UK
[4] INESC TEC - INESC Technology and Science, Campus da FEUP, Rua Dr Roberto Frias, 4200-465 Porto, Portugal

Email: fmlopes@fc.ul.pt

Abstract. The atmospheric electric field near the Earth's surface is dominated by atmospheric pollutants and natural radioactivity, with the latter directly linked to radon (^{222}Rn) gas. For a better comprehension on the temporal variability of both the atmospheric electric field and the radon concentration and its relation with local atmospheric variables, simultaneous measurements of soil-emitted gamma radiation and potential gradient (defined from the vertical component of the atmospheric electric field) were taken every minute, along with local meteorological parameters (e.g., temperature, atmospheric pressure, relative humidity and daily solar radiation). The study region is Amieira, part of the Alqueva lake in Alentejo Portugal, where an interdisciplinary meteorological campaign, ALEX2014, took place from June to August 2014. Soil gamma radiation is more sensitive to small concentrations of radon as compared with alpha particles measurements, for that reason it is more suited for sites with low radon levels, as expected in this case. Preliminary results are presented here: statistical and spectral analysis show that i) the potential gradient has a stronger daily cycle as compared with the gamma radiation, ii) most of the energy of the gamma signal is concentrated in the low frequencies (close to 0), contrary to the potential gradient that has most of the energy in frequency 1 (daily cycle) and iii) a short-term relation between gamma radiation and the potential gradient has not been found. Future work and plans are also discussed.

Keywords: atmospheric electric field, ^{222}Rn, air ionization

1. Introduction

Atmospheric electrical conductivity is mainly caused by air ions [1], as they are responsible for most of the electric charge transport in the surface atmospheric electric field. Typically, the Potential Gradient (PG) is used as a measuring means to observe such variations in the atmospheric electric field, which is defined to be positive in fair-weather (FW). The two major contributors for air ions in the lower troposphere are cosmic rays (~20 %) and natural radioactivity [2], with the latter being directly linked to radon (^{222}Rn) exhalation and mixing processes above soil surface. With an half-life of 3.82 days, radon can move from the solid grains into the air or water-filled pores of the media and further migrate via diffusion and/or advection to the atmosphere. Once in the atmosphere, the radon decay products induce air ionization through the emission of alpha particles, allowing the increase of air ions concentration and of the atmospheric electric conductivity; this causes a decrease in the local PG [3-4], while aerosols and water-droplets are the two main causes for air ions suppression and, consequently, tend to increase the PG.

In order to have a better comprehension of the temporal variability of PG and soil-emitted gamma radiation time series, the present work addresses a preliminary study on measurements of both parameters at a local scale.

2. Data and methodology

The region of study is Amieira, which is part of the largest artificial lake in West Europe as the result of the Alqueva dam, in the countryside of Portugal (Alentejo). A set of particular local features includes low pollution levels and a stable geological activity (an inactive seismic fault exists beneath the dam). All the presented data were acquired in an inter-disciplinary meteorological campaign ALEX2014, during an observational period of 3 months (June to August 2014). The campaign main focus was the study of the possible impact of the Alqueva reservoir in the local environment. Further details will be found in [5].

To have a better understanding of the daily variability in the gamma radiation and PG time series, Figure1 depicts the hourly medians of the original 1-minute time series only under FW conditions (a total of 27 days) that occurred during the whole period of observation. FW days are selected based on locally undisturbed daily solar radiation curves (not shown). This is because meteorological events like shower-clouds, fogs, strong winds and thunderstorms (which happened between 23 and 25 June) perturb dramatically the PG and shadow any possible relation with radon. In addition, one FW day, 13th June, was removed from the analysis as a PG outlier. For spectral analysis, it was used the Lomb-Scargle periodogram (LSP), which is developed for time series with irregular sampling intervals, as it is the case of the PG restricted to FW days. For that reason, LSP has been used in other PG studies [6-7] For a better view of the daily variability, a scatter plot between the daily medians of both variables is obtained along with their respective horizontal and vertical error bars (based on the Q1 and Q3 quartiles of each daily distribution).

3. Analysis and discussion

On one hand, the gamma time series shows high variability at different time scales with irregular daily cycles. This is expected since such type of complex behavior has also been found on a wide range of other locations and under diverse environmental conditions [8-11]. These patterns may likely result from nonlinear interaction of local environmental and meteorological effects whose interaction is still difficult to assess and not well known [12]. On the other hand, using only the FW days, the PG has a well-defined daily cycle (not observed in the gamma time series). This suggests that there does not seem to exist a direct relation between gamma and PG at the daily scale.

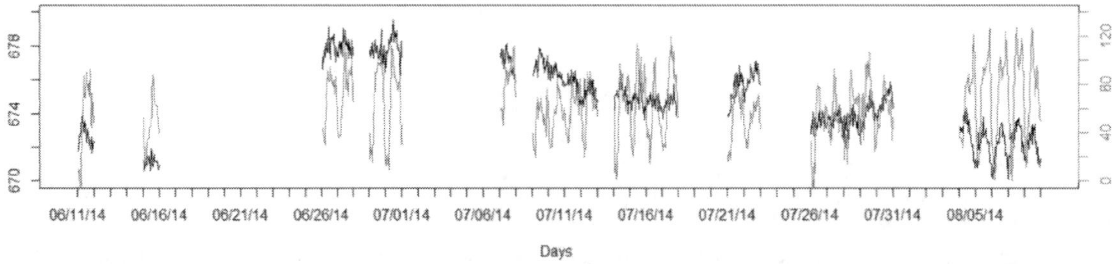

Figure 1. Hourly medians for the gamma (black) in counts/100 and the PG (red) in V/m. Only FW days, 11th June to 8th August 2014.

In fact, a closer look into the daily variability for the gamma and PG hourly time series is depicted in Figure 2. Here, the daily hourly medians are calculated by applying medians to each hour of the day for all the FW days. The distribution is uniform for the gamma (Figure 2, left), whereas the corresponding daily maximum and minimum is difficult to assess, but they seem to occur during the afternoon and during the night, respectively. As opposed to these, the PG hourly medians (Figure 2, right) show well-distributed daily pattern with the daily minimum in the early morning and daily maximum in the middle of the day, similar to the Carnegie behavior [13].

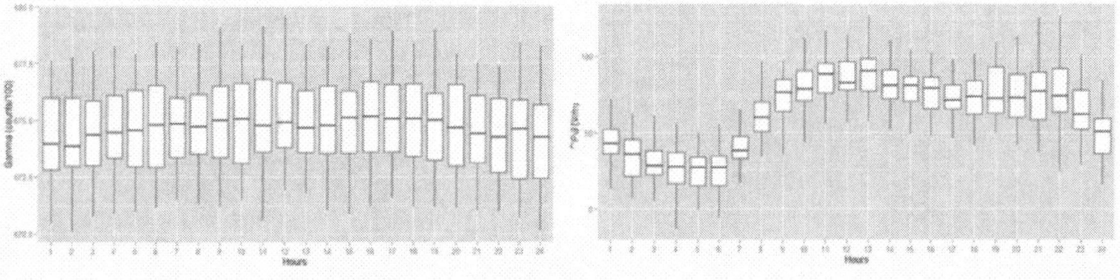

Figure 2. Boxplots of hourly medians for the Gamma (left) and PG (right) daily cycles. Only FW days, 11th June to 8th August 2014.

The energy of the depicted cycles can be seen in detail in Figure 3, where the LSP is applied to both series. Gamma radiation shows higher energies in the high period range (above 1 day) and reveals a flat response to periods below half-day. It is also observed a very weak daily period. This behavior is opposed to the PG case. The PG has a notorious and energetic daily peak, as expected [6]. For periods above 1 day the spectrum tends to flatten. A semi-diurnal cycle is also observed in the PG data, with a clear second energetic peak for period of 0.5 (two cycles per day) and a decreasing trend below this period. The cycles observed in the PG are a result of the influence of the global atmospheric electrical circuit [13] and local influences, while the physical mechanisms behind the gamma cycles are conventionally ascribed to meteorological factors such as barometric pressure and temperature [10]. It is evident that there is no spectral relation between the gamma and the PG data.

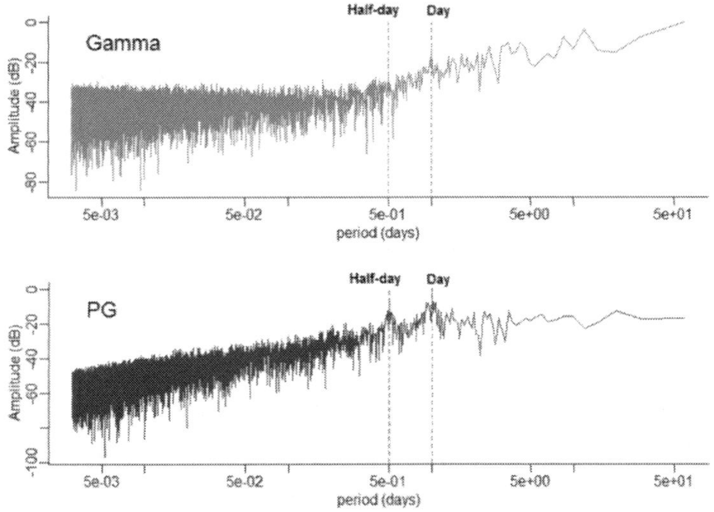

Figure 3. Lomb-Scargle Periodograms for the Gamma (top) and PG (bottom) hourly time series. Only FW days, 11th June to 8th August 2014.

Furthermore, a scatterplot (Figure 4) illustrates a uniform distribution of the daily medians (obtained through the respective hourly medians, in order to reduce the high frequencies observed previously) of the PG and gamma radiation, indicating that there is no clear correlation between variables in the present data. The high daily distribution of the PG is reflected by the vertical error bars (in grey) length that are longer than the gamma error bars (in black) which show almost the same length due to the very low change of radon concentration during the day.

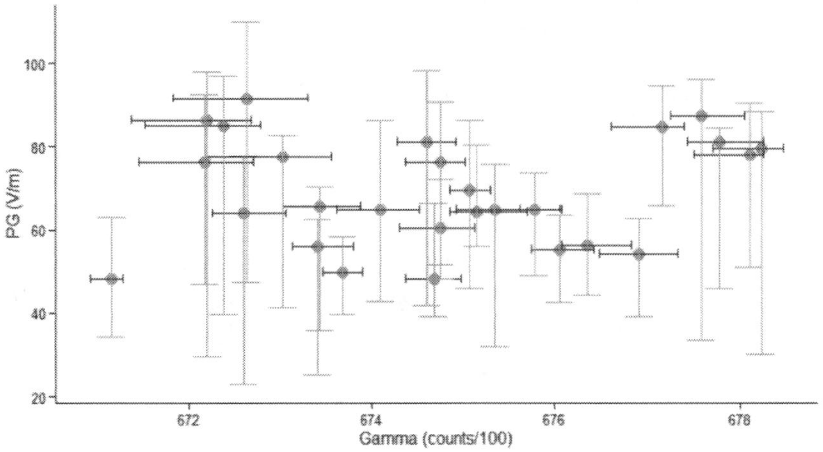

Figure 4. Scatterplot for the Gamma and PG daily medians together with their respective error bars. Grey error bars are for PG and black error bars are for the gamma. Only FW days, 11th June to 8th August 2014.

4. Final remarks

Gamma radiation showed low frequency variability contrary to PG. The results of this study point towards no obvious short-term relation between soil-emitted gamma radiation and the local potential gradient at the Amieira site. This necessarily implies future work such as performing new measurements in radon rich regions (e.g., granitic areas), acquisition of longer recording periods for long-term tendencies, measurement of airborne radon levels (which should have higher variability during the day than soil radon concentration), and measurement of air ions.

Acknowledgments

Experiments were accomplished during the campaign funded by FCT (Portuguese Science and Technology Foundation) and FEDER-COMPETE: ALEX 2014 (EXPL/GEO-MET/1422/2013) FCOMP-01-0124-FEDER-041840. The authors acknowledge the support from the projects: RADON (PTDC/CTEGIX/110325/2009); EAC (PTDC/GEO-FIQ/4178/2012) FCOMP-01-0124-FEDER-029197. H.G. Silva is grateful for the FCT post-doc fellowship: SFRH/BPD/63880/2009.

References

[1] Wright M D, Holden N K, Shallcross D E and Henshaw D L 2014 *J. Geophys. Res.* **119**
[2] Kolarz P M, Filipovi D M and Marinkovi B P 2009 *Appl. Radiat. Isot.* **67** 2062–2067
[3] Silva H G *et al* 2011 *Nat. Hazards. Earth Syst. Sci.* **11** 978-991
[4] Harrison R G, Aplin K L and Rycroft M J 2010 *J. Atmos Solar-Terr. Phys.* **72(5–6)** 376–381
[5] Salgado R *et al* 2015 The Alqueva Hydro-Meteorological Experiment (ALEX 2014) (*in preparation*)
[6] Silva H G, Conceição R, Melgão M, Nicoll K, Mendes P B, Tlemçani M, Reis A H and Harrison R G 2014 *Environ. Res. Lett.* **9** 114025
[7] Harrison R G and Märcz F 2007 *Geophys. Res. Lett.* **34** L23816
[8] Steinitz G, Piatibratova O and Barbosa S M 2007 *J. Geophys. Res.* **112** B10211
[9] Barbosa S M, Zafrir H, Malik U and Piatibratova O. 2010 *Geophys. J. Int.* **182** 829-842
[10] Zafrir H, Barbosa S M, and Malik U 2013 *Radiat. Meas.* **49** 39-56.
[11] Barbosa S M, Lopes F, Correia A D, Barbosa S, Pereira A C and Neves LF 2015 *J. Environ. Radioactivity.* **142** 14-23
[12] Barbosa S M, Steinitz G, Malik U, Piatibratova O, Silva M E and Lago P 2007 *Geophys. Res. Lett.* **34** L15309
[13] Harrison R G 2013 *Surv. Geophys.* **34(2)** 209-232

Exact closed-form solution for the electrostatic interaction of two equal-sized charged conducting spheres

Shubho Banerjee and Mason Levy
Physics Department, Rhodes College, Memphis TN 38112, USA
E-mail: banerjees@rhodes.edu

Abstract. We provide an exact closed-form solution for the electrostatic interaction of two equal-sized conducting spheres. We calculate the capacitance coefficients for the spheres in terms of the q-analogue of the digamma function. In the near limit, when the two spheres are about to touch, the closed-form exact solutions allow for much faster numerical calculations than the well-known infinite series solutions. By analyzing the exact solution in the near limit, we provide Taylor series expressions for the capacitance coefficients in terms of the surface-to-surface separation of the two spheres.

1. Introduction

The two sphere electrostatic interaction was originally analyzed by Kirchoff, Lord Kelvin, Poisson, Maxwell, and Russell [1, 2, 3] and continues to generate interest in the electrostatics community even today [4, 5, 6]. Here, we study the special case of equal-sized spheres.

The setup of the two equal sphere problem is shown in figure 1. Two equal conducting spheres A and B of radii a and b (equal to a) respectively are separated by distance s between their centers. Sphere A is held at potential V_a and sphere B at potential V_b. The capacitance coefficients C_{aa}, C_{ab}, C_{ba}, and C_{bb} of the two spheres are defined through the relations

$$\begin{aligned} Q_a &\equiv C_{aa}V_a + C_{ab}V_b, \\ Q_b &\equiv C_{ba}V_a + C_{bb}V_b, \end{aligned} \tag{1}$$

where Q_a and Q_b are the net charges on the two spheres. The coefficients $C_{ab} = C_{ba}$ are always equal. For the special case of equal-sized spheres we have $C_{aa} = C_{bb}$.

The solution for the capacitance coefficients of the two spheres can be calculated using the method of images [7, 8]. By summing up the images charges inside each sphere and substituting $a/s = \frac{1}{2}\mathrm{sech}\,\alpha$ the capacitance coefficients defined above can be written as [8]

$$\begin{aligned} c_{aa} &= \frac{C_{aa}}{4\pi\epsilon a} = \sinh\alpha \sum_{n=0}^{\infty} \mathrm{csch}\,(2n+1)\alpha, \\ c_{ab} &= \frac{C_{ab}}{4\pi\epsilon a} = -\sinh\alpha \sum_{n=1}^{\infty} \mathrm{csch}\,2n\alpha. \end{aligned} \tag{2}$$

The series in equation (2) were numerically analyzed by Pisler and Adhikari [9].

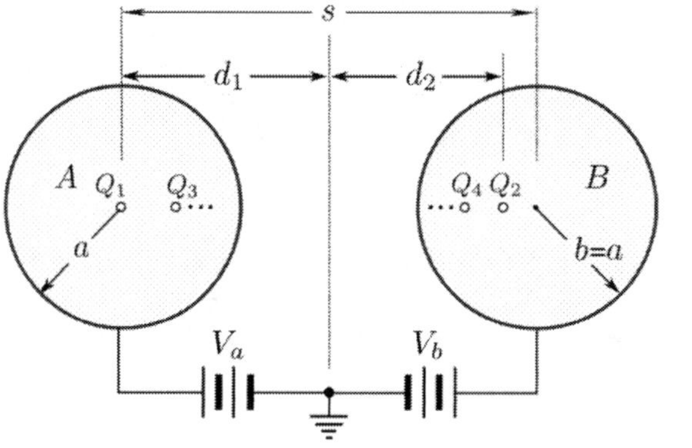

Figure 1. Two equal-sized conducting spheres are at center-to-center distance s from each other. Sphere A is maintained at voltage V_a and sphere B at voltage V_b. The image charges $Q_1, Q_2, Q_3, Q_4, \ldots$ shown are for the special case when B is grounded. The odd numbered images appear inside A and the even numbered images appear inside B.

2. Capacitance coefficients in closed form

Using the substitution $e^{-\alpha} = x$ in equation (2) and noting that $\sinh n\alpha = (e^{n\alpha} - e^{-n\alpha})/2$, the series can be rewritten as

$$c_{aa} = \left(\frac{1-x^2}{x}\right)\sum_{n=0}^{\infty}\frac{x^{2n+1}}{1-x^{4n+2}}, \tag{3}$$

$$c_{ab} = -\left(\frac{1-x^2}{x}\right)\sum_{n=1}^{\infty}\frac{x^{2n}}{1-x^{4n}}, \tag{4}$$

where the variable

$$x = \frac{s-\sqrt{s^2-4a^2}}{2a}. \tag{5}$$

In the limit of large separation between the sphere and the plane i.e. $s \to \infty$, x approaches 0, and in the limit of small separation between the sphere and the plane i.e. $s \to 2a$, we have $x \to 1$. Writing the capacitance coefficients as a series in x allows us to reexpress equations (3, 4) as

$$c_{aa} = \left(\frac{1-x^2}{x}\right)\left[\sum_{n=1}^{\infty}\frac{x^n}{1-x^n} - 2\sum_{n=1}^{\infty}\frac{x^{2n}}{1-x^{2n}} + \sum_{n=1}^{\infty}\frac{x^{4n}}{1-x^{4n}}\right], \tag{6}$$

$$c_{ab} = -\left(\frac{1-x^2}{x}\right)\left[\sum_{n=1}^{\infty}\frac{x^{2n}}{1-x^{2n}} - \sum_{n=1}^{\infty}\frac{x^{4n}}{1-x^{4n}}\right]. \tag{7}$$

Noting that the expressions within the braces above are linear combinations of Lambert series in x, x^2, and x^4 [10] , we can write the capacitance coefficients as

$$c_{aa} = \left(\frac{1-x^2}{x}\right)\left[L(x) - 2L(x^2) + L(x^4)\right], \tag{8}$$

$$c_{ab} = -\left(\frac{1-x^2}{x}\right)\left[L(x^2) - L(x^4)\right]. \tag{9}$$

The Lambert series $L(x)$ can be expressed in closed form as

$$L(x) \equiv \sum_{n=1}^{\infty} \frac{x^n}{1-x^n} = \frac{\psi_x(1) + \ln(1-x)}{\ln x}, \tag{10}$$

where $\psi_x(1)$ is the q-digamma function [11], $\psi_q(z)$, evaluated at $q = x$ and $z = 1$. For any $0 < q < 1$ and $z > 0$ the q-digamma function can be written as [12]

$$\begin{aligned} \psi_q(z) &= -\ln(1-q) + \ln q \sum_{n=0}^{\infty} \frac{q^{n+z}}{1-q^{n+z}}, &(11)\\ &= -\ln(1-q) + \ln q \sum_{n=1}^{\infty} \frac{q^{nz}}{1-q^{n}}. &(12) \end{aligned}$$

Using equation (10) the capacitance coefficients for the two sphere system can be expressed as

$$\begin{aligned} c_{aa} &= \left(\frac{1-x^2}{x}\right) \frac{4\psi_x(1) - 4\psi_{x^2}(1) + \psi_{x^4}(1) - 4\tanh^{-1}x + 2\tanh^{-1}x^2}{4\ln x}, &(13)\\ c_{ab} &= -\left(\frac{1-x^2}{x}\right) \frac{2\psi_{x^2}(1) - \psi_{x^4}(1) - 2\tanh^{-1}x^2}{4\ln x}. &(14) \end{aligned}$$

These new closed-form results exactly reproduce the numerical calculations of the infinite series solutions in equation (2) by Pisler and Adhikari (see Table I in reference [9]).

3. Capacitance coefficients in the near limit

When the separation between the spheres is relatively small it is convenient to express their surface-to-surface separation in dimensionless form as

$$\xi \equiv \frac{s-2a}{2a} = \frac{(1-x)^2}{2x}. \tag{15}$$

As the spheres are about to touch we have $\xi \to 0$ and $x \to 1$. Each term in the series in equations (6, 7) series diverges making the analysis in this limit difficult. Having the capacitances in closed form allows us to expand them in terms of ξ with $\theta = \frac{1}{4}\ln\frac{8e^{2\gamma}}{\xi}$ to obtain

$$\begin{aligned} c_{aa} &= \theta + \frac{\xi}{3}\left(\theta + \frac{1}{24}\right) - \frac{\xi^2}{45}\left(\theta + \frac{21}{160}\right) + \frac{\xi^3}{189}\left(\theta - \frac{1171}{2520}\right) &(16)\\ &- \frac{23\xi^4}{14175}\left(\theta + \frac{1402133}{1854720}\right) + \frac{263\xi^5}{467775}\left(\theta - \frac{909776807}{233291520}\right) + \dots, \end{aligned}$$

and with $\phi = \frac{1}{4}\ln\frac{e^{2\gamma}}{2\xi}$ to obtain

$$\begin{aligned} c_{ab} &= -\phi - \frac{\xi}{3}\left(\phi + \frac{7}{24}\right) + \frac{\xi^2}{45}\left(\phi - \frac{149}{160}\right) - \frac{\xi^3}{189}\left(\phi - \frac{541}{2520}\right) &(17)\\ &+ \frac{23\xi^4}{14175}\left(\phi - \frac{2619157}{1854720}\right) - \frac{263\xi^5}{467775}\left(\phi + \frac{762619303}{233291520}\right) + \dots, \end{aligned}$$

where $\gamma = 0.577216...$ is the Euler's constant. The expansions in equations (16) and (17) indicate a logarithmic divergence of the capacitance coefficients and agree with Russell's results [3].

Along with the closed-form expressions in equations (13) and (14), the series expansions in equations (16) and (17) are the main results presented in this paper. The logarithmic dependence of capacitance in the near regime is well known but expansions are available to only to order ξ^2 [3, 4]. Having the the capacitances in closed form allows us to expand the capacitance to higher orders through the analysis of the q-digamma function.

In figure 2 the capacitances in equations (13) and (14) are plotted against the relative sphere separation. Also plotted with dashed lines are the series expansions in equations (16) and (17) terminated at order ξ^5. At $\xi = 1$ (i.e. $s = 4a$), the terminated series for c_{aa} and c_{ab} are accurate to within 0.31% and 1.25% respectively.

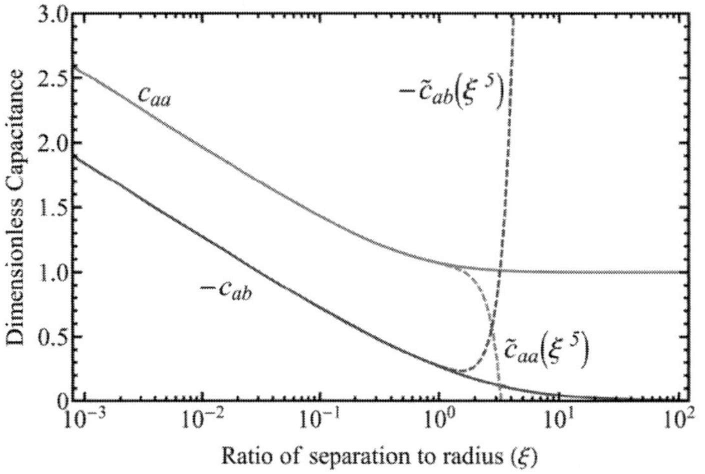

Figure 2. The capacitance coefficients c_{aa} and c_{ab} in equations (13, 14) are plotted against the relative separation between spheres. Also plotted with dashed lines are $\tilde{c}_{aa}(\xi^5)$ and $\tilde{c}_{ab}(\xi^5)$, the series expressions for capacitances in equations (16, 17) terminated at the ξ^5 term.

Acknowledgement

We thank Dr. Charles Robertson and Van Vleet Chair Support at Rhodes College for funding provided for this project.

References

[1] Thompson W 1872 *Reprint of Papers on Electrostatics and Magnetism* (London: Mcmillan) pp 86-97
[2] Maxwell J C 1954 *A Treatise on Electricity and Magnetism* 3rd ed vol 1 (New York: Dover) pp 273-276
[3] Russell A May 1927 *J. I. Electr. Eng.* **65** 517 (and references therein)
[4] Lekner J 2011 *J. Electrostat.* **69** 11-14
[5] Lekner J 2012 *Proc. R. Soc. London* A **468** 2829
[6] Murovec T and Brosseau C 2014 *Appl. Phys. Lett.* **105** 054101
[7] Pugh E M and Pugh E W 1960 *Principles of Electricity and Magnetism* (Reading: Addison-Wesley) pp 106-108
[8] Smythe W R 1968 *Static and Dynamic Electricity* 3rd ed (New York: McGraw-Hill) pp 128-132
[9] Pisler E and Adhikari T 1970 *Phys. Scr.* **2** 81-83
[10] Borwein J M and Borwein P B 1987 *Pi and the AGM: A Study in Analytic Number Theory and Computational Complexity* (New York: Wiley) pp 91-101
[11] Krattenthaler C, Srivastava H M 1996 *Comput. Math. Appl.* **32** 73-91
[12] Mansour M and Shabani A S 2009 *J. Inequal. Appl.* **10**(1) Art. 12

Simulations of the Global Electrical Circuit coupled to local Potential Gradient measurements

R Conceição, H G Silva

Departamento de Física, ECT, Instituto de Ciências da Terra, IIFA, Universidade de Évora, Rua Romão Ramalho 59, 7002-554 Évora, Portugal

Email: rfc@uevora.pt

Abstract. There are several models describing the Global Electric Circuit of the Earth's atmosphere. Here it is used the common model and parameters of Global Electric Circuit to couple it with a local circuit less studied in literature. The first objective is to test different voltage sources describing thunderstorm activity and compare the output, Potential Gradient, with the known *Carnegie Curve*. Two sets of parameters are used, the first one from values found in literature and the second one from values tweaked to get the best agreement between the simulated Potential Gradient and the *Carnegie Curve*. This study is a first step in simulations regarding the coupling of the Global Electric Circuit (primary) to local electric circuit (secondary). One of the main objectives is to estimate the aerosol load on the local resistor in case of aerosol events, e.g. fires.

1. Introduction

The existence of a Global Electric Circuit (GEC) was first recognised by the observation of the so called *Carnegie curve* based on a global daily variation of the surface Potential Gradient (PG) aboard of the Carnegie cruises [1]. For that reason, different models have been elaborated to understand the GEC proprieties [3, 4] and its relation with climate [5]. Nevertheless, scarce attempts have been made to perform simulations coupling the GEC primary circuit to a secondary circuit describing local PG measurements. Such simulations are of considerable interest because, for instance, in polluted regions the daily variation of PG differs drastically from the *Carnegie curve* [6]. Since it is expected that GEC would, in principle, impose a similar global daily variation, a deviation from the *Carnegie curve* could only be a result from local variations of the electrical components defining the secondary local circuit. If the proposed simulations were successful they would enable the separation of the global effects from local ones on real PG data. It would allow, for example, the estimation of the electric resistance load caused by atmospheric pollution from PG measurements in polluted environments like large cities [6] and severe pollution events like fires [7].

In this work, a method is described to couple a local circuit to the global one describing GEC. It is assumed that changes in the local circuit would not affect considerably the GEC. The model is presented and adjusted to reproduce the *Carnegie curve*.

2. Global Electric Circuit Simulations

Two main parts compose the electric circuit considered here. The first part englobes the components defining the primary circuit corresponding to GEC: V_s corresponds to the voltage generated by thunderstorm daily activity, R_s is the resistor associated with the thunderstorms region, R_{FW} is the resistor corresponding to fair-weather (FW) regions which closes the circuit in parallel with the Ionosphere-Earth (IE) capacitor, C_{IE}. This primary (global) circuit is coupled to a secondary circuit; which corresponds to the local circuit where the PG measurements take place. It is composed by R_{FT}, the resistor of the free-troposphere (FT) and R_{BL}, the resistor corresponding of the planetary boundary layer (BL). To account for space-charge density accumulation below the boundary layer [2] a capacitor, C_{BL}, is inserted in parallel with the R_{BL}. The diagram of the circuit is presented in Figure 1.

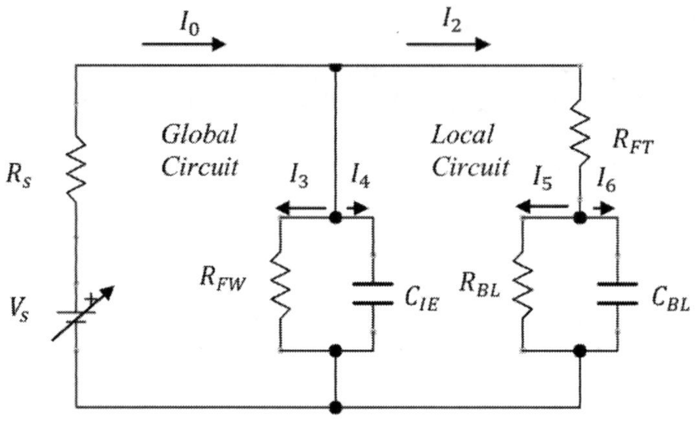

Figure 1 Circuit model and corresponding parameters.

2.1. Circuit equations

Using Kirchhoff's laws on the circuit of Figure 1 and the notation $I_n = \dot{Q}_n$ (where $\dot{Q}_n$ represents the time derivative of the *n* charge) it is obtained the following system of differential equations:

$$
\begin{aligned}
&-V_s + R_s\dot{Q}_0 + R_{FW}\dot{Q}_3 = 0;\\
&-R_{FW}\dot{Q}_3 + {Q_4}/{C_{IE}} = 0;\\
&-{Q_4}/{C_{IE}} + R_{BL}\dot{Q}_5 + R_{FT}(\dot{Q}_5 + \dot{Q}_6) = 0;\\
&{Q_6}/{C_{BL}} - R_{BL}\dot{Q}_5 = 0;\\
&\dot{Q}_0 - \dot{Q}_2 - \dot{Q}_3 - \dot{Q}_4 = 0;\\
&\dot{Q}_2 - \dot{Q}_5 - \dot{Q}_6 = 0.
\end{aligned}
\tag{1}
$$

Having V_{BL} (voltage drop at R_{BL}) as the secondary voltage output, $V_{BL} = {Q_6}/{C_{BL}}$, it is need to calculate Q_6 from Eq. (1). The numerical method used was the algorithm *ode45* from MATLAB®, which is based on an explicit Runge-Kutta (4,5) formula, with a relative and absolute tolerance of 10^{-6} [8]. A remark must be made here to explain the use of $I_n = \dot{Q}_n$; this is because such transformation converts the system of equations in (1) into a system of first order differential equations possible to integrate numerically.

3. Results and discussion

The voltage source, representing the thunderstorms, V_s, was modulated in two different ways: 1) based on the *Carnegie curve,* C_c; 2) based on the *Ionosphere Potential*, I_P, modelled in [5]. For both cases the curves were divided by its mean and multiplied by V_0, the amplitude of the estimated voltage source for the global thunderstorm activity. The expressions are:

$$V_S^{Cc} = V_0 \times {}^{C_c}\!/_{\overline{C_c}}$$

$$V_S^{Ip} = V_0 \times {}^{I_P}\!/_{\overline{Ip}}. \qquad (2)$$

Two sets of parameters were used for the different components of the circuit. Firstly, parameters according to the literature [2]: $V_0 = 100$ MV, $R_s = 100$ kΩ, $R_{FW} = 200$ Ω, $R_{BL} = 300$ PΩ, $R_{FT} = 25$ PΩ, $C_{IE} = 1$ F and $C_{BL} = 0.01$ pF. Secondly, parameters tweaked to get a closer agreement between the simulations and the observed *Carnegie Curve*: $V_0 = 110$ MV, $R_s = 80$ kΩ, $R_{FW} = 200$ Ω, $R_{BL} = 500$ PΩ, $R_{FT} = 25$ PΩ, $C_{IE} = 0.7$ F, $C_{BL} = 0.01$ pF. The values of the simulated PG were found from V_{BL} by dividing it by the height of the boundary layer, $h \sim 2000$ m. The results obtaining are presented in Figure 2, as shown:

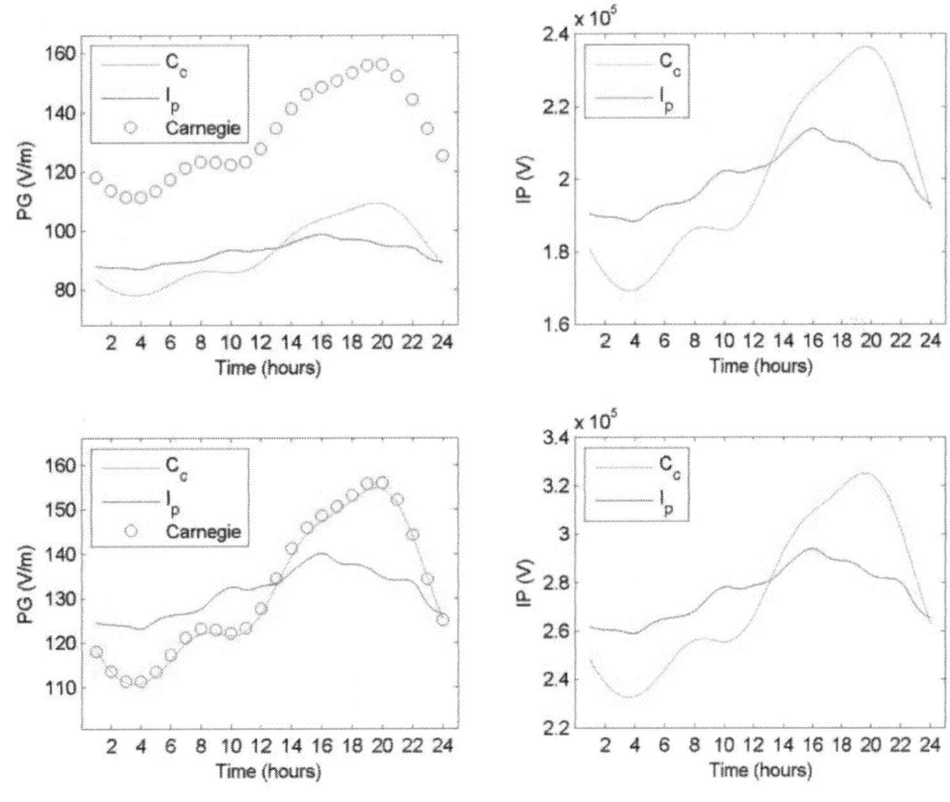

Figure 2 (Upper left): PG values simulated with the first set of parameters; (Upper right): Ionosphere Potential derived from the thunderstorms voltage source and resistance for the first set of parameters; (Bottom left): PG values simulated with the second set of parameters; (Bottom right): Ionosphere Potential derived from the thunderstorms voltage source and resistance for the second set of parameters.

It is seen that the input modulation shapes drastically the form of PG, as expected. On the one hand, using V_S^{Cc} with the typical values in the literature [3, 4] there is a remarkable difference between the observed *Carnegie Curve* and the simulated data. Using I_p modulation, V_S^{Ip}, the results show even a larger deviation because not only the maximum does not correspond to the one of daily variation, but also the PG amplitude is rather low. This results in a very poor similarity regarding *Carnegie Curve*. On the other hand, tweaking the parameters for both modulation cases, V_S^{Cc} and V_S^{Ip}, a perfect match between the simulated PG and the *Carnegie Curve* is found for the first case (as expected), but still a very poor agreement is found with the *Carnegie Curve* for the second one. The discrepancy between the PG simulated with V_S^{Ip} (expected to realistic describe the Ionosphere Potential) and the *Carnegie Curve* is noteworthy. Tentatively it can be argued that this is because of daily variation of R_{BL} associated with the daily dynamics of the boundary layer. With the present model such dynamics can be easily introduced in the simulations allowing R_{BL} to vary during the numerical integration. This is, in fact, a strong point of the present circuit simulations.

4. Conclusions

In the present work a method is described to couple the common model for the Global Electric Circuit (primary circuit) to a Local Circuit (secondary circuit) resembling real PG measurements. Though it is a preliminary approach, a significant discrepancy was found between the PG simulated with modeled Ionosphere Potential and the Carnegie Curve. Future work would involve different aspects and a major one is to test the contribution of thunderstorm activity as voltage or current source. One of the main objectives is to estimate the aerosol load on the local resistance in case of aerosol events, e.g. fires [7].

Acknowledgments

HGS acknowledges the support of Science and Technology Foundation (FCT) for the Post-Doctoral fellowship, SFRH/BPD/63880/2009. The authors acknowledge the support from the FCT/FEDER-COMPETE project EAC (PTDC/GEO-FIQ/4178/2012) FCOMP-01-0124-FEDER-029197. The authors are grateful to Heitor Reis, Mouhaydine Tlemçani, Keri Nicoll and Giles Harrison for useful discussions.

References

[1] Harrison R G 2013 Surv. Geophys. **34** 209–232
[2] Markson R, Ruhnke L H and Williams E R 1999 Atmos. Res. **51** 315–321
[3] Rycroft et al., 2007 J. Atmos. Solar-Terr. Phys. **69** 2485
[4] Rycroft M J, Israelsson S and Price C 2000 Jour. Atmosph. Solar-Ter. Phys. **62** 1563-1576
[5] Mareev E A and Volodin E M 2014 Geophys. Res. Lett. **41** 9009–9016
[6] Silva H G, Conceição R, Melgão M, Nicoll K, Mendes P B, Tlemçani M, Reis A H and Harrison R G 2014 *Env. Res. Let.* **9** 114025
[7] Conceição R, Melgão M, Silva H G, Nicoll K, Harrison R G, Reis A H 2015 *Air Qua. Atmos. Health* **8**
[8] http://www.mathworks.com/help/matlab/ref/ode45.html (retreived at 24 April 2015)

ESD in industry - present and future

Jeremy Smallwood

Electrostatic Solutions Ltd, Southampton, UK

Email: jeremys@static-sol.co.uk

Abstract. ElectroStatic Discharge (ESD) has been a recognized source of damage to unprotected electronic components and assemblies in electronic system manufacture since around 1980. Since then, manufacturers handling ESD Sensitive components (ESDS) have set up ESD Protected Areas (EPAs) for the purpose, and internationally recognized ESD control standards supported by standard test methods and procedures have been developed. Meanwhile, development of ESDS device on-chip protection has continued. This paper examines the trends in ESDS component sensitivity, and the need for development of ESD control in EPAs. It asks if ESD protection is well understood by those in industry who have to implement it, and whether there is adequate education and research in the subject. Finally, it comments on whether the electronics industry and other industry areas where reliable and effective electrostatic control is required could learn from each other's practices.

1. Introduction

Electrostatic problems have occurred and static control has been necessary for many years in many different industries and situation. In hospitals, use of flammable anaesthetics required static control to prevent ignitions. More recently lack of general static control has led to experience of shocks by personnel and malfunction of electronic systems [1]. Specific guidelines and requirements for static control are specified for various industry areas such as explosives handling [2] and electrostatic hazard avoidance in industrial processes [3]. In electronics manufacturing, sensitivity to ESD became a general concern after the adoption of metal-oxide-semiconductor (mos) transistor technology [4] exacerbated by decreasing internal physical size of semiconductor component features and the rise of integrated circuits. The first Electrical Overstress/Electrostatic Discharge Symposium in the USA was organised in 1979. The 1980 Symposium shows papers on topics as diverse as theory and practice, device failure analysis studies, failure mechanisms and modeling, design of device protective networks, and implementing ESD controls, facility evaluation and effective training. Standards [5] and technical handbooks [6] also emerged around this time. The former gave requirements for an ESD control program, while the latter gave technical data and tutorial material useful for educating the user and developing the ESD control program. ElectroStatic Attraction (ESA) [7] of contaminant particles is a problem for manufacturers of semiconductor devices and displays. For operating electronic systems, ESD provides a source of electromagnetic interference that can result in system crash, malfunction or data corruption [8]. This paper considers mainly the control of ESD damage in electronic manufacturing and handling of electronic components, assemblies and systems, the current practice and trends for the future. The practices in this industry are compared with other industry areas.

2. Electrostatic and ESD threats and component susceptibility

The main threats are the effects of electrostatic fields and electrostatic discharges. Electrostatic fields usually pose an indirect threat. An isolated conductor or device under the influence of a field may achieve an induced high voltage and become a source of ESD. In most cases it is the effects of ESD on susceptible components or assemblies (ESDS) that is the main concern. The ESD source may be a charged person or a metal object such as a tool, or a cable plugged into a connector. A device, printed circuit board or module may become charged and the source of damaging ESD to itself.

ESD susceptibility is measured using electronic circuit models that simulate ESD sources to produce standardised ESD waveforms into defined test loads. The Human Body Model (HBM) simulates a charged person discharging into a victim component. The Machine Model (MM) simulates a charged metal object, and the Charged Device Model (CDM) simulates a device itself becoming charged, and the source of damaging ESD to itself [9]. The waveform that is generated, and its effect on the device under test (DUT), depends on the particular value of the components C_{esd}, R_s and L_s as well as the characteristics of the DUT. These models have been developed to replicate known failure signatures experienced when handling devices in the past. In the test, C_{esd} is charged to a set voltage and discharged to each device pin. This is repeated for different pins and increasing voltage levels until damage occurs. The device pin ESD sensitivity is characterised as an ESD withstand voltage that is the highest test voltage that the device pin withstood without damage. Typically different pins may have different withstand voltages, and for some types of pins or devices (e.g. RF or linear input) this can be particularly low. The stray components in the tester are known to be source of irreproducibility in the test, if not carefully controlled.

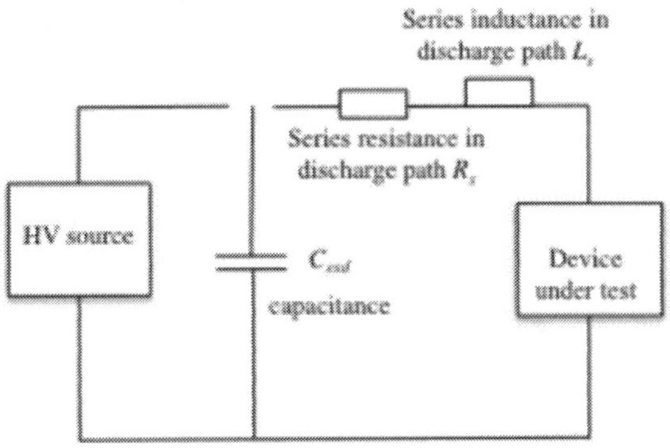

Figure 1. Generic ESD test model

Model	C_{esd}	R_s	L_s
HBM	100pF	1500Ω	stray
MM	200pF	stray	stray
CDM	Device capacitance	stray	stray

ESD model parameters

Device manufacturers have typically adopted target ESD withstand voltage levels (currently 1kV HBM, 20V MM, 250V CDM). Each device pin may require an internal ESD protection network and may go through successive design loops in order to achieve this. These networks can take up significant chip area on high pin count (500+) devices. Achieving these targets can represent a significant cost to the device manufacturer. Some pins cannot be protected, as this can compromise performance. Discrete components such as individual transistors and diodes usually have no protection. So, the range of possible device withstand voltages is very wide and can be many kV (e.g. for power junction transistors) or as little as a volt HBM (e.g. magnetoresistive heads).

Most semiconductor devices are characterised by the manufacturer for HBM ESD withstand. Many are also characterised for CDM. Unfortunately this data is often very hard to obtain. This lack of data may encourage the user to ignore or underestimate the need for ESD control and at best obstructs developing an appropriate ESD control program. Where data is available, it is still not clear how it may be applied. Many practitioners talk of HBM, MM and CDM ESD occurring in their facility, and try to evaluate their process capability against these withstand levels. In fact, what occur in the facility are ESD events from charged personnel, metal objects or devices. The source capacitance, resistance and inductance can be widely different from the models, and the device susceptibility will also vary [10]. In practice, damage can be due to exceeding energy, peak current, peak power or charge transfer thresholds and these may be quite different for the same ESD voltage. There is no guarantee that the ESD withstand voltage represents susceptibility to real ESD. While a 100V HBM device might be assumed safe if the operator's body voltage is less than 100V, this assumes a certain source capacitance and resistance. In practice the person is likely to have different body capacitance and resistance and the discharge could happen from a hand held metal tool.

ESD withstand tends to decrease as device internal component size is reduced because an ESD current passing through a smaller volume component will tend to create higher power and energy density. Thinner dielectric layers used in mosfet gates have reduced voltage breakdown strength [9], [10]. With CDM, ESD withstand is also reduced with increasing package size. Some types of device technology are also inherently more susceptible to ESD damage than others. The ESD Association have produced "roadmaps" showing the predicted ESD HBM and CDM withstand voltage range variation with time [12]. ESD withstand voltages have reduced relentlessly since the mid 1990's. In 2015 the withstand voltages of the most sensitive devices are converging with ESD control capability. Many users are handling sub-1kV or even sub 100V parts. In 2007 the Industry Council recommended reducing HBM and CDM withstand to 1kV and 250V respectively [10], [13]. It seems likely to be only a matter of time before they are reduced further. Some small package resistors and capacitors are also now showing parameter change or failure due to ESD [15].

Other trends include the use of transmission line pulse (TLP) techniques to characterize the ESD withstand of devices. While this may suit characterization of devices it is unclear how the device user could use the data generated using these techniques in specifying an ESD control program. Electrostatic attraction of particles has long been a major threat during wafer manufacture of devices. This is now also an issue in other areas, for example optical systems and displays. A dust particle can cause pixel fault and rejection of a large screen display. Even packaging of consumer goods such as mobile phones may require semi-clean conditions so that the item reaches the customer in pristine presentational condition. Some users have found that electrostatic forces can cause small components to be ejected or incorrectly placed on a PCB. One thing seems clear – electrostatics and ESD control will increasingly be of concern for the foreseeable future.

3. ESD control program development

It is difficult to know the cost of ESD failures to industry. Some companies take little care, some take great care and many fall in between. Most say they do not see ESD failures, but many just replace failed components without failure analysis. Few studies of ESD failure rates are published. The Industry Council found that the ESD failure rates were less than 0.1 per million, but their participants had probably implemented good ESD control measures. ESD failures from complacent users remain unrecorded. Some studies have shown much higher failure rates together with an impressive return on investment in ESD control. [4], [16], [9]. With wide ranges of device ESD withstand voltage and assembly processes, what constitutes adequate ESD control can be highly variable. The principle is simple: ESDS must be handled within an ESD Protected Area (EPA) in which the ESD risks are sufficiently controlled. Outside the EPA, ESD protective packaging is used. A huge range of packaging products is now available. Some protect against electrostatic fields, some against electrostatic discharges, some merely reduce electrostatic charging, and some have all these functions. Unfortunately many users do not understand the differences and are apt to select ESD packaging on price considerations. A wide variety of ESD control equipment (e.g. work surfaces, chairs, tools, footwear and flooring) is now available for use in an EPA. These replace insulators with resistive materials to provide a path to ground for charge generated during handling and processes. Equipment, materials and packaging that make contact with ESDS must have sufficiently high resistance in order to prevent high discharge currents occurring. All conducting items especially personnel must be grounded, to prevent their becoming a source of ESD. Often two or more control items (e.g. footwear and flooring) form part of a control system where both parts must be present and within specification in order to be effective. Unfortunately, lack of understanding of this system often leads to omission or poor selection of key items (e.g. floor) and can seriously impair ESD control.

Modern standards take a flexible approach that requires the user to define appropriate ESD control measures for their facility and processes [17], [18]. These aim to protect devices down to 100V HBM, and have recently incorporated CDM device withstand into their specification. MM is not included because device manufacturers no longer test according to this model. Standards for ESD protection in automated processes are still not available. Below about 500V HBM, standard basic control measures can become less effective, and more advanced techniques may be required [10].

Standards tend to lag behind real world technology needs. An example of this is in measurement methods for resistance of ESD packaging. Currently these are mainly suitable for large planar material surfaces, although a 2-pin miniature probe more suited to centimeter sized and non-planar packaging features is now available. Standard test methods have not yet been developed for packaging that has millimeter size features now in widespread use. Development of standards requires consensus, and this is not always achieved. A lack of agreed test method for ESD control garments, other than those based on material resistance, currently blocks acceptance of certain garment types. This is partly based on lack of agreement as to what parameters characterize a good ESD control garment.

In the UK, ESD seems to be almost completely ignored by Further and Higher Education establishments. Most ESD training is vocational at a level of basic awareness and use of ESD control equipment, provided internally by industrial companies or by vendors of ESD control products. Most delegates are the operators that handle or assemble electronic systems. Courses on principles and practice of ESD control are few and far between. Those who want a qualification in ESD control have few choices. I know only one UK University that has offered a short course in ESD control as part of a degree course. It seems hardly surprising that there is widespread lack of prioritisation and understanding of ESD control amongst degree qualified engineers. So, many ESD control programs are based on inadequate information and assumptions, rather than sound engineering principles.

A modern ESD control program must be flexible, in order to allow optimization for a wide range of manufacturing processes, facilities, environments climates and cultures. There is a natural desire to get the maximum benefit for the least expenditure. Unfortunately this is often interpreted as minimizing expenditure on ESD control equipment and training. Real optimization requires finding the right balance of expenditure on evaluation and planning, equipment, documentation, training and compliance verification required to make sure the program is effective in the long term [19]. Process capability and awareness that use of standard protection techniques does not guarantee effective ESD prevention are increasing concerns [20]. We currently do not have a comprehensive methodology to demonstrate that processes are capable of handling components without significant losses.

Current research seems dominated by on-device ESD protection design and test. At the 2014 ESD Symposium, 32 papers were concerned with device protection design and test. Only 6 papers were concerned with factory issues. Typically manufacturing companies will not research unless there is an issue that significantly affects production yields. Only a few companies actively try to improve their understanding of ESD control and often prefer to conceal rather than publish any knowledge gained.

In the UK, there is apparently little university research into ESD related issues. I have met only one academic with research interests in this field. The reason for this void is not clear – lack of availability and prioritization in research funding may be one possibility. Many of the knowledge gaps are practical in nature. An example is the development of device ESD susceptibility tests parameters and methodology for application in specification of ESD control measures. Development of ESD control materials, especially for packaging, is another. New test methods are also needed, for example for measurement of small feature packaging products, and for ESD control garments.

4. Other industry areas

The topic of effects of ESD on operating electronic systems can appear to be related to ESD damage in electronics manufacture. In fact they concern different communities (EMC and ESD control) and have different mechanisms and control approaches. ESD to operating systems can lead to system crash, malfunction or data corruption. Often operation may be restored by system restart. In Europe it is mandatory under EMC Directives to achieve a level of immunity before equipment is placed on the market. The system user will take no ESD control measures. This approach is mainly successful but can result in failures where ESD sources are exceptionally strong or of an unusual characteristics. A possible need is to address ESD sources other than the charged person. The confusion between these topics has led to demand for on-device ESD protection against system level ESD, and the use of the system level ESD test [8] to characterize device immunity. The Industry Council has recently proposed a System Efficient ESD Design (SEED) approach to these issues.

In ignition of flammable materials or explosives the susceptibility is measured as the material minimum ignition energy. Organisations must identify hazard zones in which the materials are handled and apply ESD control measures within that area to prevent ignition risks. In Europe it is mandatory under the ATEX Directive or other regulations to identify the risks and apply precautions.

Shocks to personnel can be due to charging of the person's body or charging of an external item of equipment, e.g. a plant item or vehicle. There is normally no regulatory obligation to control the factors leading to these shocks, and they are often tolerated to a surprising level in the workplace.

5. Conclusions

In electronics manufacture the main concern is that components will be damaged during handling. For device manufacturers, ESD withstand voltage has become a target for device qualification. For device users, this ESD withstand data can be very difficult to get and it is not clear how it can be applied in developing an ESD control program. We need to develop an improved methodology so that ESD control programs and process capability can be based on realistic ESD susceptibility data rather than assumptions. Device ESD withstand voltages are generally decreasing driven by device technology scaling, increasing performance requirements and the difficulty in achieving on-chip protection targets. With standard ESD control techniques capable of safe handling of ESDS down to 100V HBM withstand, there seems little point in including on-chip ESD protection withstand up to 1kV HBM and 250V CDM - it seems inevitable that these targets will be further reduced. ESD susceptibility of systems seems largely unconnected with ESD withstand of devices despite common assumptions.

Many companies believe they do not see ESD failures, but few analyze failures and identify ESD damage. Studies of ESD failures and their cost to industry are rare. Many invest the minimum in ESD control, often focused on control measures, with minimal training or compliance verification. ESD control is often applied with little understanding of how measures work as a system. Poor specification or equipment failures can often render them ineffective. Many engineers seem skeptical about the importance of ESD control and lack understanding of the principles and practice. Little attempt seems to be made in UK Universities to teach even basic principles and practice. Industry training tends to be mainly on awareness and use of basic equipment. There is a demand for Certification of engineers who have responsibility for ESD programs, but the options seem limited with very few providers..

The electronics industry has developed a methodology and standards for ESD control. These require that the ESD control program is planned and documented, including training and compliance verification. The user must define the specific control measures. However, it is optional whether the user concerns themselves at all with ESD control.

For flammable atmosphere hazards in industrial processes, in Europe there is a legal requirement to evaluate and control the ignition risks under ATEX. The subject is covered by guidance documents rather than standards. Ignition sensitivity data is easily available or routinely tested as part of the hazard evaluation, and easily applicable to process ESD control. There are few training courses available. The situation is similar for explosives handling. In neither of these areas is there a standard that gives requirements for developing and documenting an ESD control program with training and compliance verification. There is a general lack of guidance and standardization, training or any requirement for ESD control in prevention of nuisance shocks in workplaces or domestic environments. In Europe it is mandatory that any electronic system must pass requirements for ESD immunity. There is no requirement for control of ESD sources.

The electronics industry could benefit from learning to better characterize device ESD susceptibility and publish data and a methodology that can be applied in designing effective ESD control. As ESD withstand reduces, it will become important to base ESD protection on data, knowledge, understanding and methodology rather than assumptions. The flammable atmosphere and explosives areas could better specify and document an ESD control program with training and compliance verification, enshrined in standardization. Control of shocks to personnel would benefit from motivation or requirement to address the issue. All these areas would benefit from better training and education on the causes of ESD and principles and practice of ESD control.

References

[1] Viheriäkoski T, Kokkonen M, Tamminen P, Kärjä E, Hillberg J, and Smallwood J. 2014 Electrostatic Threats in Hospital Environment. *Proc. EOS/ESD Symp*. Paper 4B2

[2] Health and Safety Executive 2005 *Manufacture and Storage of Explosives Regulations Approved Code of Practice and Guidance*. HSE Books ISBN 0 7176 2816 7

[3] International Electrotechnical Commission 2013 *Explosive atmospheres Part 32-1. Electrostatic hazards, guidance*. PD/IEC TS 60079-32-1 ISBN 978-2-8322-1055-0

[4] McAteer O 1990 *Electrostatic discharge control*. MAC Services In. ISBN 0-07-044838-8.

[5] MIL-STD-1686A Electrostatic discharge control program for protection of electrical and electronic parts, assemblies and equipment (excluding electrically initiated explosive devices)

[6] Department of Defense. MIL-HDBK-263A 1991 *Electrostatic discharge control handbook for protection of electrical and electronic parts, assemblies and equipment (excluding electrically initiated explosive devices)*

[7] Welker RW, Nagarajan R, Newberg C E. 2006 *Contamination and ESD control in high technology manufacturing*. IEEE Press Wiley Interscience ISBN-13:978-0-471-41452-0

[8] Ind. Co. *ESD Target levels 2010 White paper 3. System level ESD Part 1:common misconceptions and recommended basic approaches*. Oct.. 2010

[9] Amerasekera A, Duvvury C 2002 *ESD in Silicon integrated circuits*. 2nd Ed. Wiley ISBN 0 470 49871 8

[10] Ind. Co. ESD Target levels 2011 *White paper 1. A case for lowering component level HBM/MM ESD specifications and requirements*. Rev. 3 Sept. 2011.

[11] Viheriakoski T, Peltoniemi T and Tamminen P et al. 2012 Low level Human Body ESD. *Proc. EOS/ESD Symp*. Paper 4A3

[12] ESD Association *Electrostatic Discharge (ESD) Technology roadmap* – revised 2013.

[13] Ind. Co. *ESD Target levels 2010 White paper 2. A case for lowering component level HBM/MM ESD specifications and requirements*. Rev. 2 Apr. 2010

[14] Paasi J, Smallwood J and Salmela H 2003 New Methods for the Assessment of ESD Threats to Electronic Components *Proc. EOS/ESD Symp* Paper 2B4.

[15] Taminnen P, Sydanheimo L and Ukkonen L 2014 ESD sensitivity of 01005 chip resistors and capacitors *Proc EOS/ESD Symp*. Paper 9A2

[16] Danglemeyer G T. 1990 *ESD Program management*. Van Nostrand Reinhold, ISBN 978-1-4615-6935-0

[17] International Electrotechnical Commission 2007 *Electrostatics – Part 5-1: Protection of electronic devices from electrostatic phenomena - General requirements*. IEC 61340-5-1: 2007

[18] ESD Association 2007 *ESD Association Standard for the Development of an Electrostatic Discharge Control Program for – Protection of Electrical and Electronic Parts, Assemblies and Equipment (excluding Electrically Initiated Explosive Devices)* ANSI/ESD S20:20-2007. ISBN 1-58537-121-1

[19] Smallwood J, Taminnen P, Viheriaekoski T. 2014 *Optimizing investment in ESD Control*. Proc. EOS/ESD Symp.

[20] Nakata Lin, Yun Liang, Wang P. 2014 *Evolution of ESD process capability in future electronics industry*. 15th Int. Conf. Elec. Packaging Tech.

Charge neutralisation from the side surface of an insulating plate

A. Ohsawa

Electrical Safety Research Group, National Institute of Occupational Safety and Health, Japan (JNIOSH), 1-4-6 Umezono, Kiyose, Tokyo 204-0024, Japan

E-mail: ohsawa@s.jniosh.go.jp

Abstract. This paper explores whether charge neutralisation from the side surface of an insulating plate is effective, on the basis of investigating the motion of positive and negative ions using 2-D numerical fluid simulations for different initial surface charge distributions and placements of an insulating plate to be neutralised.

1. Introduction

The placement of an ioniser is generally on a larger surface area of charged insulating objects to be neutralised. However, the motion of ions near the objects is influenced by not only charges on the front side surface but also charges on the opposite side when the objects are thin, where such a phenomenon was demonstrated by 2-D electrohydrodynamic (EHD) simulations of positive and negative ions emitted from an ioniser, resulting in over-neutralisation that can cause electrostatic problems when the opposite side is charged [1]. Therefore, we suggested that the situation of charged surfaces should be considered, and ionisers should be placed on charged surfaces to avoid over-neutralisation. The present work is a response to a comment regarding this suggestion at Electrostatics 2013—is the placement of an ioniser on a side surface of thin objects effective?—in which we investigated the phenomena of charge neutralisation from the side surface of an insulating plate using a 2-D EHD simulation.

2. Model

The self-consistent 2-D EHD fluid model used in this investigation consists of the mass and momentum conservation equations for incompressible fluid of airflow from an ioniser with a blower, the continuity equations for the densities of positive and negative ions emitted from the ioniser, and Poisson's equation for potential and electric field distributions in the system to be solved in addition to the charge conservation equations between the ion current and surface charge at small areas of the surfaces of an insulating plate (15 cm long and 1 cm thick) for modelling charge neutralisation on the plate surfaces, and is the same as that described in [1]. It was necessary to modify the outflow boundary conditions when modelling the neutralisation from the side surface of an insulating plate. Since we first assume that the insulating plate exists in free space with no walls except an ioniser, airflow velocities and ion densities at the outflow boundary of their computational domains were relatively higher than those that have been observed in previous simulations [1, 2] that have a wall boundary (a grounded plane in

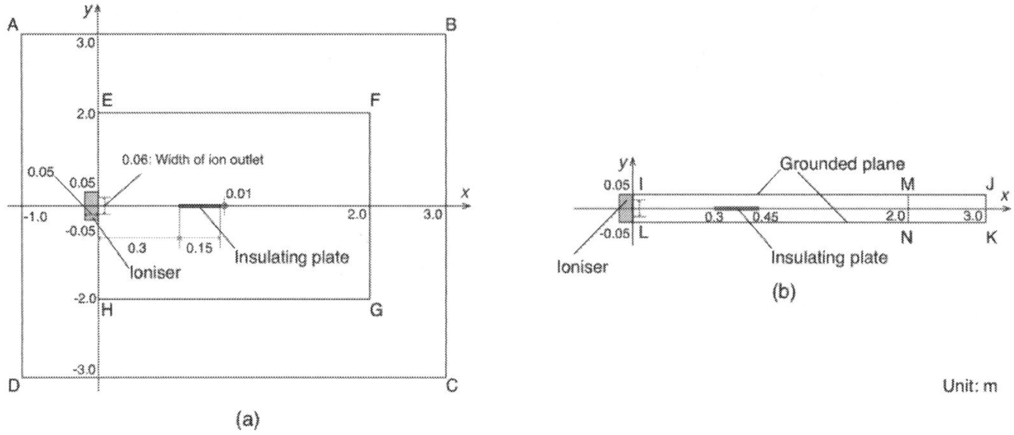

Figure 1. Geometries and computational domains when an insulating plate exists (a) in free space and (b) between grounded planes.

accordance with a standard test method [3]) existing near the plate. Therefore, a relatively larger computational domain and the convective outflow boundaries [4] for airflow velocity and densities of positive and negative ions in place of their Neumann boundary conditions were then required to solve their transient flows without unstable solutions. To investigate the influence of the charges on the opposite side surface of an insulating plate on the motion of the ions during neutralisation, three types of initial surface charge distribution (ISCD) of the insulating plate with $\pm 1.0 \times 10^{-7}$ C/m^2 were used: *Both*—the larger (upper and lower) surfaces are positively charged; *Only the upper side*—only the upper side surface is positive; and *Bipolar*—the upper side is positive and the lower is negative. We used two types of geometry for the insulating plate, where the plate to be neutralised exists in free space (figure 1a) and between grounded planes (figure 1b), because we found that when the plate is placed in free space, the main ion flows are away from the upper and lower surfaces, making neutralisation difficult, involving non-uniform charge neutralisation, as described later; placing the plate between grounded planes is expected to modify such ion flows. The densities (1.0×10^{14} m^{-3}) of positive and negative ions emitted from the ioniser are quasi-neutralised since such ions, which can lead to no space charges by ions in the entire space at a steady state, can minimise the offset voltage and can be effectively transported towards charged objects during neutralisation [2], even using a transport tube [5]. Computational domains for the plate existing in free space are ABCD for the electric field and EFGH for the air and ion flows, and those for the plate between grounded planes are IJKL for the electric field and IMNL for the air and ion flows, as shown in figure 1.

3. Results and discussion

For the plate existing in free space, in the early phase of neutralisation, an over-neutralisation on the near-side surface was observed due to charges on the other surfaces when the types of ISCDs are *Both* and *Only the upper side*, as shown in figure 2. Charge neutralisation on the near- and far-side surfaces is not easy because, as shown in figures 3 and 4, the main flow of ions separates into two flows at the near-side surface; furthermore, since they are away from the upper and lower surfaces, the charge neutralisation at the far-side surfaces is more difficult. Such ion flows involving airflow show that neutralisation from the side surface is ineffective because only parts of ions from the main flow contribute to neutralisation, as shown in figure 4. In addition, the resultant behaviour of ions causes non-uniform charge neutralisation and, especially for ISCD of *Only the upper side*, a bipolar charge distribution (the polarity is different between the upper and lower surfaces) due to the surface charge of the opposite side, as shown in figure 5. However, it is interesting that ions that can move near the plate respond to the polarity of the charge on the surface closest.

For the plate existing between grounded planes to control the air and ion flows, the neutralisation was modified so that faster charge decays and more uniform charge neutralisations were observed because the main flows of air and ions were along the plate, while charges on the near-side surfaces still remained, as shown in figure 6. Furthermore, a large bipolar charge distribution that was observed for ISCD of *Only the upper side* for the plate in a free space

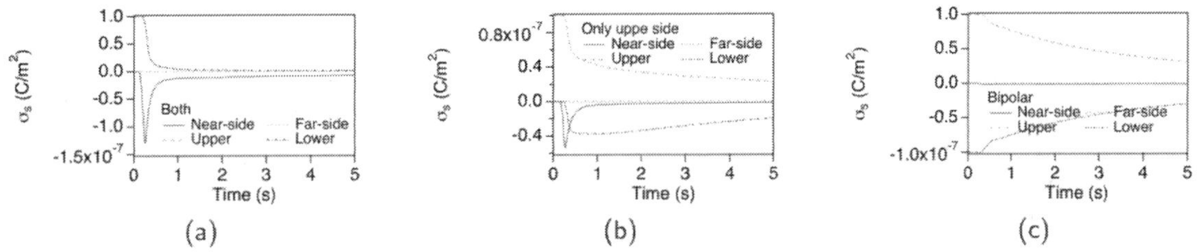

Figure 2. Time variation of the surface charge densities, σ_s, at the centre of each surface during charge neutralisation for ISCDs of (a) Both, (b) Only the upper side, and (c) Bipolar for the plate existing in free space.

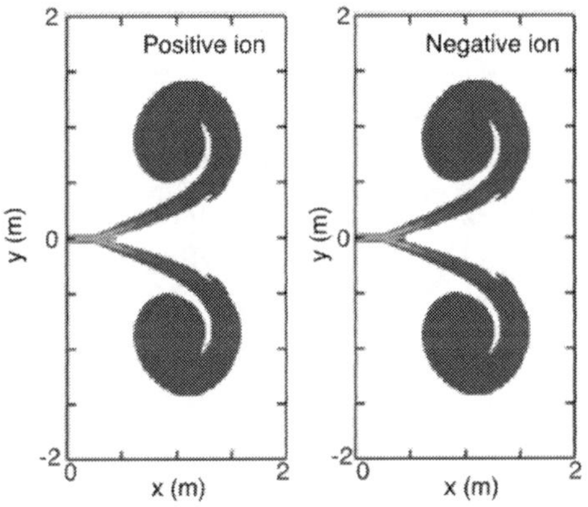

Figure 3. Density distributions of positive and negative ions at $t = 5$ s for ISCD of Both when the plate existis in free space. Contour colour maximum (red) is 1.0×10^{14} m^{-3}, and its minimum (blue) is 1.0×10^{11} m^{-3}.

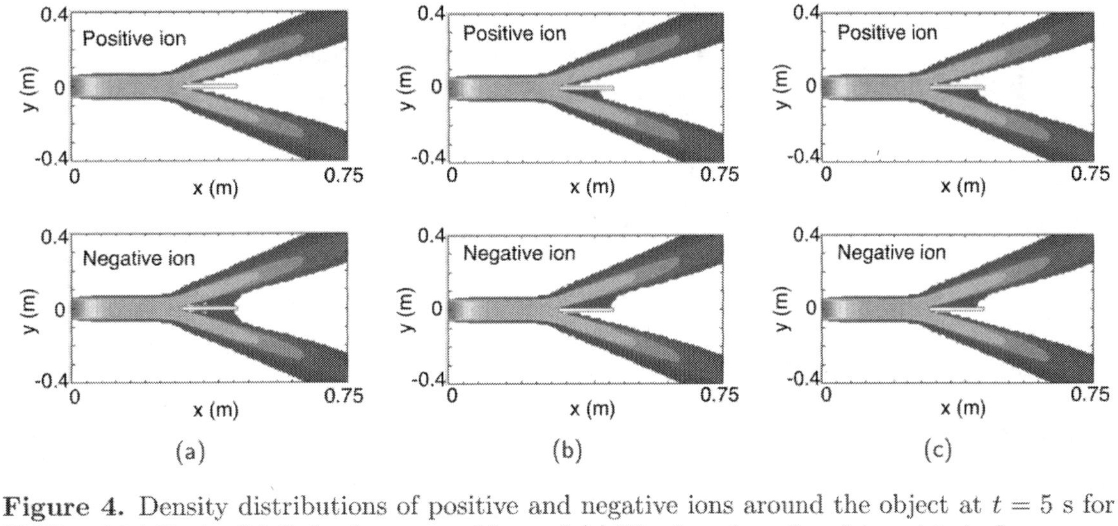

Figure 4. Density distributions of positive and negative ions around the object at $t = 5$ s for ISCDs of (a) Both, (b) Only the upper side, and (c) Bipolar when the plate exists in free space.

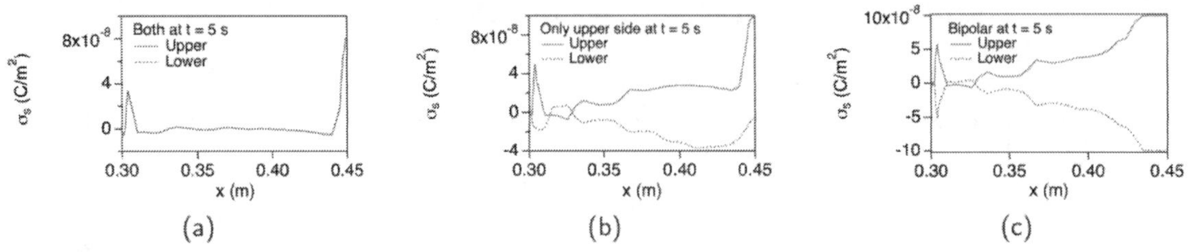

Figure 5. Distributions of the charge densities, σ_s, on the upper and lower surfaces at $t = 5$ s for ISCDs of (a) Both, (b) Only the upper side, and (c) Bipolar when the plate is in free space.

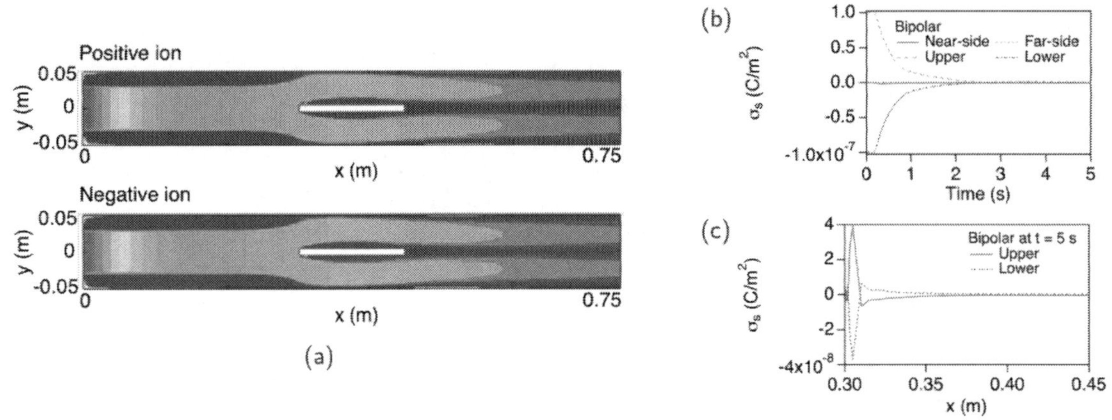

Figure 6. The phenomena of charge neutralisation for the ISCD of Bipolar when the object is placed between grounded planes, (a) density distributions of positive and negative ions at $t = 5$ s, (b) time variations of surface charge densities, σ_s, at the centre of each surface, and (c) surface charge density distributions on the upper and lower surfaces at $t = 5$ s.

can be eliminated. This implies and supports the idea that neutralisation from both larger side surfaces would not lead to a large bipolar charging that could cause problems. In addition, for a thinner plate of 1 mm thick with the ISCD of *Bipolar*, charge neutralisation became difficult because the opposite surface charges are closer; however, when the forced airflow from ionisers is normal to larger surfaces, this would be improved.

Consequently, we found that charge neutralisation from the side surface of an insulating plate is improper for practical applications; as concluded previously [1], we can suggest again that simultaneous charge neutralisation from both larger surfaces would be proper.

Acknowledgments

This work was supported by JSPS KAKENHI Grant 23510214 and motivated by the comment of Dr. Kris Kozaczek.

References

[1] Ohsawa A 2013 Computer simulations of insulator charge neutralisations with a corona ioniser — Influence of initial surface charge distribution *J. Electrostat.* **71** 287–293

[2] Ohsawa A 2013 2-D electrohydrodynamic simulations towards zero offset voltage with corona ionisers *J. Electrostat.* **71** 116–124

[3] IEC 61340-5-1 1998 *Electrostatics — Part 5-1: Protection of Electronic Devices from Electrostatic Phenomena — General Requirements*

[4] Yoshida T, Watanabe T, Nakamura I 1993 Numerical analysis of open boundary conditions for incompressible viscous flow past a square cylinder *Trans. Jpn. Soc. Mech. Eng.* B **59** 2799–2806 (in Japanese)

[5] Ohsawa A 2008 In-tube transport of ions for charge neutralization *J. Phys.: Conf. Series* **142** 012014

An Unusual Source of Multiple ESD Events in Electronic Equipment

D C Smith

University of Oxford, Department for Continuing Education, Course Tutor
Rewley House, 1 Wellington Square, Oxford, Oxfordshire, OX1 2JA, UK
US address: 1305 Arizona St. #101, Boulder City, NV, USA 89005

Email: doug@dsmith.org

Abstract. Small wall mounted power supplies that plug directly into AC power mains and furnish power, usually low voltage DC, to equipment over a low voltage cable are ubiquitous in the electronic field. It is shown that these low voltage supplies can be the source of multiple ESD events that are fed to the connected system on the low voltage cable when an ESD event is applied to the system or directly to the power supplies. The multiple ESD events produce nasty current waveforms Examples of these waveforms and the measurement method are given. The results have been duplicated in three laboratories, results are presented from two of the three.

1. Introduction

During troubleshooting of an ESD problem on a piece of equipment that was powered by a small plug type power supply connected to the equipment by a low voltage cord, strange current waveforms were noted on the low voltage cord that appeared to be multiple ESD events. This, even though only one ESD pulse was being applied to the system. The author had not observed such an effect previously.

The test setup was checked and sources of secondary ESD events were ruled out. Ultimately, the test was simplified to involve just the power supply, a current probe, and an ESD simulator. With this setup, a series of current waveforms could be quickly generated and so a number of power supplies could be checked in a short period of time.

2. Test and Measurement Setup

Figures 1 and 2 show details of the simplified test setup. In Figure 1, the AC mains connection of the supply was connected to the ground lead connection of the ESD simulator and the tip of the ESD simulator was connected to the low voltage connector from the supply. A Fischer Custom Communications F-65 current probe with a flat frequency response of 1 MHz to 1 GHz and a transfer impedance of one Ohm (one Volt/Amp) was used to measure the current.

In Figure 2, the two AC mains connections are shorted together to rule out the possibility that a protection device across the mains (such as an MOV device) could be contributing to the effects reported here. Shorting the mains connections of the supply did not change the results.

This test setup was used in three different laboratories involving different ESD simulators and different power supplies. The only common factor in the three test setups was the F-65 current probe. The performance of the current probe was verified and is reported later in this paper. The ESD simulators included a KeyTek MZ-15, a NoiseKen ESS-S3011, and one simulator at a client location.

The applied ESD level was between 500 Volts to as high as 3,000 Volts. The multiple discharges often started around 500 Volts, well below the breakdown voltage of the power supply, and did not happen above 3000 Volts. Note that the actual peak voltage across the dielectric barrier of the supply *should* be less than the setting of the ESD simulator as the charge would be split between the simulator and the capacitance across the dielectric of the supply.

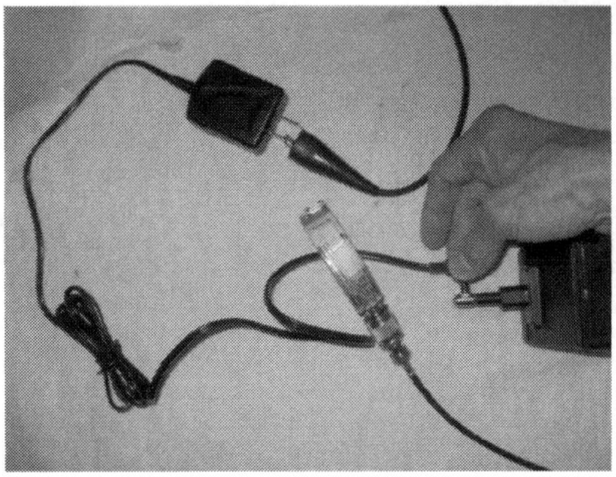

Figure 1. Test setup.

Figure 2. Shorting of the mains connections.

3. Characteristics of the Multiple Discharges

The pattern of apparent multiple discharges fell into a few types. One type is illustrated in Figures 3 and 4. The pattern of Figure 3 was a common one. The original ESD event is followed a few hundred nanoseconds later by a smaller ESD like event embedded in a low frequency discharge, itself lasting a few hundred nanoseconds. Sometimes the embedded ESD like event was not present, just the low frequency discharge was present.

Sometimes the pattern of Figure 3 was followed by a negative polarity event a few microseconds later as shown in Figure 4. A possible mechanism of the cause of the negative event will be discussed later. One thing is sure at this point, the time scales in Figures 3 and 4 are much larger than the propagation time of the current around the test setup loop composed of the power supply low voltage cord and the ground lead of the ESD simulator, which would be on the order of about 20 nanoseconds.

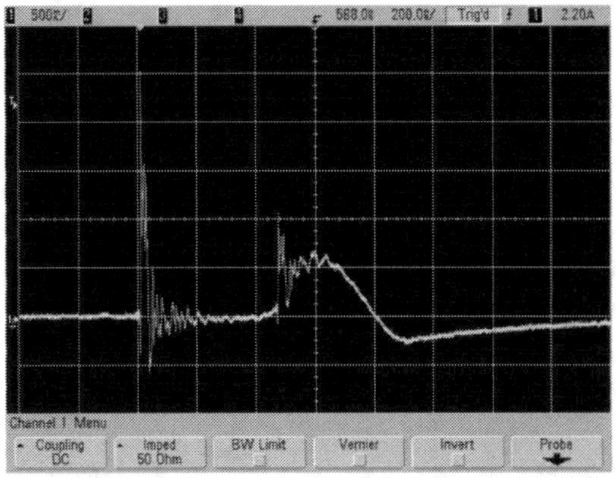

Figure 3. Secondary discharge embedded in larger low frequency event. Scale factors: 500 mA/div and 200 ns/div.

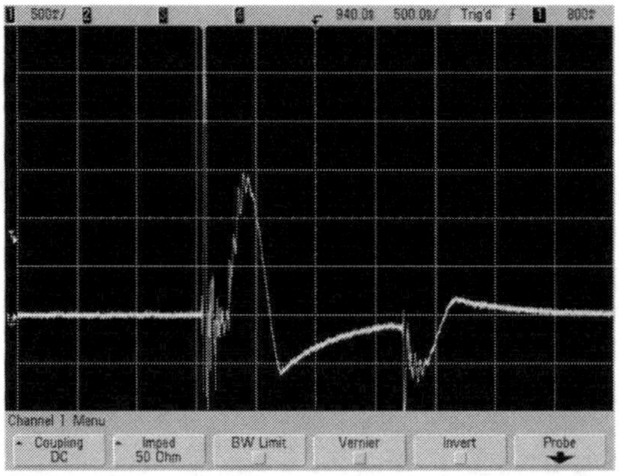

Figure 4. Additional negative polarity ESD event and low frequency event. Scale factors: 500 mA/div and 500 ns/div.

Figure 5 shows a long view at a horizontal scale of one μs/div. At this time scale, a secondary event of negative polarity can be seen about seven microseconds after the original event. This delay time is much, much greater than the approximately 20 nanosecond delay around the setup loop.

Figure 6 shows another frequent pattern. In this pattern, the secondary events come in a rapid sequence with each event 100 to 300 nanoseconds apart.

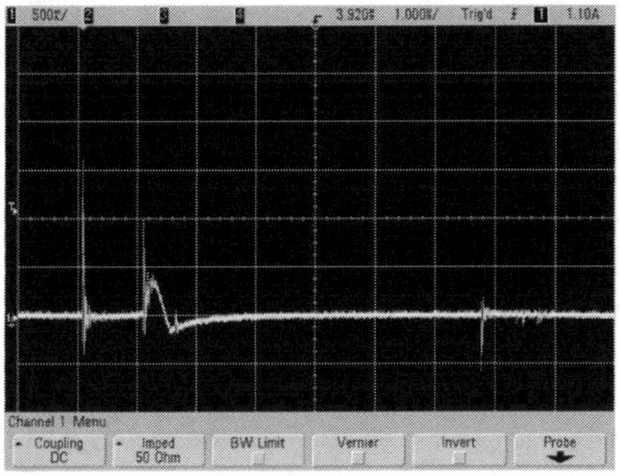

Figure 5. Long time view showing negative event. Scale factors: 500 mA/div and 1 μs/div.

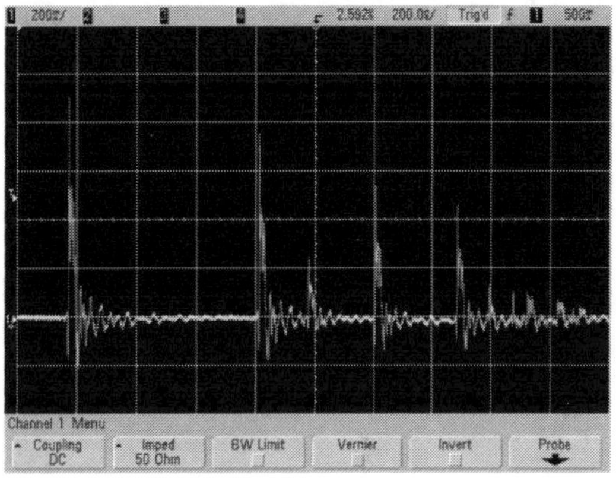

Figure 6. Multiple events in rapid secession. Scale factors: 200 mA/div and 200 ns/div.

Figure 7 shows an initial applied current waveform *and* a secondary one having the appearance of a human metal waveform, similar to that described in IEC 61000-4-2.[1] However, a secondary event shown in Figure 8 has the appearance of a cable discharge event, having a flat region of current corresponding to the discharge of a common mode transmission line.[2] Note the duration of about 20 nanoseconds. This is the expected time delay around the loop of the low voltage power cord of the power supply and the ground lead of the ESD simulator.

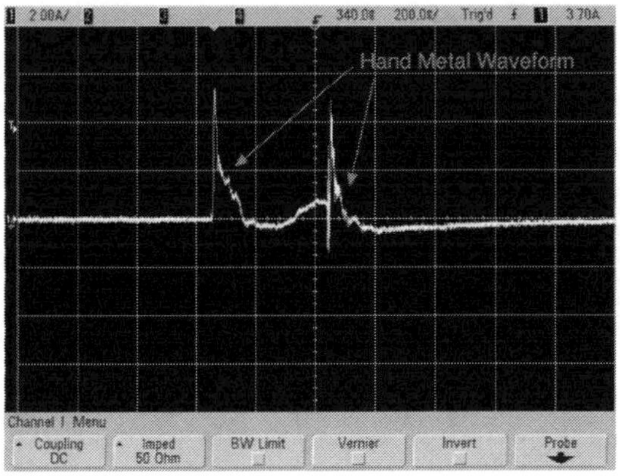

Figure 7. Secondary event showing human metal waveform characteristics. Scale factors: 2 A/div and 200 ns/div.

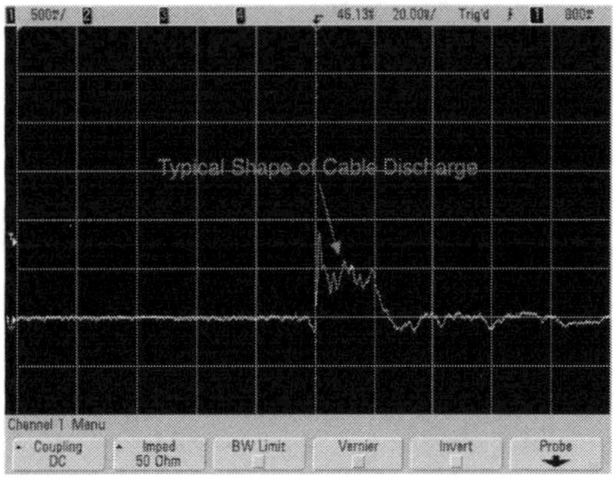

Figure 8. Secondary event showing cable discharge waveform characteristics with 20 ns width. Scale factors: 500 mA/div and 20 ns/div

Figure 9 shows a similar waveform to those above, but taken at a different laboratory using different equipment except for the current probe. To make sure the current probe or scope was not contributing significantly to the observations, Figure 10 shows the current waveform when the ESD simulator ground lead was connected to the simulator tip with the F-65 current probe measuring the current in the loop. As one can see, there is no evidence of any secondary events.

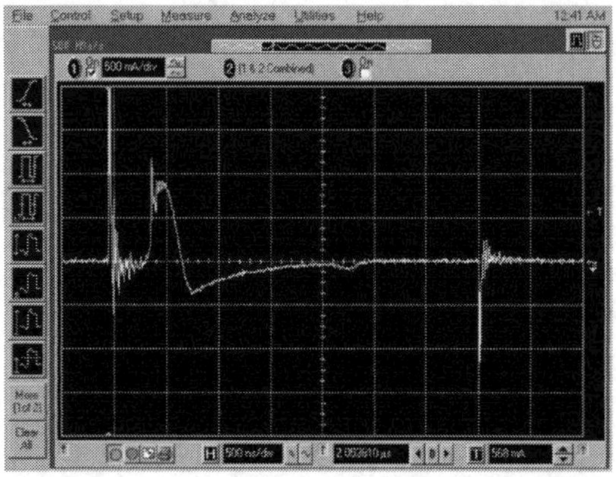

Figure 9. Secondary discharge recorded at different laboratory showing delayed negative pulse. Scale factors: 500 mA/div and 500 ns/div.

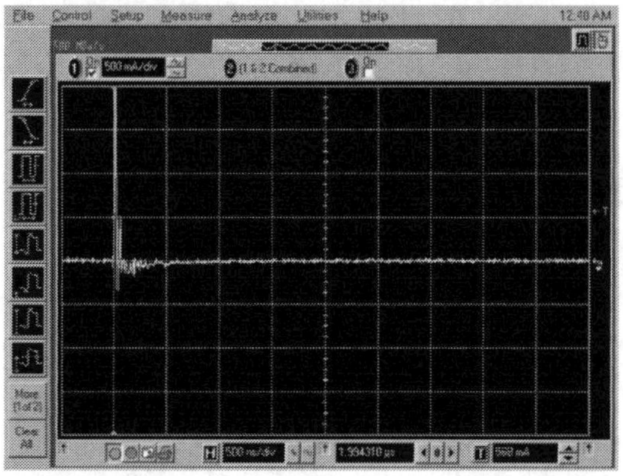

Figure 10. Current measured in ESD simulator ground lead shorted to the simulator tip showing no secondary discharges. Scale factors: 500 mA/div and 500 ns/div.

4. Analysis of Results and Possible Explanation

The long time delays in the above waveforms cannot be explained on the basis of the time delay around the current loop (about 4 meters). Likewise negative secondary events in response to a positive initial event must be explained as well a breakdown at significantly lower voltages than the barrier rating (500 V applied stress vs. >2000V barrier rating).

One possible theory is that a resonance between elements of the EMI filter and power transformer (leakage inductance, and inter-winding capacitance) causes a breakdown of the transformer insulation in both polarities.

5. Subsequent and Future Work

Similar current waveforms have been now produced using Electrical Fast Transients, IEC 61000-4-4, instead of ESD.[3] This is problematic as EFT happens frequently on AC power mains, much more so than ESD in the environment. There is the potential for power line spikes to cause system malfunctions by this mechanism!

Recently, similar results have been produced on a "brick" type power supply, such as used on laptop computers. Work to identify the mechanism within the power supply is planned.

6. Conclusion and Summary

Internally generated ESD/EFT events in system power supplies may be a cause for system upsets in normal use. Results presented in this paper may explain what may be happening. In any event, new power supply designs of all types should be checked for this problem. All that is needed is a current probe and an ESD simulator. A full ESD test area is not needed.

Reference

[1] IEC 61000-4-2, "Testing and measurement techniques – Electrostatic discharge", https://webstore.iec.ch/publication/4189

[2] "Cable Effects Part 1: Cable Discharge Events," January 2002, http://www.emcesd.com/tt2002/tt010102.htm, Douglas C. Smith

[3] IEC 61000-4-4, "Testing and measurement techniques – Electrical fast transient/burst immunity test", https://webstore.iec.ch/publication/22271

Understanding electrostatic charge behaviour in aircraft fuel systems

Jill A Ogilvy[1,3], Phil Hooker[2] and Darrell Bennett[2]

[1]Airbus Group Innovations, Quadrant House, Celtic Springs, Coedkernew, Newport, NP10 8FZ, UK
[2]Health and Safety Laboratory, Buxton, Derbyshire, SK17 9JN, UK

[3]E-mail: jill.ogilvy@eads.com

Abstract. This paper presents work on the simulation of electrostatic charge build-up and decay in aircraft fuel systems. A model (EC-Flow) has been developed by BAE Systems under contract to Airbus, to allow the user to assess the effects of changes in design or in refuel conditions. Some of the principles behind the model are outlined. The model allows for a range of system components, including metallic and non-metallic pipes, valves, filters, junctions, bends and orifices. A purpose-built experimental rig was built at the Health and Safety Laboratory in Buxton, UK, to provide comparison data. The rig comprises a fuel delivery system, a test section where different components may be introduced into the system, and a Faraday Pail for measuring generated charge. Diagnostics include wall currents, charge densities and pressure losses. This paper shows sample results from the fitting of model predictions to measurement data and shows how analysis may be used to explain some of the observed trends.

1. Introduction

An understanding of electrostatic charge in fuel systems is of interest to several industries, as it is possible that sufficient charge may be developed to lead to unwanted dielectric breakdowns. Although the topic of electrostatic discharge in fuel tanks has been studied for many decades there has been little work performed on developing models for electrostatic charge build up in the fuel prior to reaching a tank. Work has been primarily experimental or empirical. A motivation for the model presented here is to improve the understanding of measurement data at a more fundamental level. A further motivation is to provide a versatile software tool that can be used to predict the effects on charge generation of design changes, new materials or fuels, or modified refuel conditions.

Some of the theory used in the model, which is based on electrical double layer (EDL) theory, is outlined. The principles behind the measurement rig are then presented. Model data is fitted to measurement data, for a range of fuels flowing in straight pipes. This provides insight into why the charging tendencies of different fuels can be strongly dependent on fuel type.

2. Principles behind the model

Electrostatic charge build up as liquids flow through pipes has been studied for many decades, experimentally and theoretically. Some models are based on semi-empirical relationships derived from

[3] Corresponding author

measurements (see for example [1], [2]). This approach makes it difficult to extrapolate to conditions outside those used to derive the model.

More rigorous modelling, based on EDL theory, has been performed (see, for example, [3],[4]), Positive and negative ions are assumed to be present in the liquid, arising from the dissociation of ionic impurities or additives. Ions of both sign react with the pipe wall. Ions of one sign are preferentially adsorbed onto the wall, forming a layer attached to the wall. Ions of opposite sign form a 'diffuse layer' that extends some distance from the wall. If the liquid were not moving then ions would remain in these layers. When the liquid is moving then the ions in the diffuse layer will be entrained in the flow and travel down the pipe with the liquid.

The details of the modelling are dependent on the further assumptions that are made. In our model we assume that the 'small charge density' approximation is valid. This requires that the concentration of the excess ions is 'small' such that the imbalance in the concentrations of positive and negative ions is a small perturbation on these concentrations. We also assume that the flow in aircraft fuel pipes and other fuel systems components is turbulent. With these assumptions the charge density in a pipe is given by [5],[6],[7]:

$$\bar{q}(x) = q_w\left\{\frac{R\tau_w}{\rho\bar{v}^2}\left(\frac{\lambda}{a}\right)^2\left[1-\frac{\delta/\lambda}{sinh(\frac{\delta}{\lambda})}\right]+\frac{\frac{\delta}{\lambda}}{sin\,h\left(\frac{\delta}{\lambda}\right)}\frac{1}{(1+\frac{a\delta}{2\lambda^2})}\right\}\left[1-exp(-\frac{x}{L_D})\right]+\bar{q}_0\,exp\left(-\frac{x}{L_D}\right)$$
$$+\;B(\bar{q}_0-\bar{q}_\infty)\left[\frac{1-exp\left(-\frac{x}{L_D}\right)}{\frac{x}{L_D}}-exp(-x/L_D)\right] \qquad (1)$$

where $\bar{q}(x)$ is the charge per unit volume, averaged over the pipe cross section, at a distance x from the pipe inlet. The parameters are: a is the pipe radius, ρ is the fluid density, $\bar{v}$ is the fluid velocity, averaged over the pipe cross-section, R is the Reynold's number $\left(=\frac{2\rho a\bar{v}}{\mu}\right)$, with μ the fluid viscosity, τ_w is the shear stress at the wall, given by [6] $\tau_w = 0.0396R^{-1/4}\rho\bar{v}^2$, δ is the laminar sub-layer thickness, ie the radial distance from the pipe wall over which the flow is laminar, given by [6]:

$$\delta = 13v\left(\frac{\tau_w}{\rho}\right)^{-\frac{1}{2}}S^{-\frac{1}{3}},\left(S=\frac{v}{D},v=\frac{\mu}{\rho}\right) \qquad \textbf{(2)}$$

where S is the Schmidt number and v is the kinematic viscosity. The quantity λ is the Debye length $=\left(\sqrt{\frac{\varepsilon D}{\sigma}}\right)$, where ε is the fluid's permittivity, D is the average diffusion coefficient, or diffusivity, of the ionic species in the fluid, and σ is the fluid's electrical conductivity. The Debye length determines the radial distance from the pipe wall that the ions are present in the electrical double layer. The quantity q_w, the equilibrium wall charge density, is the density of ions at the wall that forms when the charging and relaxation processes are balanced. This only occurs in a pipe of sufficient length. The quantity x is the distance that the fluid has travelled down the pipe from the entrance, L_D is the charge 'development length', the distance that the fuel flows in the pipe before the equilibrium wall charge is set up. This is given by:

$$L_D = \frac{\varepsilon\bar{v}}{\sigma(1+\frac{2\lambda^2}{a\delta})} \qquad \textbf{(3)}$$

The quantity $\bar{q}_0$ is the average charge density at the start of the pipe and $\bar{q}_\infty$ is the average charge density in an 'infinite' pipe. When the pipe diameter is much larger than the Debye length, which is the regime in which we are interested for aircraft fuel system applications, the equilibrium wall charge is given by [6] $q_w = -\frac{\sigma\zeta}{D}$, where ζ is the zeta potential (see, for example, [6], [8],[12]). *B* is the ratio of the conductance per unit length of the fuel to the conductance per unit length of the fuel and pipe wall. *B=0* corresponds to a metallic pipe, *B=1* to a perfectly insulating pipe.

Other fuel system components are incorporated into the model. Micropore filters are modelled using the theory of Huber and Sonin [9],[10], taking account of charge relaxation with the theory of Washabaugh and Zahn [11]. Components such as valves, bends and junctions are modelled in terms of pressure drop [12]. The effects of temperature are incorporated into the model through a range of analytical expressions gathered from the literature. Details will be the subject of a separate paper.

3. Measurement rig

A purpose-built test rig was built at the Health and Safety Laboratory (HSL) in Buxton UK. The rig comprises; a controlled fuel delivery system operating at up to 150 m^3/hr, a test section where different components may be introduced into the system, and a Faraday Pail for measuring the charge accumulated by the fuel. Measurements include wall currents, charge densities and pressure losses. Figure 1 shows a schematic of the rig.

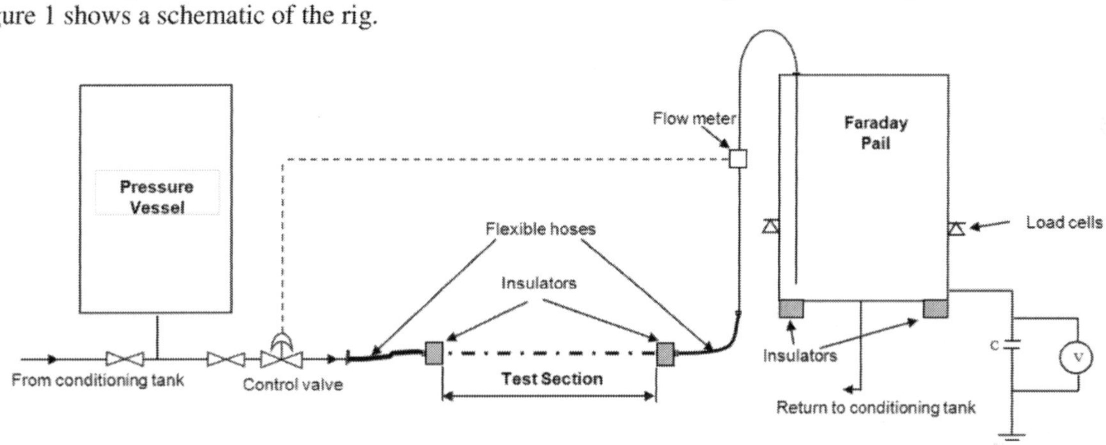

Figure 1: schematic diagram of the test rig.

4. Comparison of experimental data and model predictions

Predictions from the EC-Flow model have been fitted to measurement data, for a range of components and fuels. Fitting was carried out by varying the zeta potential and the ionic diffusivity. Sample results from the fitting are shown in figure 2. Fitted values (at 20°C) are: 35mV (3pS/m fuel) and 33mV (100pS/m fuel) for the zeta potentials and $10^{-9}\text{m}^2\text{s}^{-1}$ (both fuels) for the diffusivities.

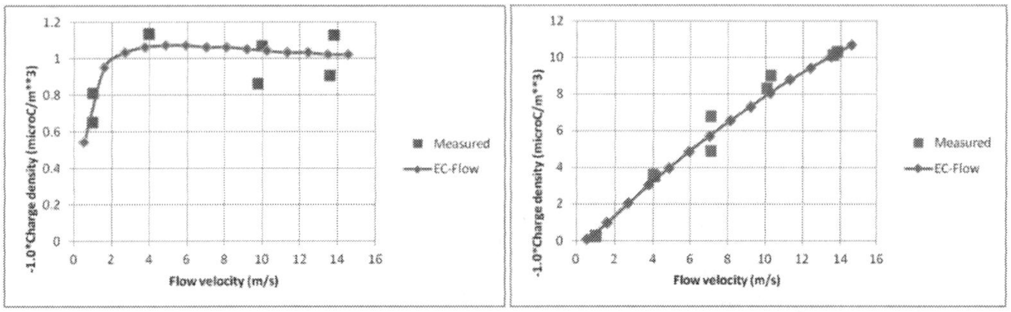

Figure 2: measured and fitted variations of fuel charge density with flow velocity, in metallic pipes of radius ~31mm, for fuel conductivities of 3pS/m (left) and 100pS/m (right).

Journal of Physics: Conference Series **646** (2015) 012021 doi:10.1088/1742-6596/646/1/012021

The very different trends seen in figure 2 for the two fuel conductivities (and for other fuel conductivities not shown here) are understandable when we consider the competing effects that arise as the flow velocity increases. The pipe becomes shorter relative to the charge development length, so less charge may build up. However, the laminar sub-layer thickness decreases so that more charge may be entrained in the flowing fuel. The balance between these competing effects depends on the ratio of the pipe length to the charge development length and on the ratio of the Debye length to the laminar sub-layer thickness. Figure 3 shows the variation with flow velocity of the ratio of the pipe length to the charge development length (solid curves) and the ratio of the Debye length to the lamina sub-layer thickness (dashed curves), for two conductivities. The competing effects are apparent, as are the different regimes for the different conductivities.

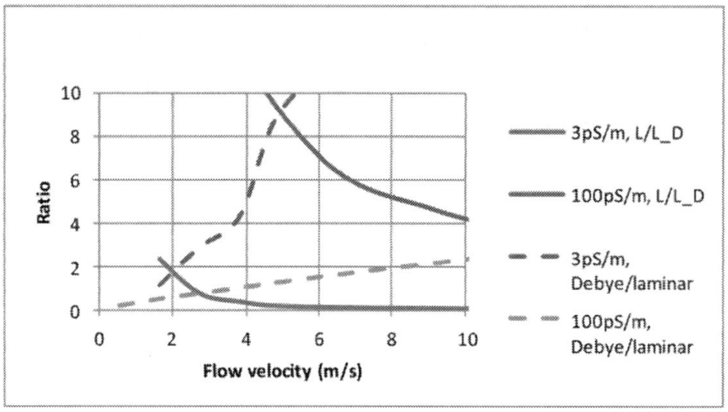

Figure 3: variations with flow velocity of the ratio of the pipe length to the charge development length (solid curves) and the ratio of the Debye length to the laminar sub-layer thickness (dashed curves), for two fuel conductivities (3 and 100pS/m).

5. Summary

A model for charge generation in fuel systems has been developed and an experimental test rig has been built and used to provide comparison data. The model helps understand measured trends in charging behaviours. This has been demonstrated for a simple component (a metal pipe) and for two fuels with different conductivities. A broader range of components, and other fuel types, can be used in the model. The model is useful to help identify primary sources of electrostatic charging, and to examine methods to mitigate against unwanted charging.

References

[1] Gavis J and Wagner J P 1968 *Chem. Eng. Sci.* **23** 381-91
[2] Gibson N and Lloyd F C 1970 *J Phys. D: Appl. Phys.* **3** 563-73
[3] El-Adawy M, Paillat T, Touchard T G and Cabaleiro J M 2011 *IEEE Trans. Diel. and El. Ins.* **18** 1463-75
[4] Zmarzly D 2013 *IEEE Trans. Diel. and El. Ins.* **20** 1497-1509
[5] Abedian B and Sonin A A 1982 *J Fluid. Mech.* **120** 199-217
[6] Abedian B and Sonin A A 1986 *Chem. Eng. Sci.* **41** 3183-89
[7] Carruthers J A and Marsh K J 1962 *J Inst. of Petroleum* **48** 169-79
[8] Gosse J P and Solofomboahangy A 1984 *J Electrostatics* **15** 351-8
[9] Huber P W and Sonin A A 1977 *J Colloid and Interface Sci.* **61** 109-25
[10] Huber P W and Sonin A A 1977 *J Colloid and Interface Sci.* **61** 126-45
[11] Washabaugh P and Zahn M 1997 *IEEE Trans. Diel. and El. Ins.* **4** 688-709
[12] Klinkenberg A and van der Minne J L 1958 *Electrostatics in the Petroleum Industry* (London: Elsevier)

Charge transferred in brush discharges

M Talarek[1], **R Kacprzyk**[2]

[1] - KOMAG Institute of Mining Technology, Gliwice, Poland
E-mail: mtalarek@komag.eu
[2] - Wrocław University of Technology, Wroclaw, Poland

Abstract. Electrostatic discharges from surfaces of plastic materials can be a source of ignition, when appear in explosive atmospheres. Incendivity of electrostatic discharges can be estimated using the transferred charge test. In the case of brush discharges not all the energy stored at the tested sample is released and the effective surface charge density (or surface potential) crater is observed after the discharge. Simplified model, enabling calculation of a charge transferred during electrostatic brush discharge, was presented. Comparison of the results obtained from the simplified model and from direct measurements of transferred charge are presented in the paper.

1. Introduction

Electrostatic discharges from plastic materials, which are mostly brush type discharges, can be incendiary if appear in an explosive atmosphere. Charge transferred test method [1], which is described in Standards [2, 3], allows assessing if the discharge will be incendiary for the specific gas mixture. The idea of the test is to measure charge Q transferred in a single discharge, as the ball electrode approaches the tested sample. Standard [2] allows using three different apparatus for charge transfer measurement (von Pidoll, Schnier, Chubb). Obtained result is compared with the criterion for the specific gas mixture, for example for the methane-air mixture acceptable transferred charge is 60 nC. However, results for charge transferred test have a strong statistical scatter [4], which is the test method disadvantage.

In the case of brush discharges, not all the energy stored in the tested sample is released and the effective surface charge density (or surface potential) crater is observed after the discharge [5,6]. Assuming the ball electrode as the closest earthed element in the vicinity of the tested sample, one can estimate the energy released W from the approximated formula (1):

$$W \cong \frac{1}{2} \cdot \frac{Q^2}{C_E} \tag{1}$$

where: Q – charge transferred during discharge, C_E – capacitance between ball electrode (on the earth potential) and surface of the sample tested just before the discharge. It is worth to emphasize, that the total energy W accumulated in the dielectric (sample) – ball electrode system changes itself with position of the electrode which determines its capacitance C_E. Total charge Q transferred in a discharge can be estimated using the following equation:

$$Q = \int_{S_{CE}} \Delta q_s ds \tag{2}$$

where: Δq_s – change of the effective surface charge density value measured for the pre- and after-discharge states; S_{CE} – part of investigated dielectric surface covered by brush discharges (surface exhibiting change of Δq_s value due to discharges).

The main task of this work was to observe real distributions of surface charge density after the discharge and to propose a simplified model describing it and allowing to assess the total charge Q transferred in brush discharges. Distribution of surface charge density was determined indirectly, using the measurements of surface potential, U_Z which depends on the local surface charge density as follows:

$$U_Z = \frac{q_s \cdot d}{\varepsilon_0 \cdot \varepsilon_r} \quad (3)$$

where: q_s – local value of the surface charge density, d – thickness of tested sample, ε_0 – dielectric permittivity of a free space and ε_r – relative electric permittivity of the tested sample.

Determination of the C_E capacitance, which changes with a distance between the ball electrode and tested sample is a different problem, not considered in the paper.

2. Test set-up and experimental procedure

Test set-up using in the experiments is shown in Figure 1. It consists of: HV source, 100-needles fakir-electrode, non-contact electrostatic voltmeter with compensative probe, capacitor, discharge probe and voltmeter. Electrostatic voltmeter, Trek Model 341B, with compensative no-contact probe, operating in the measurement range ±20 kV was the main instrument in this test stand.

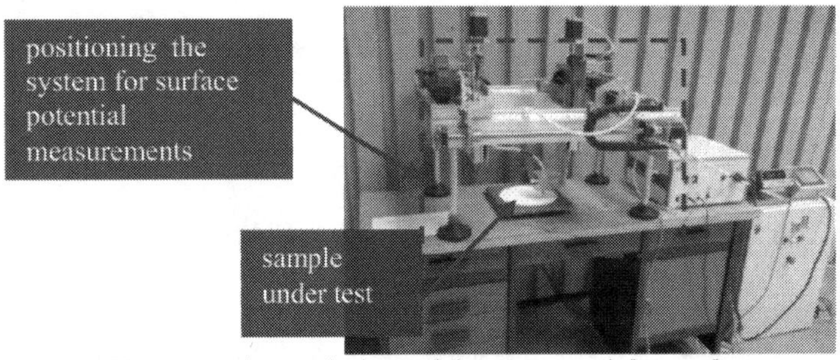

Figure 1. General view of the test stand for surface potential measurements.

Plastic, high resistivity samples of dimensions (150 x 150 x 6) mm and PTFE disc of diameter 250 mm and 6 mm thickness were the test objects. Before the measurements the samples were cleaned with isopropyl alcohol. Climatic conditions of the tests were as follows: temperature T=23°C and relative humidity RH≤ 30%. Multi-needle electrode was used for corona charging, to have satisfactory repeatability of the electrification process. Corona electrode was supplied with negative high voltage -30 kV dc. Freshly charged samples were submitted to transferred charge measurements. Electrostatic discharge was provoked by approaching ball electrode to the tested sample, which was lifted about 100 mm above the metal earthed plate. Ball electrode was connected to the standard capacitor, across which the voltage was measured to determine the transferred charge [3]. Surface potential, U_Z was measured using contact-less method with compensative probe placed in a distance of 5 mm above the tested sample surface.

3. Results and discussion

Typical result of surface potential distributions observed after discharge provoked on the tested PTFE sample is shown in Figure 2. Some results for other samples are presented in Figure 3, with the information about transferred charge obtained from the simplified model.

Journal of Physics: Conference Series **646** (2015) 012022 doi:10.1088/1742-6596/646/1/012022

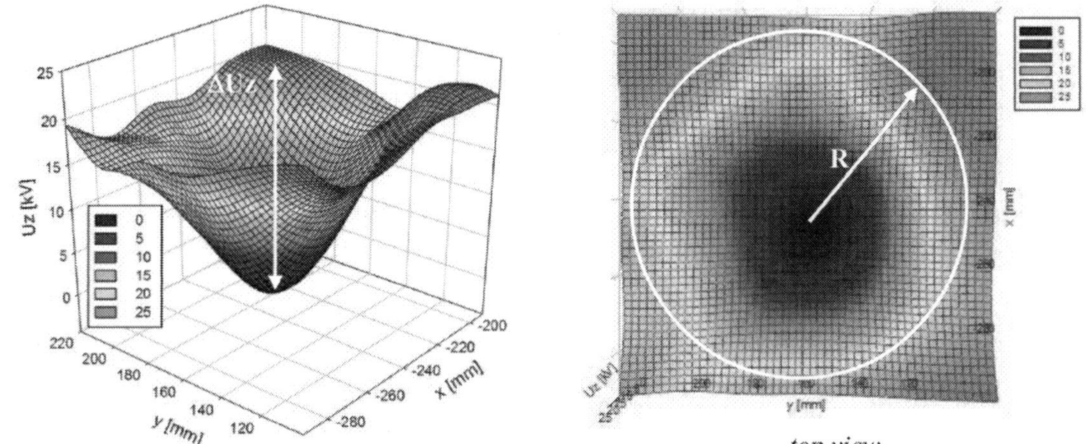

top view

Figure 2. Example of surface potential distribution after discharge at PTFE sample.

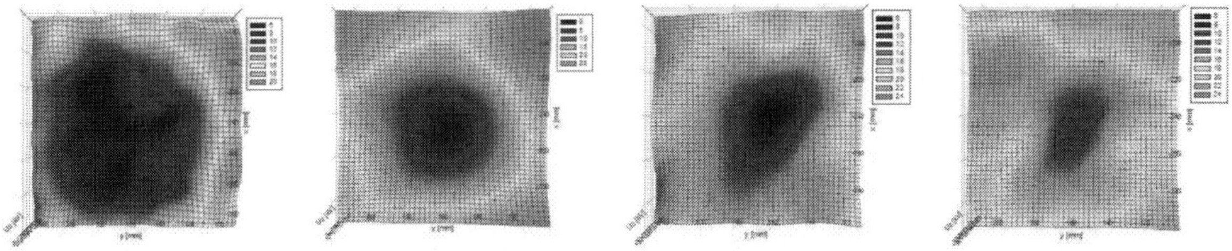

Q=359nC Q=181nC Q=104nC Q=61nC

Figure 3. Some results of surface potential distribution after discharge.

The real potential distributions were modelled by a cone–type distributions using ΔU_Z and R parameters shown in Figure 2 and defined as follows:

$$\Delta U_Z = U_{Z\max} - U_{Z\min} \tag{4}$$

where: $U_{Zmax.}$ and $U_{Zmin.}$ - maximum and minimum values of surface potential, U_Z measured for the same "crater", respectively - determining "crater" depth. The "crater" equivalent radius R was estimated from the following equation:

$$R = \sqrt{\frac{S_C}{\pi}} \tag{5}$$

where: S_C was area of a cross-section (limited by an equipotential line) of the voltage "crater" for the level of 0,8 $U_{Zmax,}$. It should be noted, that $S_C \le S_{CE}$.

The proposed simplified model has the following assumptions:

1) The sample is electrically and geometrically uniform – exhibits constant electrical permittivity and thickness;

2) Uniform distribution of surface charge density along the surface of the whole sample before the discharge;

3) Cone type distribution of effective surface charge density occurs in the "crater" region - after the discharge;

4) Probe-to-surface separation distance does not influence the surface potential results.

Transferred charge Q value was calculated on the basis of a surface charge density change in the "crater" region using (in the first approximation) eq. (2).

The equation can be transformed, for the assumptions given above, to the following form:

$$Q = \frac{\pi}{3} \cdot \frac{\Delta U_Z \cdot \varepsilon_0 \cdot \varepsilon_r}{d} \cdot R^2 \qquad (6)$$

Some results of comparison between measured transferred charge and the charge obtained from the simplified model are presented in Table 1.

Table 1. Transferred charge. Values measured and estimated from the model

	***Q* [nC]**		***k***
Tested sample	Simplified model (SM)	Direct measurements (DM)	k=SM/DM
Plastic sample 1	61	66	0.93
Plastic sample 2	104	108	0.96
PTFE disc	133	164	0.81
Plastic sample 3	184	217	0.85
PTFE disc	181	260	0.70
Plastic sample 4	359	397	0.90

4. Conclusions

Surface charge density craters observed after discharge were previously reported for thin dielectric layers [4,5]. Similar phenomenon was observed for dielectric (plastic) samples, 6 mm thick and with total area equal to 225 cm^2 (according to the requirements of samples for charge transferred test method [2]).

Simplified model, which is used to assess the value of transferred charge, assumes conical shape of the charge density crater. Real surface charge density craters have more complex shape than the conical one. However, relationship between value of transferred charge obtained from the model and the measured value, *k* is about 0.7 to 0.9 which gives an error in the range from 10% to 30%.

The model simplicity is its advantage. This might be useful in the test method, which is to be developed in future research work.

References

[1] U. von Pidoll, E. Brzostek, H. R. Froechtenigt, Determining the incendivity of electrostatic discharges without explosive gas mixtures, IEEE Transactions on Industry Application, 40 (2004), 1467-1475

[2] IEC 60079-32-2:2015 Explosive atmospheres – Part 32-2: Electrostatics hazards – Tests

[3] EN 13463-1:2010 Non-electrical equipment for use in potentially explosive atmospheres – Part 1: Basic method and requirements

[4] M. Talarek, L. Orzech, Results of an interlaboratory comparison on the measuring of transferred charges, Electrostatics 2013 proceedings

[5] T. Oda, Y. Ito, Studies of electrostatic surface discharges on corona-charged polymer surfaces, IEEE Trans. Ind. Appl. 26 (4) (1990) 22-27

[6] J.L. Davidson, T.J. Williams, A.G. Bailey, Electrostatic discharges between charged insulators and grounded spheres, Journal of Electrostatics 56 (2002) 29-42

2.2. Target values and safety margins
IEC 60079-32-1 [1] indicates that to avoid incendive discharges from an insulating liquid surface its potential must be less than 25 kV. Given the likely slurry movement in reality, compared with the inevitable flat surface which would be used for modelling, as well as some other uncertainties, it was somewhat arbitrarily decided to aim for an order of magnitude margin of safety. In other words, the target maximum surface potential as determined by modelling would be 2.5 kV.

To avoid spontaneous discharges across a liquid surface (as opposed to discharges initiated by objects approaching the surface) the maximum electric field must not exceed the breakdown strength of the atmosphere in the vessel head space. Taking the breakdown strength to be that of air under normal atmospheric conditions ($3x10^6$ V.m^{-1}) and applying the larger but no less arbitrary safety margin of two orders of magnitude (mainly allowing for the numerically higher values), a target maximum electric field of $3x10^4$ V.m^{-1} was specified.

2.3. Initial Modelling
A sketch of the high shear blender vessel and a simplified version which might be used for modelling, where only the potentials and fields at the liquid surface are of interest, are shown in Figure 1. Essentially, the simplified version sets the bottom cone of the model vessel approximately at the position of the top of the blender.

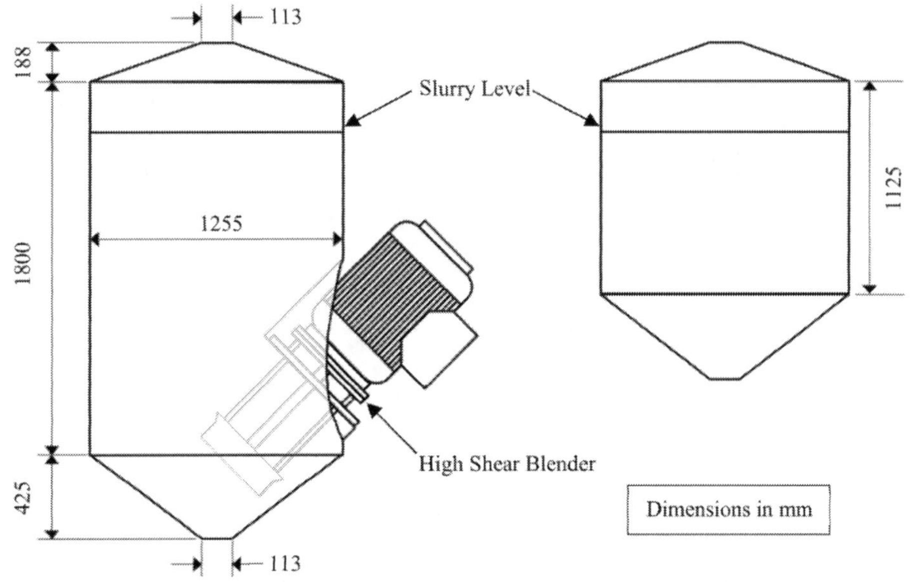

a) High shear blender vessel b) Simplified version for modelling

Figure 1. Sketch showing model dimensions.

The first modelling trials used both of the geometries shown in Figure 1, as well as a charge density in the slurry of $1x10^{-4}$ C.m^{-3}. This value is towards to top end of the $5x10^{-6}$ to $4x10^{-4}$ C.m^{-3} range given in IEC/TS 60079-32-1 for pumped flow in pipes, thereby making some allowance for the expected high charging due to particulates and high shear.

The results showed similar values for electric fields and potentials between the two geometries (better than 10%), such that later trials used only the easier to model simplified version. However, at about $1x10^6$ V.m^{-1} the maximum electric field was very much higher than the target value and, at 340 kV, the maximum potential was well over the safe limit even without applying a safety margin.

3. The detailed assessment

3.1. Measured and modelled fields and potentials for laboratory scale equipment

Laboratory scale trials were carried out using low concentration model slurries and a commercial slurry. The latter was the commercial slurry known to have the lowest conductivity. The laboratory scale blender had a nominal 2 litre capacity with a normal stirrer and a high shear blender. It had a bottom valve through which slurry samples were dropped into a Faraday Pail to allow direct charge to mass ratio measurement. A field meter (Simco Model FMX-003) was used to monitor the electric field above the liquid. Figure 2 shows the blender geometry and the potential distributions determined by the computer model on two planes through the field meter. Table 1 shows a comparison of results.

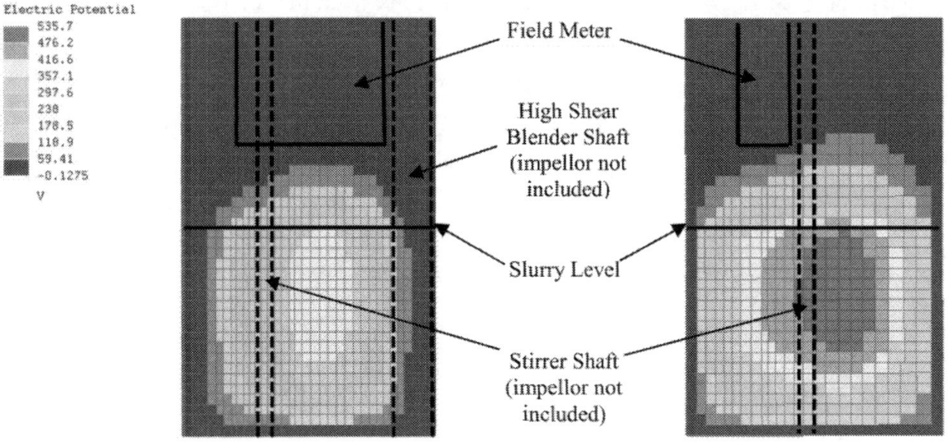

Figure 2. Potential distributions on perpendicular planes through field meter in laboratory blender.

Table 1. Experimental and computer model results for laboratory scale blender.

Formulation	Experimental			Computer Model	
	Slurry Charge:Mass Ratio ($nC.kg^{-1}$)	Measured Field* ($kV.m^{-1}$)	Slurry Surface Potential* (kV)	Field (at meter) ($kV.m^{-1}$)	Slurry Surface Potential (kV)
Model Formulation 1	41	10	0.4	6.6	0.4
Model Formulation 2	820	180	7.2	161	11
	1200	260	10	252	17
Model Formulation 3	6	2	0.08	0.6	0.04

*Measure field calculated from displayed surface potential. Surface potential adjusted for measuring distance.

The good agreement between experimentally determined and modelled values provided additional verification for the computer model and supported the slurry charge-to-mass ratio measurements.

3.2. Estimation of maximum commercial slurry charge to mass ratio

The least conducting commercial slurry tested in the laboratory scale blender was still much more conducting than the model slurries of Table 1. Also, the much higher solids concentration compared with the model slurries meant samples could not be taken for direct charge to mass ratio measurements. However, field meter measurements were taken. In fact, the field was too low to register a sensible reading, but that alone was enough to indicate the electric field at the field meter must have been $<400\ V.m^{-1}$. With confidence from the previous comparisons, running the model with values selected by trial and error showed the charge to mass ratio must have been $<1.3x10^{-9}\ C.kg^{-1}$.

It is well known that charge generation increases with specific power of the blender [1]. The laboratory scale blender provided specific power of 75 $W.kg^{-1}$, compared with 15 $W.kg^{-1}$ for the full scale blender. Hence, using the above determined maximum charge to mass ratio for modelling the full scale blender fields and potentials would be erring on the side of safety.

3.3. Final modelling results for the full scale blender

Final modelling of the full scale blender used a charge to mass ratio of $1.3\text{x}10^{-9}$ C.kg^{-1} (determined above), together with the simplified geometry model referred to in Section 2.3. Graphical and numerical results are shown in Figure 3 and immediately afterwards, respectively.

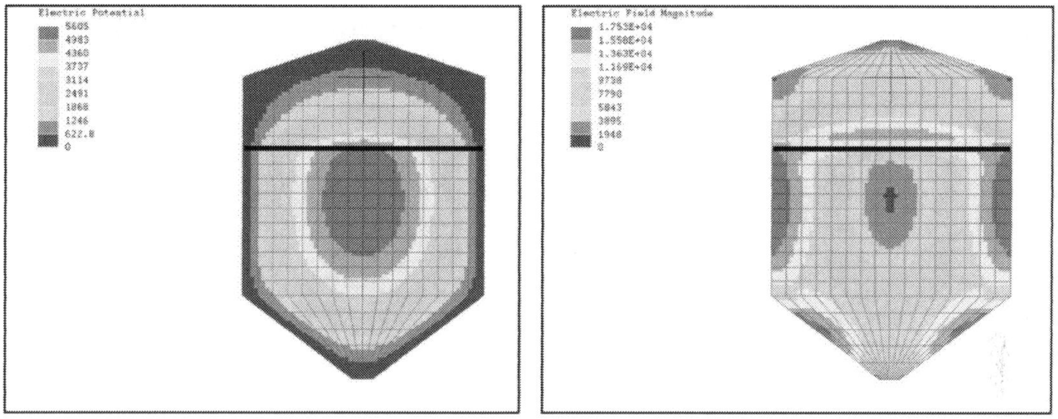

Figure 3. Final graphical results of modelling the full scale blender.

From the results shown graphically in Figure 3 the following values were obtained:

- Maximum potential at the slurry surface: $5.2\text{x}10^3$ V
- Maximum electric field at the slurry surface: 1.6×10^4 V.m^{-1}

The maximum electric field is below the selected target value and, although the surface potential is not as low as its target value, it is well below the 25 kV cited in IEC/TS 60079-32-1.

4. Conclusions

By combining computer modelling with experimental measurements as described it was concluded that incendive electrostatic discharges in the full scale high shear blender vessel are so unlikely that no additional protective measures (such as inerting the vessel headspace) would be required.

Inevitably this conclusion came with a number of limitations based on assumptions made along the way. Set out as recommendations these were: all other ignition sources must be properly assessed and addressed; no fixed or moveable items should protrude into the vessel headspace; reduced batch sizes must not be permitted if the blender specific power exceeds 50% (arbitrary margin) of the laboratory scale blender power, and; a new assessment must be undertaken for new formulations of lower conductivity than the current worst case commercial slurry.

Overall, an approach for combining computer modelling of electric fields and potentials in process plant with experimental measurements has been successfully demonstrated and used for a particular situation. However, it is considered this general approach could represent a useful starting point for undertaking many similar assessments for a wide range of other process equipment.

References

[1] IEC/TS 60079-32-1 2013 Explosive atmosphere Part 32-1 Electrostatic hazards, guidance
[2] Glor M and Pey A 2013 *J. Electrostatics: Electrostatics 2013, Budapest* **71** pp362-367
[3] Britton L G 1999 *Avoiding Static Ignition Hazards in Chemical Operations* (New York: AIChE/CCPS)

Charge generation associated with liquid spraying in tank cleaning and comparable processes - preliminary experiments

Carsten Blum[1] and Oswald F J Losert[2,3]

[1]DEKRA EXAM GmbH, Abteilung Explosionsschutz, Dinnendahlstr. 9, D-44809 Bochum, Germany,

[2]Berufsgenossenschaft Rohstoffe und chemische Industrie, Präventionszentrum Nürnberg, Südwestpark 2-4, D-90449 Nürnberg, Germany

E-mail: oswald.losert@bgrci.de

Abstract. The BG RCI has initiated investigations in order to improve the data basis for assessing the ignition hazard by electrostatic charging processes associated with the spraying of liquids. On the base of preliminary experiments, we established procedures for measurements of electric field strength and charging current in the presence of aerosol particles. Results obtained with three different nozzle types, variation of pressure and with built-in deflecting plate are presented.

1. Introduction

1.1. Motivation

In Germany the technical guidelines of TRBS 2153 [1] regulate avoiding ignition hazards resulting from electrostatic charging. This document was established and is updated by the Deutsche Gesetzliche Unfallversicherung (DGUV)[4] under the direction of the Berufsgenossenschaft Rohstoffe und chemische Industrie (BG RCI)[5].

Many industries, e. g. chemical, pharmaceutical, petrochemical, and food industry, perform processes in which liquids are sprayed. Therefore those responsible must assess the danger of ignition by electrostatic discharges. In marked contrast to this requirement the guidance given by TRBS 2153 is scarce. One example is the maximum pressure that is still acceptable when cleaning containers with different solvents in order to avoid dangerous charging. The lack of clear guidance is due to limited experimental data for the actual conditions that result in hazardous charging. For the existing statutory provisions it is not known how far the "safe range" of parameters is apart from dangerous conditions.

This is the reason why BG RCI has decided to initiate investigations in order to improve the data basis for assessing the ignition hazard by electrostatic charging processes associated with the spraying of liquids. Two objectives of the preliminary experiments are presented here. First we establish procedures for measurements in the presence of aerosol particles, which is essential prior to

[3] Author to whom any correspondence should be addressed
[4] German Social Accident Insurance
[5] German Social Accident Insurance Institution for the raw materials and chemical industry

quantitative investigations. Secondly, we target the most critical parameters among the many that impact charge generation in technical applications.

1.2. Physical context and previous studies
The charging of small droplets in spraying processes is well known as Lenard effect and commonly referred to as "waterfall electricity", a nickname, which reflects that this effect is not confined to liquids of low conductivity. The fact that discharges resulting from spraying liquids can act as an ignition source was brought to mind in 1969, when within a couple of weeks three supertankers suffered heavy damages due to explosions during the cleaning of empty tanks with seawater. These incidents stimulated extensive studies researching electric fields and potentials generated by liquid jets at 12 bar in tanks of some 20.000 m³ size, e. g. [2]. The findings were that most likely water slugs discharging charged clouds of droplets were the cause of the explosions. While this setup is unlikely to occur in smaller tanks and containers, the question arose as to whether in smaller vessels brush discharges capable of igniting explosive atmospheres can be produced involuntarily. In 1989 Post et al. [3] published results from experiments with combustible liquids and water in a 1 m³ container at up to 50 bar and with water in a 24 m³ container at pressures from 100 bar to 500 bar. They gave safe conditions for operations in the investigated pressure range, which have been implemented in the German regulations. With the advance of technologies the call has intensified for answers extending the scope of permissible conditions.

2. Experimental setup
The tests were performed by spraying industrial and purified water with a commercial high pressure cleaner (HD9/50-4, Kärcher) through the manhole into a modified cylindrical stainless steel 1 m³-IBC. The outlet valve was left open during spraying and the water was collected in an insulated container (not shown in the experimental setup, figure 1). The electric field strength was measured with a field meter (JCI 131, JCI Chilworth). In addition, the temporal progress of charge accumulation on the wall

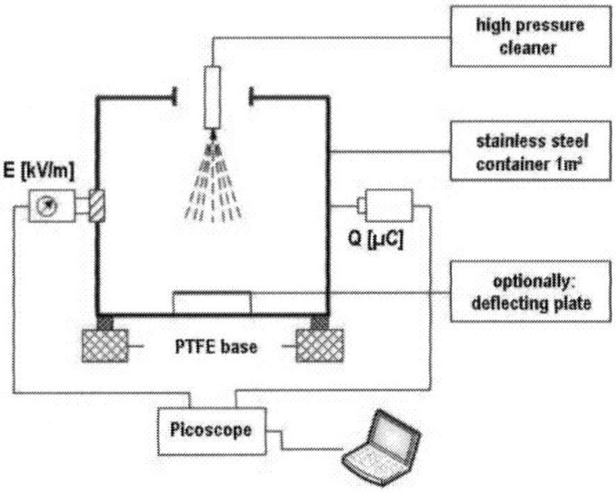

Figure 1. Experimental Setup.

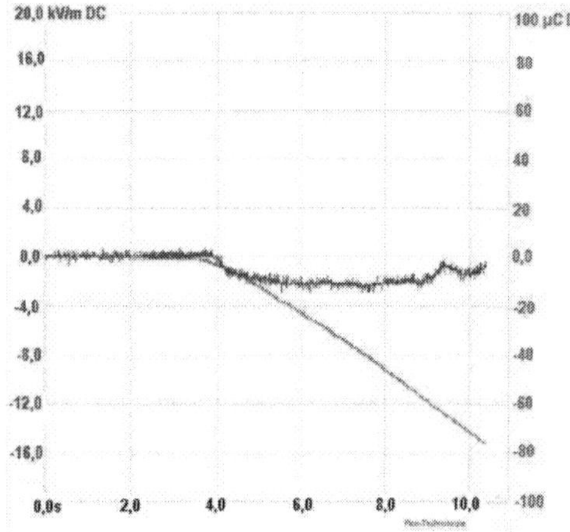

Figure 2. Simultaneous record of field strength (top) and charging current (bottom writing).

was recorded with a charge amplifier (5011B, Kistler). In all experiments the observed increase of charge during the spraying time was linear (e. g. bottom writing in figure 2). Therefore, we evaluated the gradient dq/dt, which yields the charging current. The nozzle types used were a pencil jet nozzle (.3 mm), a flat spray nozzle (15°, .3 mm), and a rotating nozzle (only 300 bar). The pressure was varied from 100 bar to 500 bar. The IBC could be fitted with a deflecting plate mounted close to the bottom but electrically isolated from the IBC. Two different plates were used, an insulating plate of polyethylene and a metal plate earthed by a cable laid through the outlet valve.

3. Results

3.1. Procedures for measurements of electric field strength and charging current

The objective of the experiments was to establish a procedure to measure simultaneously i. the electric field strength of the spray cloud using the induction field meter and ii. the charge generated by droplets arriving at the wall of the IBC – the charging current – using the charge amplifier and the oscilloscope (figure 2). Both methods were suited. The induction field meter worked reliably in the presence of aerosol particles. Furthermore, the measurements of charging currents produced correct results. Since charge accumulation on the wall can only be determined if the vessel is electrically isolated against earth, simultaneous measurements could not be continued, because in the course of the measurements the water fog shorted the insulating installation of the field meter, draining the charge via the casing. So the electric field strength and the charging current had to be measured consecutively maintaining constant parameters.

3.2. Pressure dependence and influence of installations

As anticipated, charging currents and electric field strength increased with pressure (figure 3). The differences in charging between the pencil jet nozzle and the flat spray nozzle were less marked (figure 4). Introducing the deflecting plate lead to a lower charge on the wall. The currents measured at the wall and from the earthed metal plate had similar values (figure 4 bottom lines), summing appropriate values almost exactly equals the currents measured for the polyethylene plate.

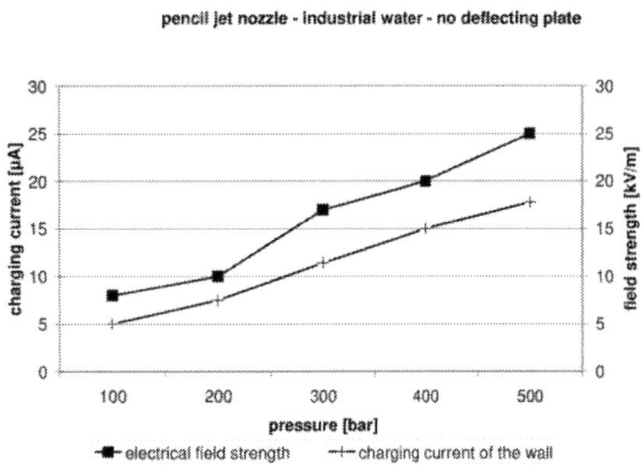

Figure 3. Pressure dependence, el. isol. container

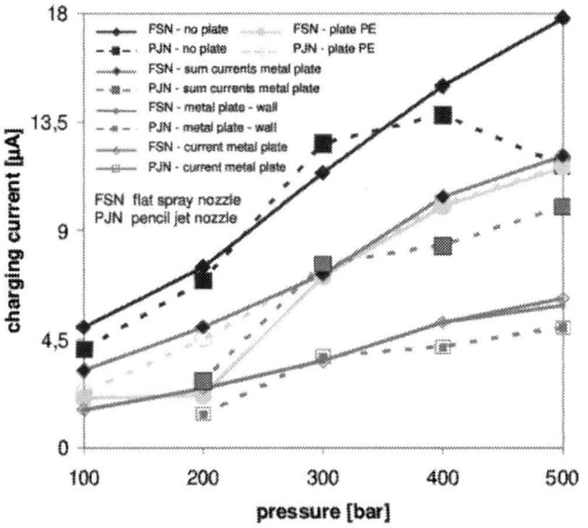

Figure 4. Charging current, industrial water.

3.3. Conductivity of the sprayed water

The purified water and the industrial water used differed in conductivity by approximately one order of magnitude. This was the most important parameter giving the most significant differences in measured field strengths and currents. At identical conditions, the charging current using purified water was found to be up to eight times that of the value measured using industrial water. When spraying into the grounded container only spraying with purified water yielded a measurable electric field of the spray cloud.

3.4. Grounded container

In practical applications containers are grounded rather than isolated against earth. With respect to the objective of assessing the danger of ignition by electrostatic charging the electric field strength was also monitored when spraying into a grounded container. Table 1 gives the results obtained for spraying purified water with a pressure of 500 bar (flat spray nozzle and pencil jet nozzle) and 300 bar (pencil jet nozzle and rotating nozzle) without deflecting plate and using a polyethylene deflecting plate. This shows that the field strength increased from flat spray to pencil jet and was highest employing the rotating nozzle. Also the introduction of a polyethylene deflecting plate increased the field strength considerably.

Table 1. Electric field strength – container grounded.

Nozzle	Pressure [bar]	Deflecting plate	E [kV/m]
flat spray	500	no	4
flat spray	500	polyethylene	16
pencil jet	500	no	6
pencil jet	500	polyethylene	25
pencil jet	300	no	0
pencil jet	300	polyethylene	16
rotating	300	no	12
rotating	300	polyethylene	20

4. Discussion and conclusions

When a liquid is sprayed into a container charging occurs at the phase boundaries in the feed pipe and the nozzle bringing charged liquid into the container and also if the jet breaks up and when it hits the wall of the container. As we collected the water draining from the container in an insulated vessel, while the container itself was electrically isolated against ground, our measurements of the charging current of the wall give the total net charge reduced by the charge of the cloud of droplets. Thus a total charge of the liquid of 71 μC/kg is calculated from the measured current of 17.8 μA and a flow rate of 15 l/min (values for industrial water, 500 bar, flat spray nozzle, no deflecting plate).

The charging currents we measured for flat spray resp. pencil jet nozzles at 500 bar are one order of magnitude higher than the value found by [3] for industrial water sprayed at 50 bar into a 1 m^3 container. Taking into account different nozzle geometries and sizes the values match fairly well.

In order to arrange conditions close to practice and to allow validation by calculations, we measured the electric field strength with the container grounded. Using the same parameters as given in the first paragraph, we obtained a field strength of 4 kV/m. Assuming a homogeneous space charge density in the cylindrical container and using Gauss law, we can derive a space charge density of 133 nC/m^3. This estimate is about a factor 2 smaller than the value of 240 nC/m^3 given by [3] for their experiments with a pair of nozzles in a 24 m^3 container.

The measured field strengths in the grounded container were far from the physical limit given by the uniform breakdown field of air of 3 MV/m. For this reason there is no ignition hazard from electrostatic discharges under conditions comparable to our tests.

Installations reduced the measured charge of the liquid and increased the measured field strength. It appears that fewer droplets arrive at the wall resulting in a higher charge of the cloud of droplets. Since the geometry is more complex, further model calculations of the potential and the resulting electric field could help in clarifying the contributions of the different charging mechanisms.

The preliminary tests have shown that the procedure outlined above procures valid information on charging in spraying and related processes. It can be used to obtain experimental data in application-oriented setups, e. g. bigger containers and higher pressures. Thus the conclusion of [3] that the space charge density is converging at pressures above 100 bar to an upper limit of approx. 240 nC/m^3 can be put to the test. Eventually this is a step on the way to safe operation conditions based on a better understanding of the underlying physical processes.

The authors gratefully acknowledge technical support, esp. the modified container, provided by Merck KGaA, Darmstadt/Germany.

References

[1] 2009 *TRBS 2153 „Vermeiden von Zündgefahren infolge elektrostatischer Aufladungen“*
[2] Bright A W, Hughes J F and Makin B 1975 *Journal of Electrostatics* **1** 37-46
[3] Post L, Glor M, Lüttgens G and Maurer B 1989 *Journal of Electrostatics* **23** 99-109

Electrostatic Ignition Hazards Associated with Dissipative Dust Filters

M Glor, S Forestier and P Meier
Swissi Process Safety GmbH, Basel, Switzerland

E-mail: martin.glor@swissips.com

Abstract. In present guidelines and codes of practice it is strongly recommended to use dissipative filter media in dust separators, if the minimum ignition energy of the powder is below 3 mJ and/or a hybrid mixture may be present. Despite being in compliance with this rule and despite the filter media passed the relevant tests, burnt product was found in a dust separator of pharmaceutical production. Extensive investigations have shown that the demands made in the present standards on dissipative filter clothes are not sufficient to prevent ignition of deposited powder and ignition of an explosive atmosphere. Proposals for additional tests and requirements are presented in the paper.

1. Introduction

In the relevant guidelines to avoid ignition hazards due to static electricity TRBS 2153 [1] and IEC/TS 60079-32-1 [2] the use of dissipative filter media is strongly recommended in dust separators, as soon as the minimum ignition energy of the powder is below 3 mJ and/or hybrid mixtures cannot be excluded. The dissipative properties of textile filter media are usually determined according to the standard DIN 54345-5 [3]. Sometimes the measurements are also performed according to the standard DIN EN 1149-Part 1 [4] though this standard is really intended for personal protective clothing.

Despite being in compliance with the present guidelines and despite the filter media installed in the dust separator passed the relevant tests, burnt product was found in a dust separator used in pharmaceutical production. The product being processed had a high resistivity of more than $10^{13}\,\Omega{\cdot}m$ and low minimum ignition energy of less than 3 mJ. Solvent vapors or hybrid mixtures have not been present. Extensive investigations outlined in the following have shown that the demands made in the present standards on dissipative filter clothes are not sufficient to prevent ignition of deposited powder and/or ignition of an explosive atmosphere in dust separators.

2. Investigations and results

In DIN 54345-5 [3] the resistance between two metal electrodes of length 50 mm and variable distance between 10 and 300 mm put on top of the textile filter media is measured. In DIN EN 1149-Part 1 [4] the resistance between to concentric circular electrodes with a gap of about 10 mm put on top of the textile filter media is measured. In both cases the homogeneity of electrical conduction through the whole filter media with a large contact surface of the electrodes with the filter media is determined. Most dissipative or conductive filter media are made dissipative or conductive by incorporation of very thin metal or carbon loaded filaments. Due to the large contact area of the electrodes in the test arrangements the electrical current between the electrodes flows along a huge number of conductive filaments as explained above. Thus, the electric current flowing through one single filament remains rather moderate, particularly if a voltage of 100 V is applied as recommended in the standards. In practice the current is accumulated all over the area of the filter cloth and flows to earth through the contact points with earth. Depending on the geometry and installation of the filter cloth the contact area with earth may be small. Particularly if the filter is purged the filter cloth may become detached from the supporting earthed metal grid and the contact area with earth may become very small. Exactly at this moment the current density all over the filter area will be highest due to the separation process between powder particles and filter cloth surface.

Based on new investigations [5] the charging current density due to contact and separation of powder particles in pneumatic transfer systems and dust separators may reach 1 mA/m^2. Thus, from a filter cloth area of 0.1 m^2 a total current of 100 μA has to be discharged to earth.

If in a filter assembly the charging current generated over the whole surface is lead to earth through a few small contact points, a high current will flow through a few very thin conductive filaments. Under these conditions the thin conductive filter filaments may start to glow. This phenomenon was observed in a dust separator in pharmaceutical production. Fortunately this phenomenon was observed before leading to a dust explosion. The resulting scorch marks on the filter cloth are shown in figure 1. The scorch marks are exactly in the region, where the filter cloth was in contact with the supporting earthed structure of the filter housing.

Figure 1. Scorch marks on the filter cloth of a dissipative filter material from a dust separator in pharmaceutical production

In further investigations, the filter media were earthed at the rim and contacted at different locations on the surface with the pointed electrode of a megohmmeter. Immediately after the contact the filter medium surface began to glow at the contact point with the pointed electrode. In addition glowing was observed at the earthing contact, which was achieved with an alligator clip. With an applied voltage of 1000 V the total electrical current flowing from the megohmmeter to the earthing alligator clip was typically between 70 μA and 100 μA. Photographs of this experiment are shown in figures 2 and 3. In case the pointed electrode was slid across the surface a luminous tracer could be observed, which left behind a trace with burnt surface spots. Several video clips showing such luminous tracers have been recorded.

Figure 2. Glowing spot on the filter medium when contacted by the pointed electrode of a megohmmeter.

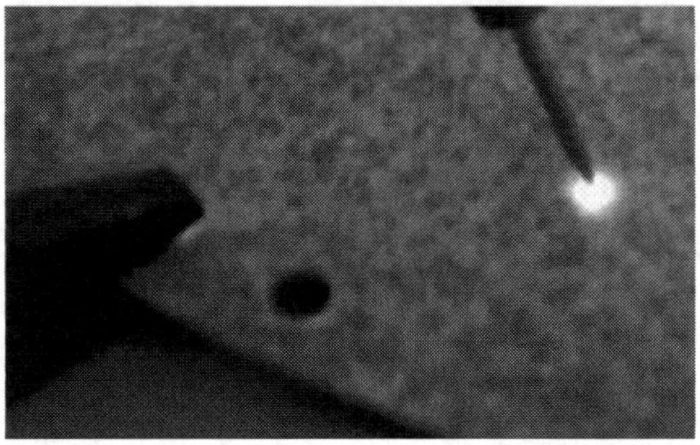

Figure 3. Zoomed in region of figure 2 with the glowing spot. The glowing spot around the earthing contact at the location of the alligator clip becomes also visible.

Since already at moderate electrical currents of 70 µA to 100 µA such glowing spots have been observed and former experiments [5] have shown that such currents are realistic in dust separators, the hypothesis, that the total electrical current integrated over the whole area of the filter cloth during the separation process between the powder and the filter cloth surface has been responsible for the scorch marks on the filter cloth as shown in figure 1, was strongly supported.

In additional experiments it was shown, that neither glowing spots nor traces of burnt material could be observed, if the electrical current in the circuit with the megohmmeter was below 30 µA.

3. Proposal for additional tests

To prevent phenomena as described above the density and interconnection of the conductive filaments in filter media must be increased and improved.

In order to identify and prevent the phenomena described in this paper the following tests have proven to be very useful and reliable in addition to a measurement of the surface resistance or surface resistivity according to the standards mentioned above:

The filter cloth is laid on an insulating surface. An alligator clip is fixed to the rim of the filter cloth. The alligator clip is connected via a µA-meter to the earth port of a megohmmeter. A test tip is connected to voltage port of the megohmmeter. The measuring voltage is set to 1000 V. An electrical contact between the test tip and the filter cloth surface is applied at 5 different locations. In addition the test tip is slid across the surface of the filter cloth. These measurements are repeated with 4 different earthing points.

The tests must be performed in the dark and the following observations and measurements must be made:

- Occurrence of glowing spots at the test tip
- Occurrence of glowing spots at the earthing clip
- Measurement of the electrical current

Based on the observations and measurements listed above the following two different safe kinds of filter media construction types have to be distinguished:

1. The surface resistance or surface resistivity of the dissipative filaments in the filter medium is rather high. The surface resistance or surface resistivity meets however the requirements specified in the relevant guidelines [1, 2]. In this case the electrical current stays rather low - also in case of a very local contact of the earthing point. In this case all of the following requirements must be met:
   a) The surface resistance is below $10^8\ \Omega$.
   b) No glowing spots are observed at the test tip. The very weak spark, which may occur due to the contact of the test tip at 1000 V and the dissipative structure of the filter medium, is tolerated.
   c) No glowing spots are observed at the earthing clip.
   d) The current stays below 20 μA.

2. The resistance of the dissipative filaments in the filter medium is rather low. The density of the dissipative filaments in the filter medium is rather high or the conductive cross section of the dissipative filaments is sufficiently high that also in case of a punctual contact now glowing of a single filament occurs. In this case all of the following requirements must be met:
   a) The surface resistance is below $10^8\ \Omega$.
   b) No glowing spots are observed at the test tip. The very weak spark, which may occur due to the contact of the test tip at 1000 V and the dissipative structure of the filter medium, is tolerated.
   c) No glowing spots are observed at the earthing clip.
   d) The current is not relevant.

Note: If it would only be stated that the resistance to earth from any point on the filter material shall be less than $10^6\ \Omega$ as mentioned in IEC/TS 60079-32-1 [2], sub-clause 13.2.2 - where it is explained that if the charging current is 100 μA or higher, the resistance to earth should be less than $10^6\ \Omega$ - glowing filter filaments could still occur, since already one or a few very thin metallic filter filaments will accomplish a resistance below $10^6\ \Omega$ but may start to glow at a current of 100 μA or higher.

Though the tests described in this paper may appear to be rather empirical lacking any quantitative scientific background they have proven to be very useful and applicable to many different kinds of filter clothes and - most important - filter clothes, which have passed these tests, no longer provoked burned product.

4. References

[1] TRBS 2153 Technische Regel Betriebssicherheit 2153 „Vermeidung von Zündgefahren infolge elektrostatischer Aufladungen", Germany, June 2009.
[2] IEC/TS 60079-32-1, Technical Specification EXPLOSIVE ATMOSPHERES – Part 32-1: Electrostatic hazards, Guidance, Edition 1.0, 2013-08.
[3] DIN 54345-5 Elektrostatisches Verhalten: Bestimmung des elektrischen Widerstandes von Streifen aus textilen Flächengebilden (Juli 1985).
[4] DIN EN 1149-Part 1 Schutzkleidung - Prüfverfahren für die Messung des Oberflächenwiderstandes.
[5] Fath W, Blum C, Glor M and Walther C D *Journal of Electrostatics* **71** 377-382.

Effects of magnetic fields during high voltage live-line maintenance

Gábor Göcsei, Dr. István Kiss, Bálint Németh
Budapest University of Technology and Economics
High Voltage Laboratory

E-mail: gocsei.gabor@vet.bme.hu

Abstract. In case of transmission and distribution networks, extra low frequency (typically 50 or 60 Hz) electric and magnetic fields have to be taken into consideration separately from each other. Health effects have been documented from exposures to both types of fields. Magnetic fields are qualified as possibly carcinogenic to humans (category "2B") by WHO's cancer research institute, International Agency for Research on Cancer (IARC), so it is essential to protect the workers against their harmful effects. During live-line maintenance (LLM) electric fields can be shielded effectively by different kinds of conductive clothing, which are enclosed metal surfaces acting as a Faraday-cage. In practice laboratory measurements also prove their efficiency, the required shielding ratio is above 99% by the related standard.. A set of measurements have proved that regular conductive clothing used against the electric fields cannot shield the magnetic fields effectively at all. This paper introduces the possible risks of LLM from the aspect of the health effects of magnetic fields. Although in this case the principle of shielding the electric fields cannot be applied, new considerations in equipment design and technology can be used as a possible solution. Calculations and simulations based on the data of the Hungarian transmission network – which represents the European grid as a part of ENTSO-E – and high-current laboratory measurement results also prove the importance of the topic.

1. Introduction

Live-line maintenance can guarantee the execution of various kinds of activities on the different equipment of the low, medium and high voltage grid while the system is energized. This way of work has numerous technical and economic benefits without any consumer disturbance. The method – often called as the "future of maintenance" – is beneficial from both the side of the consumer and the transmission system operator (TSO) or distribution system operator (DSO) [1]. Lately different methods of LLM became popular because of the advantages introduced above. Besides its well-known benefits, from the other side all risks related to the LLM personnel have to be taken into consideration. The main aim of this paper is to clarify the non-visible sources of danger from the aspect of both short and long-term effects of electric, but especially magnetic fields. With the growth of TSOs and DSOs applying different LLM methods, this mostly unclarified question become up-to-date, important, practical and more and more urgent to handle in a proper way.

2. Electric and magnetic fields (EMF)

Both extra low frequency (ELF) electric and magnetic fields (below 100 kHz) can have several short- and long term health effects which have to be taken into consideration separately from each other. There are presently two sets of International Commission on Non-Ionizing Radiation Protection (ICNIRP) guidelines in use. For ELF electric fields, the guideline values from ICNIRP (2010) and ICNIRP (1998) both stipulate 5 kV/m maximum permissible exposures for the general public and 10 kV/m for occupational exposures [2], [3], [4]. The proposed maximum magnetic flux density values given in ICNIRP (2010) are twice as high as those in ICNIRP (1998), with permitted vales of 200 μT for the general public and 1 mT for occupational exposures. 19 European Union members have adopted exposure guideline values of 100 μT for power frequency exposures for the general public, sometimes with special provisions included [13].

2.1. *Effects of ELF electric fields*

It is generally recognised that short-term above-the-limit exposure to electric fields values may cause discharges on the surface of the skin and generate eddy currents in the human body. This is as a result of drift currents, which have similar long-term effects to static magnetic fields. These are a major reason why electric fields also have specified limits as prescribed by ICNIRP and elsewhere.

2.2. *Effects of ELF magnetic fields*

Short-term effects of above-the-limit ELF magnetic fields have several effects on body organs especially with high moisture rate (e.g. human eyes). Decrease in precision of movements is also reported in various articles based on the results of different in-vitro and in-vivo experiments inspecting bio-currents [2], [3]. Numerous epidemiological inspections have been executed lately to inspect the possibly harmful long-term effects of magnetic fields. The main reason is that ELF magnetic fields are often referred to as a possible cause of childhood leukaemia [3]. As a result of the studies, WHO's cancer research institute, International Agency for Research on Cancer has qualified ELF magnetic fields as possibly carcinogenic to humans (category "2B") [5]. As it can be seen, proper shielding of ELF magnetic fields is important. For the proper simulation of a typical high voltage power line, a cross-border Hungarian interconnection – as a part of ENSTO-E's European grid – has been selected. Based on the data of the Hungarian TSO, typical peak current load of this given power line is about 2 kA per phase.

In case of localized sources (in a distance of less than 20 cm) limits cannot be applied directly [2], [3], [6]; dosimetric calculations have to be executed case by case instead. For this case the reference current density value is 10 mA/m^2 inside the human body. Model of IEC 62233 [7] is shown in Figure 1 with the results of the finite element calculations for current densities. The distance from the phase conductor was 10 cm (3.94") which is typical in case of barehand method – widely applied as a technology of high voltage live-line maintenance worldwide. As it can be seen form the results, current density in the head – which is one of the most critical part of the human body – may exceed the limit during normal operational conditions. The main question about the topic is whether ELF magnetic fields can be shielded by the conductive clothing similarly to ELF electric fields or not. To clarify this question different types of conductive clothing have been inspected in the High Voltage Laboratory of Budapest University of Technology and Economics (BUTE). As results show none of the conventional clothing is effective from the shielding of ELF magnetic fields at all [8]. Because of the exposure values (induced currents) above the limits and the ineffective way of current shielding it is especially urgent and important to find a solution to prevent the workers from the possibly carcinogenic effects of these kind of fields.

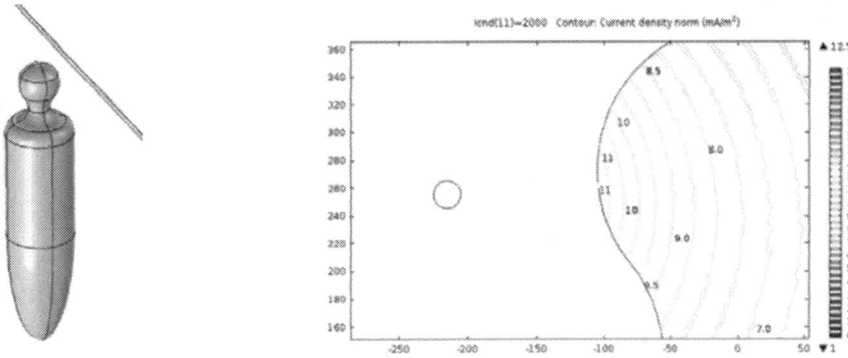

Figure 1. 3D human body model and calculated induced current densities in the head.

3. Ways of EMF shielding EMF

Although generation of both ELF electric and magnetic fields are regardless from voltage level or current load, in case of most low and medium voltage overhead lines, relatively low voltage and current (or relatively high distances) can guarantee the safety of the worker without any additional need of any kind of shielding. In case of high voltage LLM, both electric and magnetic fields might be above their valid limits [9], [10], [11]. Electric fields can be shielded effectively by a conductive clothing acting as a Faraday cage. Inside an ideal Faraday cage (an enclosed conductive surface) electric field is zero from outer source. However ideal Faraday cages cannot been applied practically during any kind of live-line work, conductive clothing – with a face screen [12] – usually act as an effective way of shielding, with screening efficiencies above 99%. However shielding of ELF electric fields is solved practically during LLM, in the current practice magnetic fields cannot be reduced significantly at all during any kind of live-line work. The main aim of this paper is to present a possible (currently mainly theoretical) way of shielding ELF magnetic fields during high voltage live-line maintenance. Development of practical way of application is under development in the High Voltage Laboratory of Budapest University of Technology and Economics as a part of an international research and development project in co-operation with European TSOs.

4. Principle of shielding ELF magnetic fields

It is possible to "shield" ELF magnetic fields by the application of basic physical laws. If an additional current path is ensured in parallel with the original current-carrying conductor, two effects of the distributed currents will decrease magnetic flux density simultaneously. Original magnetic field of phase conductors is generated by the phase current flowing through them. In case of parallel current path(s) ensured at the site of the work, summarized current is divided by a ratio depending on the number of branches; current of each branch – so the magnetic flux density generated by them – will be lower than the original value. If there are two or more conductors with a current in the same direction a given area between the conductors will be "shielded" by the cancellation of the magnetic field between the conductors. The size and the shape of this area depends on the currents of each branch and the geometry of the arrangement. As a result of the reasons above a so-called "protected area" is being generated between the conductors. If the worker remains inside this area, this kind of shielding can guarantee the safety of the work at any time. An example for the shielding in case of 2 conductors (1 original phase conductor and 1 additional conductor installed at the working site) is shown in Figure 2. Protected area marked with "P" is also shown both in the magnetic flux density distribution map and the magnetic flux density strength curve as a function of distance.

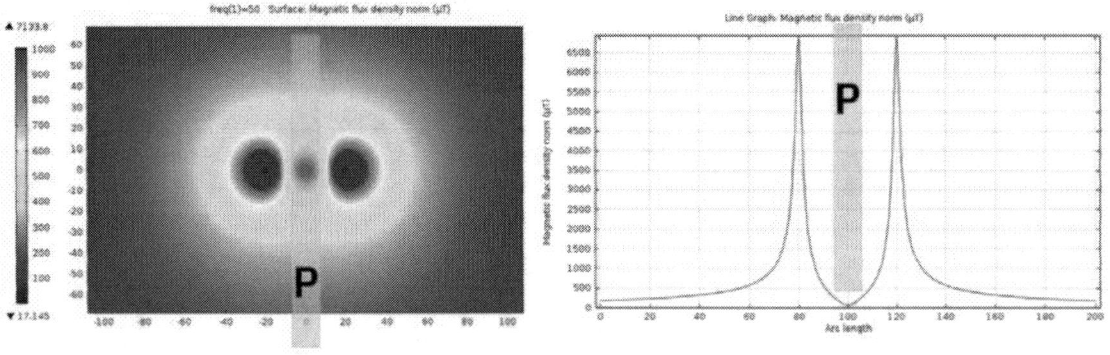

Figure 2. Principle of shielding ELF magnetic fields.

5. Conclusion

Based on the latest research of WHO IARC and others, it is recognised that extra-low frequency magnetic field exposure can cause several long-term health effects. During high voltage live-line maintenance magnetic field might be higher than current limits defined by ICNIRP (there are a number of countries within the European Union that apply limits different from those given in ICNIRP guidelines [13], [14]. None of conventional conductive clothing have significant shielding effect on ELF magnetic fields. ELF magnetic fields can be shielded effectively by a parallel conductor as a summarized result of two basic physical laws. In case of conductive live-line equipment with more than one potential clamps, an undefined amount of current may flow through the structure of the equipment itself. The risks related to these currents have to be handled during any kind of live-line work. In case of proper design guidelines the conductive structure itself can be used as an alternate current path for ELF magnetic field shielding. Principle of a possible way to shield ELF magnetic fields has been introduced in this paper. Practical application is under development in the High Voltage Laboratory of Budapest University of Technology and Economics.

References

[1] CIGRÉ Joint Working Group, B2/B3.27 Report 561 2013 *Live Work – A Management Perspective* (Paris, France, December 2013)

[2] International Commission On Non-ionizing Radiation Protection 2002 ICNIRP Statement, General Approach to Protection Against Non-Ionizing Radiation Protection *Health Physics* **82**(4) 540-548

[3] International Commission On Non-ionizing Radiation Protection 2010 ICNIRP Guidelines for Limiting Exposure to Time-Varying Electric and Magnetic Fields (1 Hz – 100 KHz) *Health Physics* **99**(6) 818-836

[4] Directive 2013/35/EU of the European Parliament and of the Council of 26 June 2013 on the minimum health and safety requirements regarding the exposure of workers to the risks arising from physical agents (electromagnetic fields) (20th individual Directive within the meaning of Article 16(1) of Directive 89/391/EEC) and repealing Directive 2004/40/EC

[5] World Health Organization International Agency For Research On Cancer 2002 *IARC Monographs On The Evaluation Of Carcinogenic Risks To Humans, Non-Ionizing Radiation, Part 1: Static And Extremely Low-Frequency (ELF) Electric And Magnetic Fields* Vol 80 (Lyon: IARCPress)

[6] Krajewski W 2008 BEM evaluation of eddy currents induced in a live-line worker body *9th Int. Conf. on Live Maintenance* (Toruń, Poland, 4 – 6 June 2008)

[7] International Electrotechnical Commission *IEC 62233 ed1.0* 2005 Measurement methods for electromagnetic fields of household appliances and similar apparatus with regard to human exposure

[8] Göcsei G, Németh B, Kiss I and Berta I 2013 Shielding efficiency of conductive clothing in magnetic field *Journal of Electrostatics* **71**(3) 392-395

[9] Göcsei G, Németh B and Tarcsa D 2013 Extra low frequency electric and magnetic fields during live-line maintenance *IEEE EIC Conference* (Ottawa, Ontario, Canada) pp. 100-104

[10] Göcsei G, Németh B, Kiss I and Berta I 2014 Health effects of magnetic fields during live-line maintenance *Int. Conf. on Live Maintenance* (Budapest, Hungary, 2014) 235 p. 6, ISBN: 978-1-4799-5993-8

[11] Göcsei G, B Németh B, Tamus Á, Kiss, I and Meixner J 2014 Shielding efficiency of conductive clothing during live-line maintenance *International Conference on Live Maintenance* (Budapest, Hungary, 2014), 306 p. 5 ISBN: 978-1-4799-5993-8

[12] Göcsei G, Németh B, Tamus Z A and Kiss I 2012 Face Protection Investigation Against Electric Field On Live Line Workers (keynote) *IEEE International Symposium of Electrical Insulation* (San Juan, USA) pp. 535-539 ISBN: 978-1-4673-0486-3

[13] http://www.emfs.info/wp-content/uploads/2014/07/standards-table-revision-5l-July-2014.pdf, date of visit: 18 May, 2015

[14] Grelliera J, Ravazzanid P and Cardisa E 2014 Potential health impacts of residential exposures to extremely low frequency magnetic fields in Europe *Environment International* **62** 55–63

Interaction of a particle bed with a droplet under an applied electric field

P M Ireland , G B Webber, E D Jarrett, K P Galvin

Centre for Advanced Particle Processing and Transportation, University of Newcastle, University Drive, Callaghan NSW 2308, Australia

E-mail: Peter.Ireland@newcastle.edu.au

Abstract. Transport of dry solid particles to a liquid is relevant to a number of emerging applications, including 'liquid marbles'. We report experiments where the transport of dry particles to a pendent water droplet is driven by an external electric field. Both hydrophilic and hydrophobic materials (silica, PMMA) were studied. For silica particles (hydrophilic, poorly conductive), a critical applied voltage initiated transfer, in the form of a rapid 'avalanche' of a large number of particles. The particle-loaded drop then detached, producing a metastable spherical agglomerate. Pure PMMA particles did not display this 'avalanche' behaviour, and when added to silica particles, appeared to cause aggregation and change the nature of the transfer mechanism. This paper is largely devoted to the avalanche process, in which deformation of the drop and radial compaction of the particle bed due to the electric field are thought to have played a central role. Since no direct contact is required between the bed and the drop, we hope to produce liquid marble-type aggregates with layered structures incorporating hydrophilic particles, which has not previously been possible.

1. Introduction

The formation of liquid-hydrophobic particle agglomerates via an electrostatically-driven 'avalanche' process was first described by our group in 2013 [1], and is shown in Figure 1. This phenomenon was interesting for several reasons. First of these was the 'avalanche' itself, which had not been observed before. The second source of interest was the resemblance of the resulting metastable agglomerate to the 'liquid marbles' – liquid drops encased in a shell of non-wetting particles – first observed over a decade ago [2], which have remarkable properties and a variety of current and potential applications [3,4]. Unlike these 'traditional' liquid marbles, our agglomerates consisted of a water droplet entirely filled with hydrophilic particles – a configuration which cannot be achieved using the conventional marble formation method of rolling the drop on a particle bed (as the drop simply soaks in). Our electrostatic process raises the prospect of a new class of liquid marble complexes with layered structures, for a variety of new applications.

In this paper, we briefly recapitulate the main details of the original experiments with hydrophobic glass ballotini particles and briefly describe more recent experiments with admixtures of hydrophobic PMMA particles. The rest of the paper is devoted to a simple model of the electrostatic stress in the bed, which provides a possible explanation for many features of the remarkable particle 'avalanche' process.

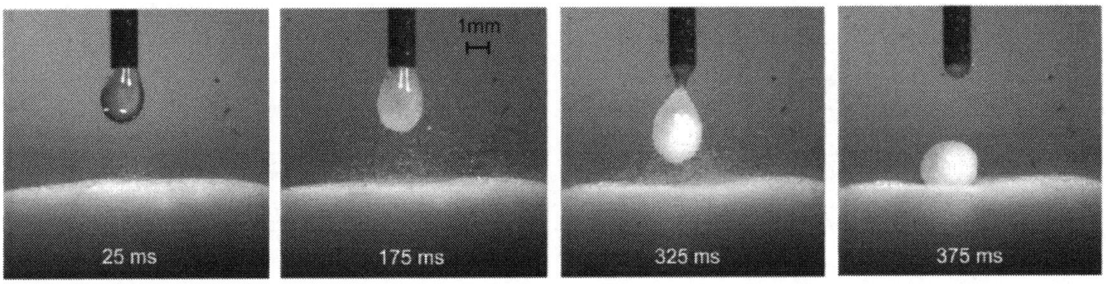

Figure 1. Sequence showing the particle 'avalanche', detachment and metastable agglomerate for 100μm ballotini particles.

2. Experimental observations

Figure 2 shows the experimental configuration used in these experiments. Full details are given elsewhere [1]. The particle bed was laid on either a film of pre-tribocharged PTFE, or a glass slide resting on a metal plate connected to an HV power supply. In the initial experiments, silica in the form of 100μm diameter ballotini (glass beads) was used, and the substrate gradually brought closer to the pendent drop. Particles jumped intermittently to the surface of the drop as it approached. Eventually, at some point, a sudden 'avalanche' of particles burst upward and into the drop. In around 400 milliseconds, the drop filled up like a 'sack of marbles' before detaching and forming a pearl-like metastable agglomerate. Figure 3 shows the magnitude of the charge measured flowing from the drop to earth, at each stage of the approach, avalanche and detachment. The behaviour was altered by an admixture of PMMA particles of the same size as the silica. For a mass fraction of 10% PMMA, the 'avalanche' process proceeded as with pure silica. For 30% and 50% PMMA, the particles jumped irregularly in large 'clumps', and no 'avalanche' occurred. Pure PMMA particles did not jump to the drop at all.

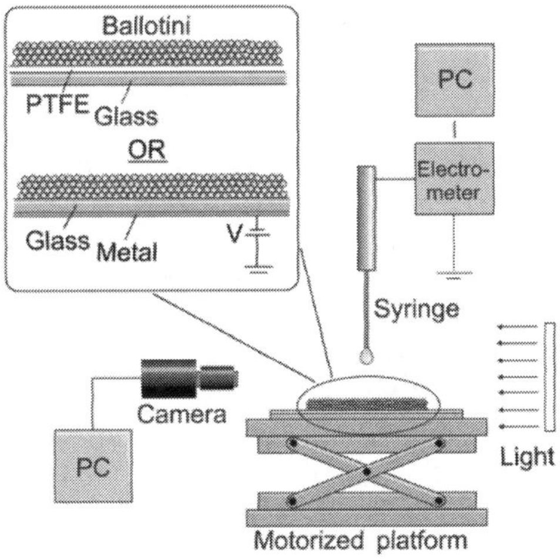

Figure 2. Experimental configuration.

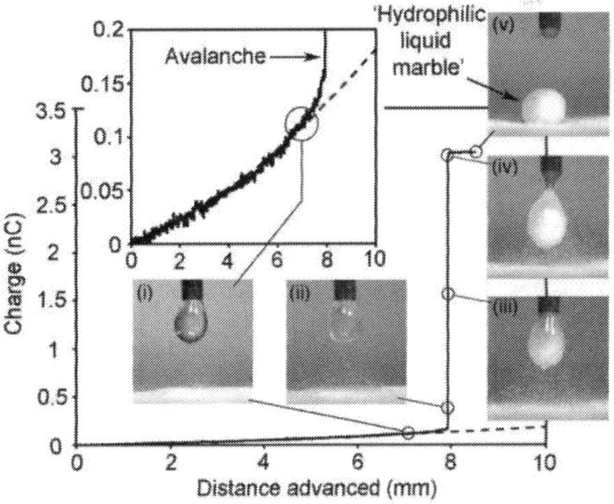

Figure 3. Charge leaving drop vs. distance advanced by the substrate.

Our current understanding is that the silica particles, being poorly conductive but also poorly insulating, were negatively charged by the tribocharged PTFE substrate or the electrically-biased plate, as evidenced by the charge transferred to the drop during the 'avalanche'. The PMMA particles, as insulators, were unable to charge sufficiently over the relevant timescale. When the bed was composed of a mixture of charged silica and uncharged PMMA particles, the two particle types aggregated electrostatically, producing a more cohesive bed and transferring to the drop in clumps.

3. Frictional cohesion model

It is clear that initiation of the avalanche does not simply correspond to the upward electrostatic force component exceeding the particle weight. High-speed video footage confirms that, at first, single isolated particles detach from the bed and accelerate upwards at 50-200 $m.s^{-2}$ while the surrounding particles remain in place. Furthermore, as reported in [1], instances were observed where the approach was halted as soon as these precursor transfer events were observed, followed by a delay, *then* an avalanche. This implies the bed is behaving cohesively, undergoing structural rearrangement, and sometimes releasing individual particles, and that these rearrangements persist as long as the bed is subject to the concentrated electric field between the substrate and drop. Bed cohesion can be explained in terms of interparticle friction. The force on the particles above the substrate on either side of the axis of symmetry should have a radial inward component parallel to the substrate, in addition to the upward component, due to convergence of the electric field lines on the image charge on the drop (Figure 4a). Both field components will increase as the drop-bed distance decreases. The radial inward component places the particles beneath the drop under a radial compressive stress. This stress increases interparticle friction, opposing the vertical component of the force and preventing vertical transport of the particles. As the stage is driven upward, the compressive stress increases, and the particles undergo structural rearrangements in order to pack more closely. Some of these release a few particles.

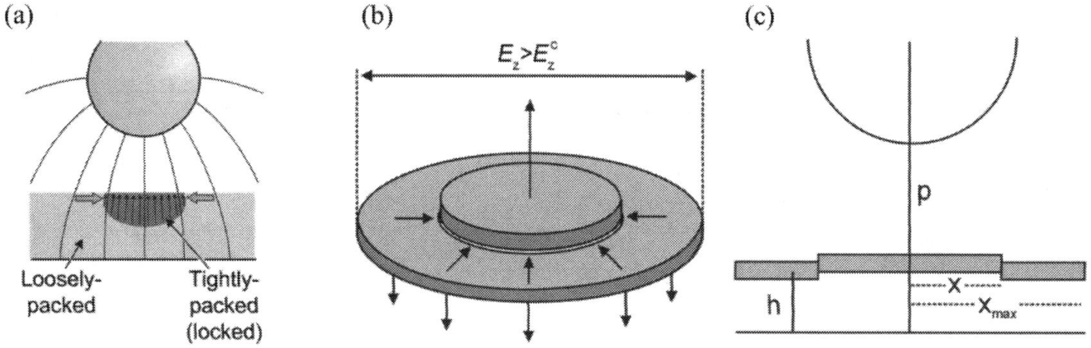

Figure 4. Bed compaction mechanism and frictional cohesion model.

We investigate this hypothesis using a much-simplified physical model. Let us assume that the drop is a sphere with radius a, that there is a concentrated, thin layer of charge on the substrate, and that the distance from the centre of the drop to the substrate is p. The topmost layer of particles in the bed is treated as a thin layer, distance h above the substrate (Figures 4b, 4c). This layer has a uniform area charge density, which is assumed to be too small to affect the electric field. The system has rotational symmetry. The radial in-plane and vertical components of the electric field, given by the model of Berry and Higginbotham [5] for a charged insulating plane and conducting sphere, are shown in dimensionless form in Figure 5 (with h/a=1). The model itself is too involved to be reproduced here. Let x_{max} be the radial distance at which the vertical electric field is sufficient to overcome the gravitational force on the particles. The net vertical force on the layer outside this radius will be zero.

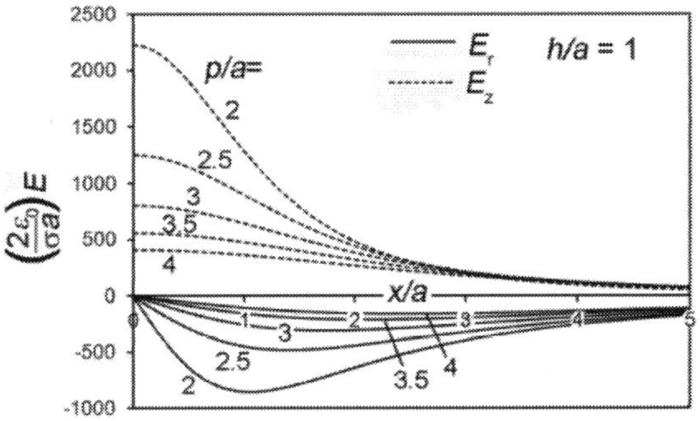

Figure 5. Radial in-plane and vertical electric field components [5].

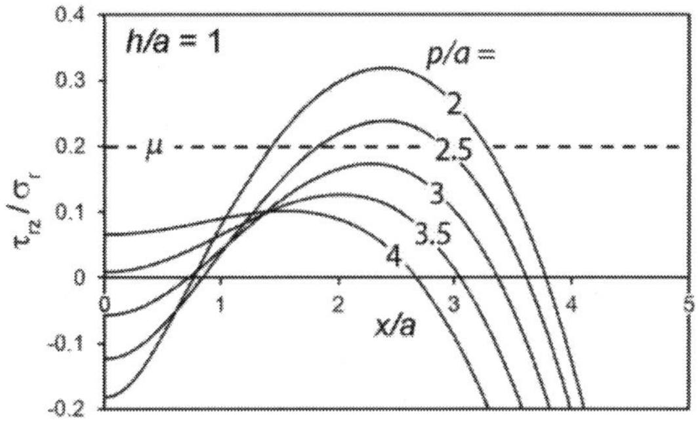

Figure 6. Ratio of shear to radial compressive stress in the bed.

Consider a circular 'slab' of radius $x \le x_{max}$ at the centre of the layer, as shown in Figure 4b. If the shear stress at x is sufficient to overcome internal friction, the whole central slab will be able to lift off. For simplicity, to avoid introducing bed densities and so on, let us denote the vertical electric field required to overcome the gravitational force on the top layer of particles as E_z^c. The internal shear resistance at radius x will be equal to the radial compressive stress at that radius, multiplied by some friction coefficient μ. The internal friction will be overcome by the shear stress when

$$\frac{\int_0^{x}\left[E_z(\xi)-E_z^c\right]\xi\,d\xi-\int_x^{x_{max}}\left[E_z(\xi)-E_z^c\right]\xi\,d\xi}{-\int_x^{x_{bed}}E_r(\xi)\xi\,d\xi} > \mu \qquad (1)$$

The parameter x_{bed} is the radial extent of the particle bed and ξ is a dummy integration variable that stands in for x. Figure 6 shows the ratio of shear to compressive stress (the left-hand side of equation 1) as a function of radial distance for $h/a = 1$. As the distance between the drop and the substrate decreases, the peak value of the stress ratio also increases. If (for instance) the internal friction coefficient is 0.2, as shown, the peak stress ratio will pass this when p/a is somewhere between 2.5 and 3.0, and an avalanche will occur. Note that negative values of the shear stress do *not* indicate parts of the layer that are unable to lift off the substrate. This quantitative argument appears to indicate that the proposed release mechanism is at least plausible, and is worth modelling in more detail using COMSOL or a similar numerical package. A more sophisticated model would include deformation of the drop, a bed of finite thickness, a non-uniform distribution of charge on the bed, and so on.

References

[1] Liyanaarachchi K R, Ireland P M, Webber G B, Galvin K P 2013 *Appl. Phys. Lett.* **103,** 054105
[2] Aussillous P, Quere D 2001 *Nature* **411**, 924
[3] McHale G, Newton M I 2011 *Soft Matter* **7**, 5473
[4] Bormashenko E 2011 *Current Opinion in Colloid and Interface Science* **16**, 266
[5] Berry J S, Higginbotham I G 1975 *J. Phys. A - Mathematical and General* **8**, 1842

Resonant vibration of a droplet located on a super-hydrophobic surface under the vertical and horizontal ac field

Y Higashiyama[1], T Ohuchi[1], and T Sugimoto[1]

Graduate School of Science and Engineering, Yamagata University
4-3-16, Jonan, Yonezawa 992-8410, Japan

higashi@yz.yamagata-u.ac.jp

Abstract. A water droplet under an ac electric field with resonant frequency changes drastically its shape repeating extension and shrinkage, alternatively. To develop an electrostatic mixing method of small amount of liquid, resonant vibrating motion of a water droplet was investigated. Both horizontal and vertical fields were applied to the droplet placed on a super-hydrophobic plate with a contact angle of 150 degrees. From the video images of the droplet, the degree of deformation of the droplet shape was evaluated by deformation rate. Under the two-directional electric field, the deformation ratio at shrinkage was increased significantly. The height of the droplet varies from 1.2 to 1.8 times larger than that of the original droplet during vibrating motion. Furthermore, the deformation rate at shrinkage varies with time periodically due to rotating motion of the droplet. The vertical electric field might be effective to cause the turbulent flow inside of the droplet.

1. Introduction

A sessile droplet located on the surface of an insulating sheet under the strong electric field forms a cone to cause corona discharge or to form filamentary channel along the field direction. Under the relatively low ac electric field, a droplet vibrates more or less depending on the frequency of the electric field, especially, it vibrates strongly at a particular resonant frequency [1]. The researches on resonant vibration of a droplet have been widely carried and effects of the droplet size, surface tension, and liquid density on natural frequency were investigated theoretically and experimentally [2, 3, 4].

Resonant vibration of a droplet affects the electrical discharge aspect. For example, resonant vibration of a droplet placed on an insulating plate affects flashover voltage via the droplet due to large deformation [5]. Furthermore, occurrence frequency of dc corona discharge from a fine droplet is governed by the resonant frequency depending on the droplet volume [6].

An application to an electro-hydrodynamic mixer for a millimeter-size droplet using resonant vibration is a promising subject, although the flow field generated during droplet oscillation under AC frequencies above 10 kHz can be also used to enhance the mixing in a droplet [7, 8]. In this paper, resonant vibration of a water droplet placed on a super-hydrophobic surface was investigated experimentally to develop a non-contact electrostatic stirrer. To cause the larger deformation of the droplet or irregular motion of the droplet, both horizontal and vertical electric fields were employed and the effect on the deformation under the resonant state was examined.

2. Experimental method

The experimental set up for applying two-directional electric field to a water droplet placed an insulating plate is shown in figure 1. An insulating pate to place a water droplet is a 20 mm x 20 mm PMMA plate with 3 mm thick was coated with super-hydrophobic paint, HIREC 1450 (NTT AT), to form micro

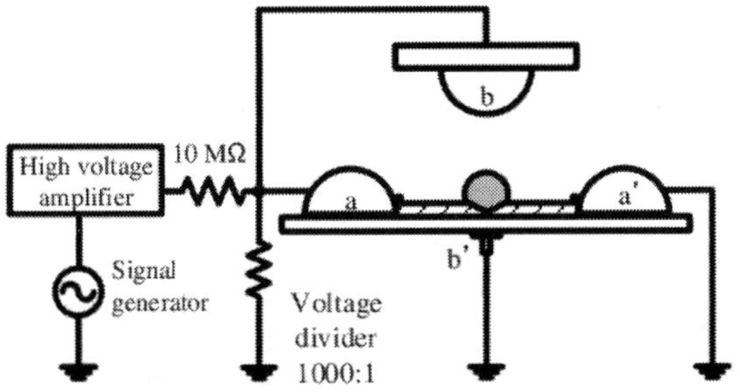

Figure 1. Arrangement of the electrodes and droplet for forming a horizontal and vertical electric field.

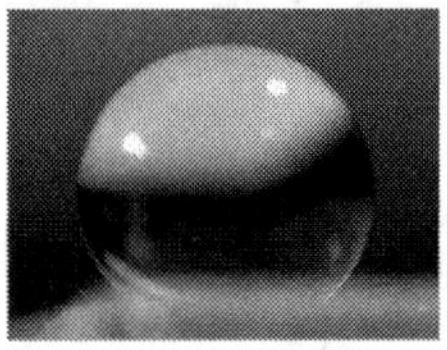

Figure 2. 7 μL water droplet placed on the super-hydrophobic plate.

structure over the surface. The insulating plate used here is referred as HIREC. An example of the shape of a 7 μL droplet on HIREC is shown in figure 2. A contact angle of the droplet is 150 degrees.

A 100 μL droplet placed on the super hydrophobic plate moves easily towards the either electrode at application of high voltage. In order to prevent a droplet from contacting with the electrodes during resonant vibration, the surface of the square plate was scraped in circularly with 18 mm diameter and 1mm depth. Moreover, to fix the position of vibrating motion of a droplet, a conical pit with a vertical angle of 120 degrees and a depth of 1mm was formed at the centre of the plate as shown in figure 1.

A pair of semispherical brass electrodes, a-a', with a radius of 5 mm was arranged for forming a horizontal field along the insulating plate. To form a vertical field, an upper semispherical and a lower-plate electrodes, b-b', were used. The upper electrode was set at 10mm above the HIREC plate.

To confirm the effect of viscosity on a motion of a droplet, water solution of polyethylene glycol, PEG, with a molecular number of around 20,000 was used as a viscous liquid. The value of a viscosity of the droplet was adjusted from 0.9 mPa·s for deionized water to 29 mPa·s for 15 % PEG water.

The sinusoidal voltage with a frequency of around 10Hz was applied to the electrodes at time zero, thus droplet motion was taken with a high-speed video camera (Photoron, RGB Rabit, 250 frames/s) synchronized with time variation of the magnitude of an applied voltage. Under resonant state, a droplet extends toward the horizontal and vertical direction, alternatively. To quantitatively evaluate the degree of deformation of the droplet with time, The deformation rates were defined for each directions: The width-deformation rate, W/W_0, is a ratio of the width of the droplet, W, to the initial width of the standstill droplet placed calmly on the plate, W_0, and the height-deformation rate, H/H_0, is a ratio of the height of the droplet to the initial height, H_0. From the deformation rates, a resonant frequency of a droplet was obtained and the behavior of a droplet under resonant vibration was analyzed in detail.

3. Results and discussion

3.1 Resonant vibration of a droplet

Figure 3(a) shows an example of the time variation of deformation of a 100 μL droplet in the resonant state during a half cycle of vibrating motion driven by only horizontal field without a pair of vertical electrodes, b-b' and figure 3(b) shows that by both horizontal and vertical field. The droplet repeats extension and shrinkage in the horizontal direction. Deformation of a droplet under the resonant state is considerably different depending on the applied field. The degree of deformation for combination of two directional field is much larger than that for the horizontal field. The vertical field effectively acts to increase the height of the droplet at the shrinkage state.

3.2 Time variation of deformation rate

Figure 4 shows the time variations of the height- and width-deformation rate during resonant vibration during two cycles of the applied field at an elapsed time of around 5 s under the two-directional field for deionized water and viscous droplet. A droplet deforms clearly twice during one cycle of the applied field, although the motion of the droplet always delayed to the applied field. Since the droplet shows

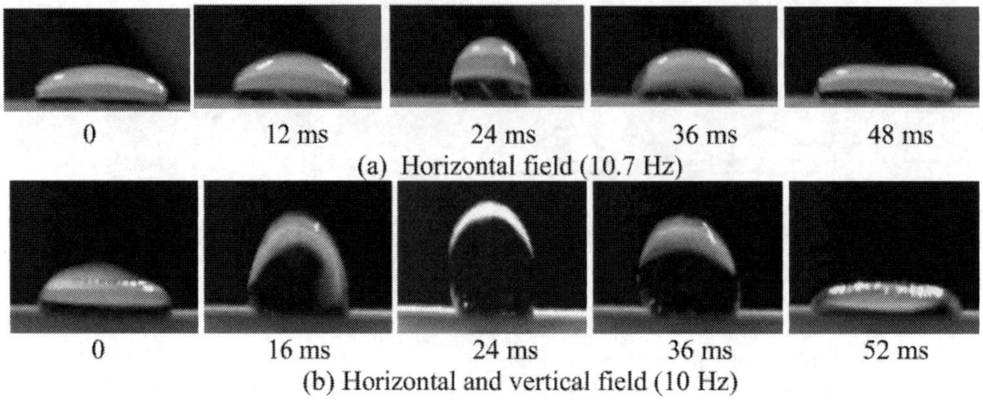

Figure 3. Resonant vibration of a 100 μL water droplet placed in the horizontal and vertical field.

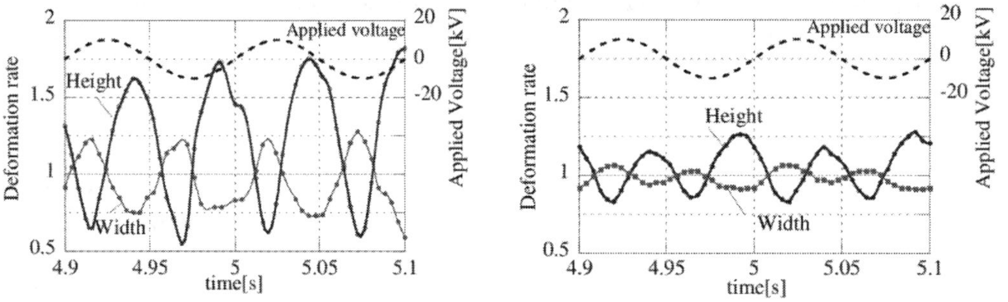

(a) Deionized water (b) PEG water solution with 29 mPa·s

Figure 4. Time variation of the width and height of a 100 μL droplet during resonant vibration.

resonant vibration under the sinusoidal field of 10 Hz, the natural resonant frequency of a 100 μL droplet should be 20 Hz.

The height-deformation rate is much larger than that the width-deformation one. The maximum of either deformation rate appears alternatively, corresponding to the shrinkage and extension. The viscous liquid also shows resonant vibration more or less. The magnitude of the height-deformation rate reached 1.8 for deionized water and 1.2 for viscus liquid.

Figure 5 shows the time variation of the maximum and minimum height-deformation rate for a deionized water droplet obtained at the shrinkage state appeared every one cycle. The height-deformation rate varies with time from 1.2 to 1.8 in a roughly regular manner. This means the maximum height of the droplet repeats an increase and decrease periodically. The maximum height differs for every vibrating motion. Thus, a droplet extends up to 1.8 times higher than original one every one second.

A droplet placed on a super-hydrophobic plate moves and slides smoothly. The droplet decreases the width or increases its height by the vertical electric field. Irregular liquid flow inside the droplet could cause some instability by occasion. Thus, the droplet might start to turn around. As a result, a droplet at resonant state seems to cause irregular motion periodically.

3.3 Deformation of viscous liquid

A viscous droplet vibrates by ac field more or less as shown in figure 4(b). Figure 6 shows the maximum and minimum values of two deformation rates appeared at the most elongated and shrunk state for a viscous droplet. At the shrinkage state, the deformation rate approach to unity for both state as the viscosity is increased. The width of the droplet or deformation in the horizontal direction varies ranging from 0.6 to 1.3 times larger than the original width for an ionized water with 0.9 mPa·s. The

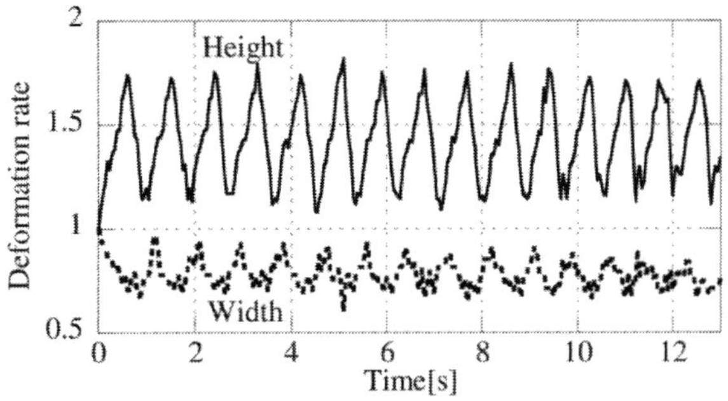

Figure 5. Time variation of the maximum height- and the minimum width-deformation rate of a droplet at shrinking state during resonant vibration under the two directional fields.

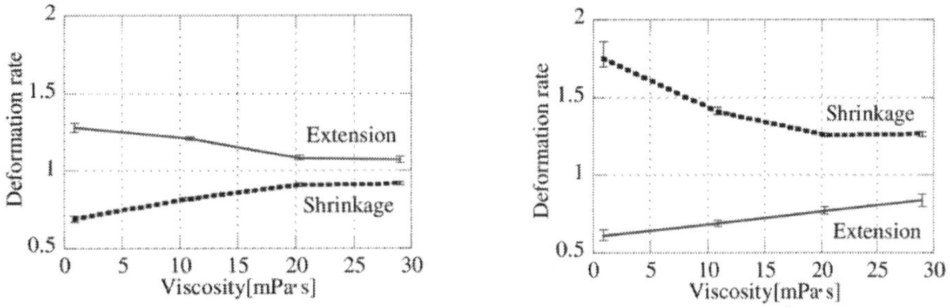

(a) Width (b) Height

Figure 6. The width- and height-deformation rate for a viscous droplet during resonant vibration under horizontal and vertical field.

height varies ranging from 0.6 to 1.7 times of the original shape. Since the deformation rate of viscous liquid up to 10 mPa·s keeps relatively higher value, it might be possible to mix viscous droplet.

4. Summary

The resonant vibration of a 100 μL water droplet placed on a super-hydrophobic plate under the horizontal and vertical electric field was investigated. The droplet varies its shape under resonant state. The height of the droplet becomes 1.8 time higher than original one and extra rotational motion appeared in two-directional extending motion. To stir a droplet, it would be effective to use resonant vibration on super-hydrophobic plate. It necessary to confirm the mixing performance using this system.

Acknowledgements

This research was supported by Grant-in Aid for Scientific Research of Ministry of Educational, Culture, Sports, Science and Technology (23560313)

References

[1] Yamada T, Higashiyama Y, Sugimoto T, Takeishi M and Aoki T, 2003 *IEEE Trans. on Industry Application*, **39**, 59-65
[2] Sample S B, Raghupathy B and Hendricks C D, 1970 *Int. J. Eng. Sci.*, **8**, 97-109
[3] Strani M, Sabata F, 1984 *J. Fluid Mech*, **141**, 233-247
[4] Tsukada T, Sato M, Imaishi N, Hozawa M, Fujinawa K, 1987 J. *Chem. Eng. Japan*, **29**, 388-93
[5] Higashiyama Y, Takada T, Sugimoto T, 2005 *J. Electrostatics*, **63**, 883–889
[6] Higashiyama Y and Saito S: 2013 *J. Electrostatics*, **71**, 499-503
[7] Lee H, Yun S, Ko S H, and Kang K H, 2009 *Biomicrofluidics*, **3**, 044113
[8] García-Sánchez P, Ramos A and Mugele F, 2010 *Phys. Rev. Lett.*, **E 81**, 015303(R)

Electrostatic adhesion of polymer particles to a foil electrode

Li Ziteng, Matthew Praeger, Jeremy Smallwood, Paul Lewin,

Tony Davies High Voltage Laboratory, University of Southampton, UK.
Email: J.smallwood@soton.ac.uk

Abstract. The SPABRINK EU project requires temporary adhesion of coloured solid "ink" particles to a surface, for later recovery and reuse. This is achieved through the use of dielectrophoretic force under the control of a voltage applied to an interdigitated electrode pattern on the polymer foil. One concern is the ability to hold particles under vibration conditions. In this paper we present an experimental study of the adhesion of 50-300 μm polymer particles to an experimental interdigitated electrode structure on flexible polymer foil. Powder loss as a function of calibrated displacement and applied voltage to the electrodes are presented. This is compared with theoretical results obtained by modelling adhesion using Pohl's equation in terms of an "adhesion factor". Some difficulties in directly comparing experimental and modelling results are discussed.

1. Introduction

The SPABRINK project (www.spabrink.eu) requires that coloured "ink" particles are attached to a planar surface "foil" to form a temporary image that can be held or released under electronic control. This provides novel printed image enabling reuse of the ink particles for low cost environmentally friendly advertising display applications. The image is required to be held for hours or days. Initial experiments had used charged particles or surfaces, but highly insulating particles are needed to hold charge for long periods, and it is difficult to control charge on the particles, especially over long periods of time and humid conditions. It is not simple to switch electrophoretic forces under electronic control, and high particle charge levels and voltages would be required for adhesion. Gravitational force increases rapidly compared to electrostatic force as the particle size increases, and so large particles will not easily adhere. In contrast, dielectrophoretic theory predicts that dielectrophoretic force is proportional to particle volume [1].

$$F = 0.5(\frac{4}{3}\pi a^3)\varepsilon_1\varepsilon_0\left(\frac{3(\varepsilon_2 - \varepsilon_1)}{\varepsilon_2 + 2\varepsilon_1}\right)\nabla[E]^2$$

where F is the force on the particle, a the particle radius, ε_2 and ε_1 the relative permittivity of the particle and surrounding medium (air, $\varepsilon_1 = 1$) respectively, ε_0 is the permittivity of free space and E the electrostatic field. The dielectrophoretic force and gravitational forces both increase with particle volume, so the ability to suspend particles should be nearly independent of particle size. It is possible to suspend surprisingly large particles using highly divergent surface fields produced by closely spaced electrodes at modest voltages. Fields up to breakdown field strength can be produced by a closely spaced electrode pattern on a surface under modest voltages (up to 1000 V) under electronic control. Particle dielectric properties rather than conductivity are the key parameters. It should be possible to hold particles for as long as power is supplied to the electrodes. With d.c. excitation low power consumption would be expected. The system would be expected to be insensitive to atmospheric humidity. Despite these advantages, dielectrophoretic manipulation appears to have been mainly used with micro and nano particles, with the exception of Calle et al.'s work [2] on moon dust control.

In order to create strong divergent fields at the surface of the foil, an interdigitated electrode structure having electrode and gap widths of 500 μm has been used [3]. Particles were shown to adhere successfully not only to the electrode surface, but also to the reverse (non-electrode) side of the polymer foil, at least for foil thicknesses up to 100 μm. It was then of interest to test the strength of adhesion to the vertical surface against vibration and the effect of reducing supply voltage. This paper explores the effects of these using polyester particles supplied by a partner in the SPABRINK project.

2. Methodology

2.1. Experimental

The electrode system used was an interdigitated copper structure with electrode and gap widths of 500 μm, inkjet printed on a 115 μm thick polyester foil. The foil was mounted in a rigid plastic frame supported from an electromagnetic vibrator that gave a displacement of 0.52 mm per volt applied. This was driven using 133 ms rectangular pulses to give a controlled displacement with each rising and falling edge of the waveform. The vibrator and foil were mounted on a rotating support so that powder could be applied in a horizontal position and then rotated to vertical.

Figure 1. Experimental arrangement . In this photograph the foil has been turned to the vertical position and fallen powder may be seen on the paper beneath it. The electrodes are not visible as they are on the rear surface of the foil.

The foil was energized from a d.c. high voltage power supply via 20 MΩ resistor for protection against sparking or accidental short circuits. The d.c. current consumption of the electrode structure was <1 mA. A multimeter directly measured the voltage across the electrode structure.

The foil was cleaned with methanol and neutralised using an ioniser before each application of powder. A leveled spatula was used to measure repeatable quantities of powder, which was passed through a 90 μm mesh sieve using a brush to produce a 20 mm diameter uniform thin layer of 63-90 μm diameter material on the foil. With the powder and voltage applied, the vibrator and foil were turned to vertical (Figure 1). Fallen powder was collected and weighed. The foil was subjected to "there and back" displacement of various distances using the vibrator. Fallen powder was again collected and weighed. This was repeated until all the powder had been removed by vibration. The experiment was repeated five times for each powder loss data point and the average of the results taken.

2.2. Modelling

A 2-dimensional electrostatic model was developed of the foil and electrode arrangement [3]. We defined an "adhesion factor" as the ratio of dielectrophoretic to gravitational force normal to the foil surface, taking into account material dielectric constant and density. An adhesion factor >1 indicated that adhesion would be expected. The forces were plotted as shading on a model, giving indication of where regions of strong and weak adhesion would occur (Figure 2). The amount of powder that could be expected to adhere to the surface could then be estimated by finding the area of the model for which the adhesion factor was >1.

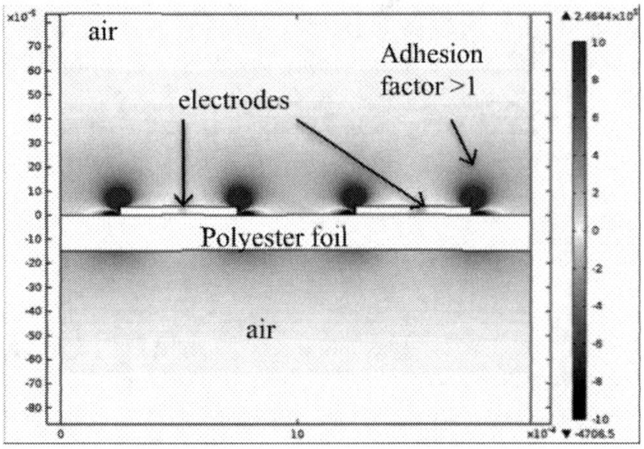

Figure 2. 2-dimensional electrostatic model showing regions of strong adhesion force (adhesion factor >1) around the edges of the electrodes

3. Results and discussion

The loss of powder from the foil due to foil displacement is given for PET powder in Figure 3. Negligible powder was lost with foil displacement less than 0.5mm, showing surprisingly strong adhesion. All the powder was lost for displacement greater than 3 mm.

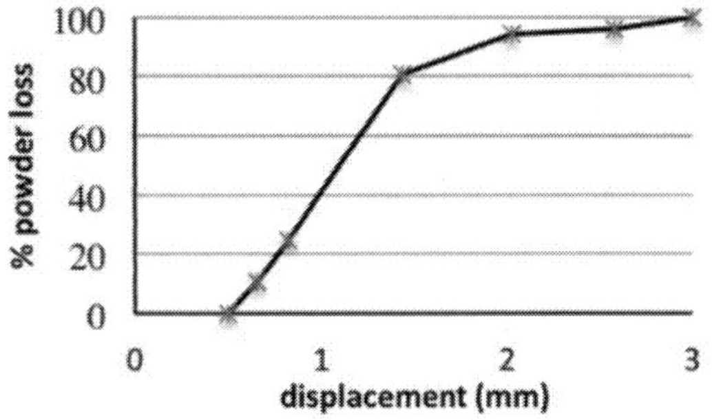

Figure 3. Powder loss due to vibration at various displacements

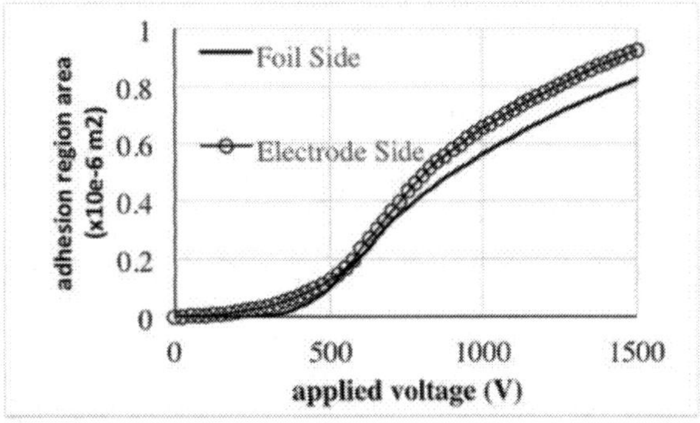

Figure 4. Adhesion region area modeled for a 115 μm foil as a function of applied voltage for electrode and non-electrode (foil) sides

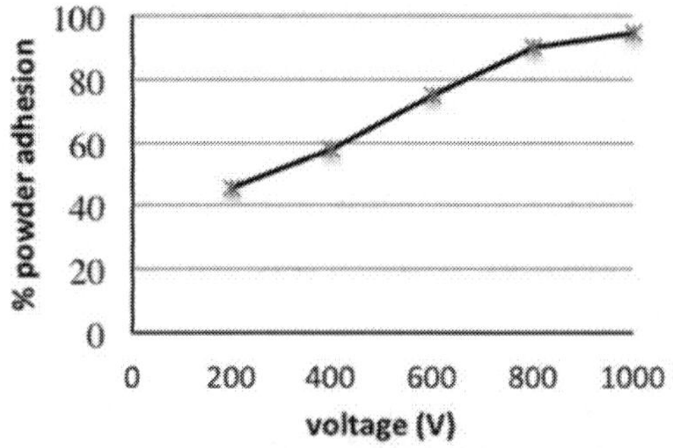

Figure 5. Powder loss from the non-electrode side of foil as a function of supply voltage for a 115 μm foil

The electrostatic model predicts that the area having adhesion factor >1 is reduced as the foil applied voltage is reduced. For a 50 μm foil, the area of the region of adhesion on the non-electrode side of the foil is similar to the electrode side. For a 115 μm foil, the adhesion region area is reduced on the non-electrode side (Figure 4) as the adhesion force is spread and attenuated due to the foil thickness. A threshold of adhesion appears, in this case around 500 V for a 115 μm foil. In our practical experiments, using a 115 μm foil and polyester powder we found that an average of 94% of the powder adhered at the maximum applied voltage of 1000 V, and this fell to 54% at 200 V.

The experimental results are broadly in line with our predictions although adhesion at low voltages was much greater than expected. It is difficult to directly predict the powder mass adhesion for a variety of reasons. The packing fraction of a powder is not known and can be expected to be variable with applied force. Secondly, Pohl's equation for dielectrophoretic force assumes the presence of one spherical particles. In practice there are many non-spherical particles in a packed layer. This can be expected to affect the strength and divergence of the field, and hence the dielectrophoretic adhesion forces. The model also assumes uncharged particles. In practice this is unlikely to be the case with a highly insulating material like polyester, and the method of handling the powder may have a large effect on particle charging. The powder was found to adhere preferentially to the positive electrode, and it seems likely the powder became negatively charged by brushing it through the sieve. So, it seems likely that the adhesion force observed here is due to a mixture of dielectrophoretic and electrophoretic force.

4. Conclusions

We have shown that 2-d electrostatic modeling can be used to make general predictions of the behavior of dielectric particles adhering to an interdigitated electrode with d.c. voltage applied. Particle adhesion to the electrode and non-electrode sides of the foil has been shown to withstand shaking with displacement up to 0.5mm. Higher displacements up to 3 mm led to increasing powder loss, with 100% loss for over 3mm displacement.

Powder adhered well to the non-electrode side of a 115 μm foil with 1000 V applied, losing only around 6% of material when shaken. Material loss rose to 10% at 800 V and 54% at 200 V applied. The adhesion forces in this experimental work are likely to include electrophoretic force due to negative charge on the polymer particles as well as dielectrophoretic forces.

Acknowledgement

This project has received funding from the European Union's Seventh Framework Programme for research, technological development and demonstration under grant agreement no 605299.

References

[1] Pohl H A 1978 *Dielectrophoresis*. Cambridge University Press. ISBN 0 521 21657 5
[2] Calle C I, Buhler, C R, McFall J L and Snyder S J 2009 *J. Electrost.* **67** 89-92
[3] Smallwood J, Praeger M, Chippendale R and Lewin P 2014 Dielectrophoretic adhesion of 50-300 μm particles under ambient atmospheric conditions. Dielectrophoresis 2014, Institute of Physics London

Electrostatic separator for micronized mixtures of metals and plastics originating from waste electric and electronic equipment

Sara Messal[1], Razvan Corondan[1], Ionut Chetan[1], Rabah Ouiddir[2], Karim Medles[3], Lucian Dascalescu[1]

[1]PPRIME Institute, UPR 3346, CNRS-University of Poitiers-ENSMA, France.
[2]University of Sciences and Technology Mohamed Boudiaf, El-M'naouar, Oran, Algeria.
[3]University Djillali Labes, Sidi-bel-Abbes, Algeria.

E-mails: sara.messal@univ-poitiers.fr lucian.dascalescu@univ-poitiers.fr

Abstract. In spite of their extensive use for processing mixtures of granules exceeding 1 mm in size, very few industrial electrostatic separators are capable of handling micronized metals and plastics originating from waste electric and electronic equipment. The aim of the present work is to validate the possibility of using a novel belt-type electrostatic separator for the selective sorting of such particulate mixtures, the dimensions of which are in the order of 0.1 mm. In this type of separator, the metal particles get charged by electrostatic induction in contact with the grounded metal belt electrode, while the plastics remain uncharged in the electric field and are collected separately. The experiments are performed with 2-g samples of a mixture composed in equal proportions (50 % - 50 %) of Aluminium and Acrylonitrile Butadiene Styrene (ABS) particles of average diameter ranging between 125 μm and 250 μm. They enabled the evaluation of the effects and the interaction of two control variables of the process: the angle of inclination of the roll-type electrode and the high voltage applied to it.

1. Introduction

Electrostatic separation is a method extensively employed in industry for processing mixtures of granules typically exceeding 1 mm in size [1]-[3]. However, very few electrostatic separators are capable of handling smaller sized particulate mixtures [4]–[7]. The micronized mixtures of metals and plastics originating from waste electric and electronic equipment (WEEE) are characterized by dimensions in the order of 0.1 to 0.5 mm.

The aim of the present paper is to demonstrate the feasibility of the electrostatic sorting of such mixtures, using a novel multifunctional belt-type electrostatic separator. The experiments are carried out with micronized mixtures of Al and ABS, two materials frequently encountered in the composition of WEEE. The study is performed according to a full factorial experimental design [8], [9] that enables the evaluation of the effects of the two major control variables of the process.

2. Experimental set up

The novel belt-type electrostatic separator (figure 1), designed by the PPRIME Institute in Poitiers, France, is a multifunctional equipment that is able to sort granular or micronized materials through the use of three different physical mechanisms of electrical charging : (1) triboelectric effect [10]-[13] involving the charge transfer at the impact and/or friction of the particles between themselves or against the walls of dedicated devices; (2) DC corona discharge [14], [15] ; (3) electrostatic induction [16]-[18]. The particles are disposed as a monolayer onto the surface of a metallic belt-type conveyor (6) (length: 700 mm; width: 70 mm), which is connected to the ground and entrained by a variable speed drive (5).

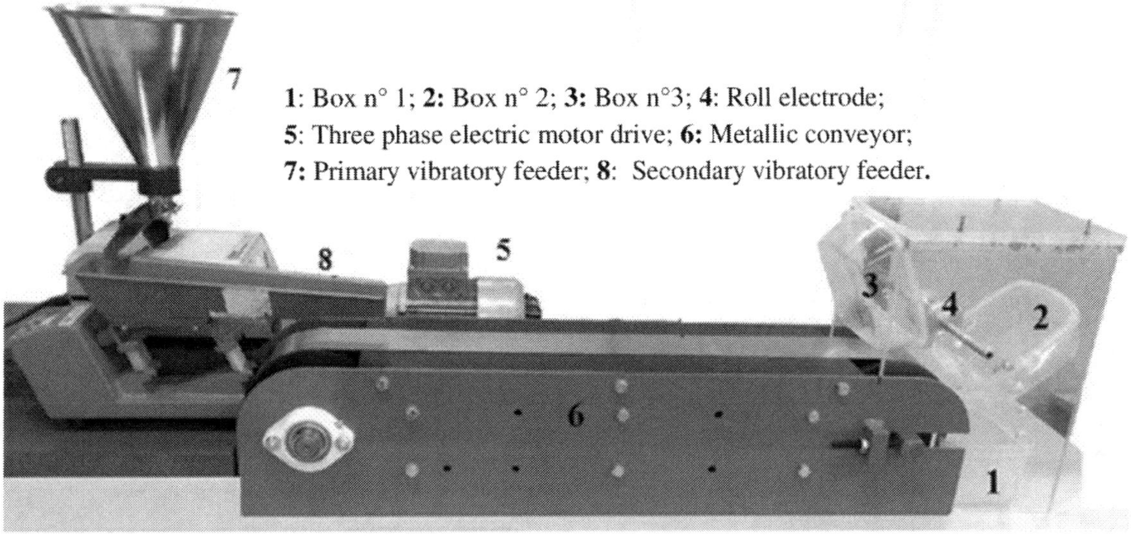

Figure 1. Electrostatic separator for micronized mixtures.

The primary vibratory feeder (8) (model PDB 4/30, VIBRA France) may be used in association with a custom-designed particle-dispersion device and has a dedicated control unit. The speed of the particles at the surface of this tray can be thus correlated with the speed of the metallic conveyor belt. A secondary vibratory feeder (7) (model DR 100, RETSCH) is employed for adjusting the mass of material processed in a time unit. The trays of both vibratory feeders are connected to the ground.

A roll electrode (4) (stainless steel; diameter: 30 mm; length: 70 mm) is connected to a reversible (positive or negative) high voltage power supply (not shown in figure 1; model SPL 300, 100 kV; 3 mA; SPELLMAN). A zone of intense electric field is generated between this electrode and the grounded metallic belt-conveyor. The uncharged insulating particles are not affected by this electric field, and are collected in the box (1). The conductive particles charge by conductive induction in contact with the grounded belt (6) and are attracted to the roll electrode of opposite polarity. Most of them are collected in the box (2). Some of the conductive particles attracted to the roll electrode bounce back after contact with it and end up in the box (3) of the collector.

The distance between the roll electrode and the surface of the metallic belt conveyor is 25 mm. Its angular position is expressed by the angle α [°] between the vertical direction and the plan defined by the axes of the roll and of the conveyor drum. This angle can be set in the range from 0° to 90°, by increments of 15°.

3. Materials and method

The experiments are carried out on a 50% Al + 50% ABS mixture of particles originating from WEEE. The average diameter of the particles of the mixture ranges between 125 μm and 250 μm.

Based on the results of a preliminary electrostatic separation test, the angular position α [°] of the roll electrode and the high-voltage U [kV] applied to it have been designated as the most significant factors of the electrostatic separation process and their domain of variation has been fixed as shown in Table 1.

Table 1. The domain of variation of the control variables: angle α and high-voltage U.

	Minimum	Maximum
α (°)	15	45
U (kV)	14	20

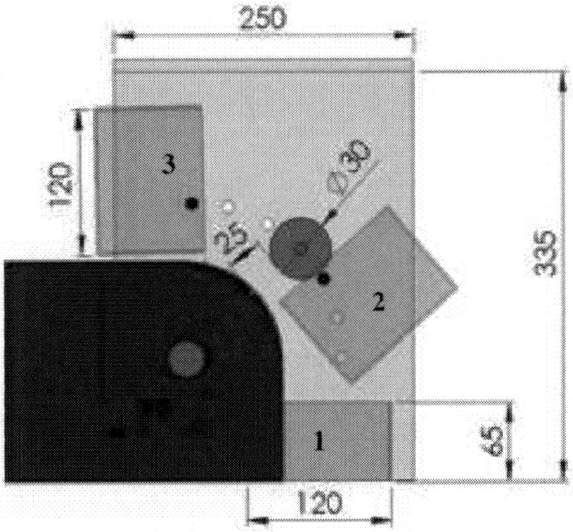

Figure 2. Schematic of the collecting system.

A full-factorial experimental design is adopted for the present study, as it enables the evaluation of the effects of these two control variables and the interaction between them. These experiments are all made in stable climatic conditions of temperature (between 23.9 ° and 25.6 °) and relative humidity (between 56.2 % and 62.4 %). Each of the 2^2 experiments is duplicated, to make possible the estimate of the experimental error.

The masses collected in the three boxes (figure 2) are measured with a precision electronic scale (resolution: 0.01 g). The values recorded for the boxes 2 and 3 are summed up to give the total mass of the conductor product. The roll electrode and the conveyor belt are cleaned after each experiment. The data are analysed with commercial software (MODDE 5.0, Umetrics, Sweeden).

4. Results and discussion

The results of the eight experiments are given in Table 2, where m_{Al} and m_{ABS} are respectively the masses of the conducting and insulating separated products, expressed in percentage of the feed.

Table 2. Electrostatic separation results.

Run	U [kV]	α [°]	m_{Al} [%]	m_{ABS} [%]	Loss [%]
1	14	15	53.27	44.86	1.87
2	20	15	58.06	40.32	1.61
3	14	45	33.17	66.35	0.48
4	20	45	49.72	49.73	0.55
5	14	15	47.32	51.71	0.98
6	20	15	60.99	38.46	0.55
7	14	45	36.04	62.94	1.02
8	20	45	42.42	57.07	0.51

The material losses for these experiments are negligible because they represent less than 2% of the feed (an average of 0.94%). Using MODDE 5.0, a linear model is obtained for of each of the output variables (responses) m_{Al} and m_{ABS}, in function of the normalized centred values U^* and α^* of the two control variables (figure 3):

$$m_{Al} = 47.62 + 5.17\, U^* - 7.28\, \alpha^* \quad (1)$$

$$m_{ABS} = 51.43 - 5.03\, U^* + 7.59\, \alpha^* \quad (2)$$

The average mass of the conductive product m_{Al} collected in boxes 2 and 3 increases from about 42% at an applied high voltage $U = 14$ kV (which means that the electric is too weak for all the Al particles to be attracted to the roll electrode), to roughly 52% at $U = 20$ kV. Indeed, by increasing the high voltage, the electric field becomes strong enough to drive to the roll electrode not only the 50 g of Al, but also some of the ABS particles that carry a positive charge, acquired by tribocharging effect during transportation. Neutralization of the materials prior to electrostatic separation might solve this problem.

The angular position α of the roll electrode has also a significant effect on the mass of conductive product. The average mass collected with the roll electrode at α = 15° represents more than 55% of the feed, which means that the electric field drives to the boxes 2 and 3 not only the Al particles, but also a relatively large percentage of insulating particles. This is due to both the Coulomb force acting on the insulating particles carrying a residual positive tribo-charge, and the dielectrophoretic force that acts on neutral, but polarized ABS particles. At α = 45°, the mechanical forces (centrifugal and gravitational) have a stronger influence on both Al and ABS particles, and the average mass of the conductive product represents only 35% of the feed, if $U = 14$ kV. According to the linear model (1), the best results (i.e., m_{Al} = 50%) are expected to be obtained at $U^* = 1$ (i.e., $U = 20$ kV) and $\alpha^* = 0.4$ (i.e., $\alpha = 33°$).

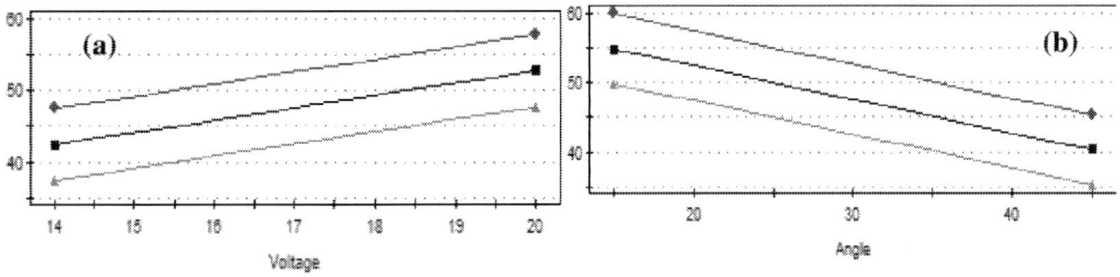

Figure 3. MODDE 5.0- predicted variation of the collected mass of aluminum, expressed in [%] of the total mass of the particulate mixture fed to the separator, as function of the applied voltage U [kV] (a), and of the angle α [°] (b). The upper and lower curves indicate the limits of the confidence interval.

However, a composite factorial experimental design should be performed in order to rigorously establish the optimal operating conditions. An appropriate technique for evaluating the purity of the products would be necessary for validating that optimum.

5. Conclusions

The belt-type electrostatic separator presented described in this paper can successfully perform the selective sorting of metallic and insulating particles from micronized WEEE. Within the limits of the experimental domain considered in the present study, the mass of the particles recovered as conductive product increases with the applied high-voltage and decreases with the angle that gives the position of the axis of the high-voltage electrode.

Further research is needed for a more accurate modelling and optimization of the separation process, as well as for analyzing the purity of the products.

References

[1] Dascalescu L, Morar R, Iuga A, Samuila A and Neamtu V 1998 *Part. Sci. Technol.* **16** 25-42
[2] Higashiyama Y and Asano K 1998 *Part. Sci. Technol.* **16** 77-90
[3] Tilmatine A, Bendimered S, Boukhoulda F, Medles K and Dascalescu L 2009 *Waste Management* **29**, 228–232.
[4] Ban H, Li T X, Hower J C, Schaofer J L and Stencel J M 1997 *Fuel* **76** 80-85
[5] Trigwell S, Tennal K B, Mazumder M K and Lindquist D A 2003 *Part. Sci. Technol.* **21** 353-364
[6] Das S, Kohnlechner R, Aman F and Dascalescu L 2010 *IEEE Trans. Ind. Appl.* **46** 2157-2164
[7] Dascalescu L, Dragan C, Bilici M, Beleca R, Hemery Y and Rouau X 2010 *IEEE Trans. Ind. Appl.* **46** 659 - 665.
[8] Eriksson L, Johansson E, Kettaneh-Wold N, Wikström C and Wold S 2000 Design of Experiments. Principles and Applications (Stockholm: Learnways AB)
[9] Dascalescu L, Medles K, Das S, Younes M, Caliap L and Mihalcioiu A 2008 *IEEE Trans. Ind. Appl.* **44** 3-8
[10] Harper W R 1998 Contact and Frictional Electrification (Morgan Hill, CA : Laplacian Press)
[11] Iuga A, Calin L, Neamtu V, Mihalcioiu A and Dascalescu L 2005 *J. Electrostat.*63, 937-942
[12] Blajan M, Beleca R, Iuga A and Dascalescu L 2010 *IEEE Trans. Ind. Appl.* **46** 1558-1563
[13] Buda G, Samuila A., Atroune S, Bilici M and Dascalescu L 2013 *J. Electrostat.* **71** 407-412
[14] Dascalescu L, Iuga A, Morar R, Samuila A, Neamtu V and Suarasan I 1994 *J. Phys. D: Appl. Phys.* **27** 1242-1251
[15] Dascalescu L, Samuila A, Rafiroiu D, Iuga A and Morar R 1999 *IEEE Trans. Ind. Appl.* **35** 543-548
[16] Crowley J M 1999 Fundamentals of Applied Electrostatics (Morgan Hill, CA: Laplacian Press)
[17] Iuga A, Vlad S, Mihailescu M and Dascalescu L 2004 *Part. Sci. Technol.* **22** 275-283
[18] Das S, Medles K, Younes M, Dragan C and Dascalescu L 2007 *IEEE Trans. Ind. Appl.* **43** 1137-1143

Dust particles precipitation in AC/DC electrostatic precipitator

A Jaworek[1)], A Marchewicz[1)], A Krupa[1)], A T Sobczyk[1)], T Czech[1)], T Antes[2)], Ł Śliwiński[2)], M Kurz[2)], M Szudyga[2)], W Rożnowski[2)]

[1] *Institute of Fluid Flow Machinery, Polish Academy of Sciences, Fiszera 14, Gdańsk, POLAND, jaworek@imp.gda.pl*
[2] *RAFAKO S.A., ESP Division, Górnośląska 3A, Pszczyna, Poland*

Abstract. Submicron and nanoparticles removal from flue or exhaust gases remain still a challenge for engineers. The most effective device used for gas cleaning in power plants or industry is electrostatic precipitator, but its collection efficiency steeply decreases for particles smaller than 1 micron. In this paper, fractional collection efficiency of two-stage electrostatic precipitator comprising of alternating electric field charger and DC supplied parallel-plate collection stage has been investigated. The total number collection efficiency for PM2.5 particles was higher than 95% and mass collection efficiency >99%. Fractional collection efficiency for particles between 300 nm and 1 μm was >95%.

1. Introduction

The efficiency of removal of dust particles smaller than a few micrometers in diameter by conventional electrostatic precipitators decreases with decreasing particle size. The electric charge on such particles is too small for the electric force to overcome air drag force that causes the particles to leave the precipitator. In order to avoid difficulties with the removal of submicron fly ash particles from exhaust gases, the charging and collection processes can be separated and accomplished in two different stages. Two types of such devices can be distinguished: electrostatic scrubber and two-stage electrostatic precipitator. In electrostatic scrubbers [1], the charged particles are deposited onto oppositely charged water droplets. In two-stage electrostatic precipitators [2-7], the particles pass through the charging stage, where they are charged by ionic current, and next, enter the precipitation stage, which is usually formed by two parallel plates at opposite potential, where they are collected. Additional advantage of such device is that it is free of back discharge due to lack of ionic current in the precipitation stage.

Two-stage electrostatic precipitator was first investigated by White [2] and applied by Masuda et al. [3,4], which used the boxer charger [5] to charging and control the emission of high-resistivity dust. Later, nozzle charger equipped with corona electrodes, corona-triode charger, quadrupole precharger, and alternating electric field charger were used as the charging stage by various authors [6-8].

In this paper, fractional collection efficiency in PM2.5 range of laboratory scale, two-stage AC/DC electrostatic precipitator, with alternating electric field charger (AC stage) and DC supplied precipitation stage, is investigated. In alternating electric field charger [9,10], the particles are charged by ionic current in alternating electric field, and the charge imparted to the particles can be higher than for DC chargers. The precipitation stage has no discharge electrode, and the precipitation occurs due only to electrostatic field between the electrodes that reduces power consumption by precipitator.

2. Experimental

The measurements were carried out in experimental stand shown schematically in Figure 1. The charger consisted of two discharge electrodes comprising of 104 stainless steel pins 0.6 mm in diameter and 5 mm long, mounted to brass sheet. Between the discharge electrodes, two parallel grids 140 mm high and 160 mm long, made of 9 stainless steel rods, 2 mm in diameter spaced 20 mm from each other were placed. The distance between the grids was 50 mm, and between each grid and the tip of pins - 45 mm. The space between the grids formed the charging zone. The electrodes were supplied from two high voltage transformers through the circuit made of diodes and resistors [7]. The ionic current emitted by discharge electrode at negative potential flow through the nearby grid, which was at ground potential, and through the charging zone between the grids, to the opposite grid at positive potential. The role of discharge electrodes and potential of the grids changed alternately in every half-cycle of supply voltage. The particles flowing through the charging zone are charged, and leave the charger as charged particles. The alternating electric field existing within the charging zone causes charged particles to oscillate, and partially to agglomerate during their motion, but prevents them from migrating to the charger electrodes. Two parallel-plates electrodes were used in the collection stage. The electrodes, 140 mm high and 340 mm long, were made of brass sheet. The distance between them was 80 mm. One of these plates was connected to high voltage power supply (SPELMANN HV model SL600W/30kV/P), and the other was grounded. The alternating electric field charger and collection electrodes were placed in a channel of square cross section of 160x160 mm, made of PMMA. A flow straightener was mounted at the inlet of the channel to eliminate air vortices. The airflow through the channel was forced by a suction fan placed at its outlet. The experiments were carried out for MgO test particles of mass mean diameter of about 2.8 μm. The particles were dissolved in methanol forming a colloidal suspension and stirred for at least 1 hour. The solution was sprayed by Aerosol Generator ATM 226 (TOPAS) and injected to the channel at its inlet. The concentration of particles was measured at the outlet of the channel using Aerosol Particle Size Spectrometer LAP 322 (TOPAS). The air flow velocity in the channel was measured by hot wire anemometer TSI model 8455.

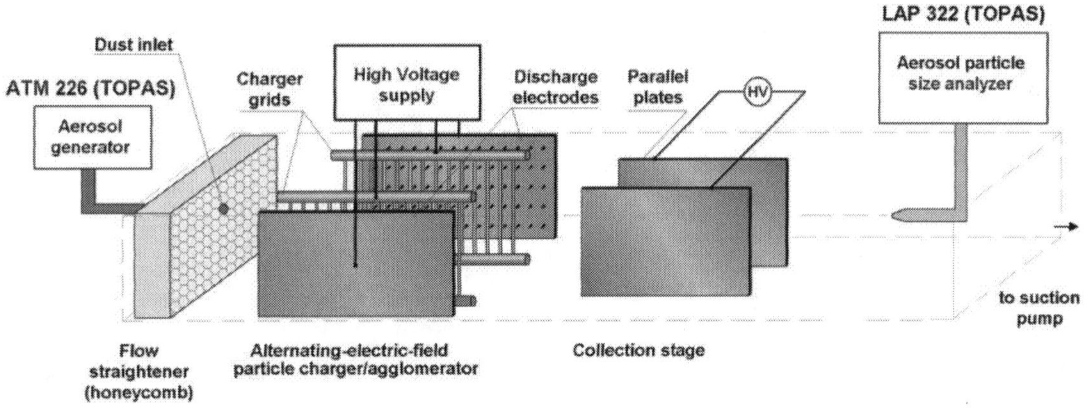

Figure 1. Schematic of experimental set-up of two-stage electrostatic precipitator with alternating electric field charger.

3. Results

Size distributions of particles at the outlet of AC/DC precipitator for AC and DC voltages switched OFF and ON are shown in Figure 2. The number concentration of particles for switched-off voltages varied between 8000 and 10000 $\#/cm^3$. When both AC and DC voltages were ON, the total concentration decreased below 500 $\#/cm^3$. The maximum in particle number size distribution for voltages OFF was about 350 nm and it decreased to about 250 nm during precipitator operation. The volume mean diameter decreased from 2.82 μm to 1.44 μm.

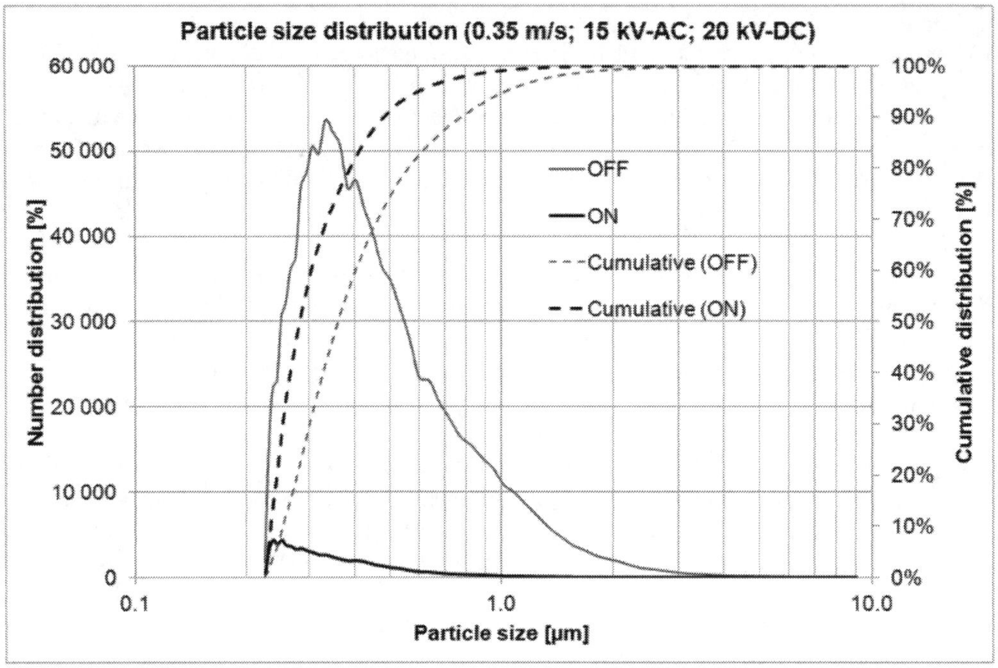

Figure 2. Number size distribution of particles leaving AC/DC precipitator for AC and DC voltages switched OFF and ON.

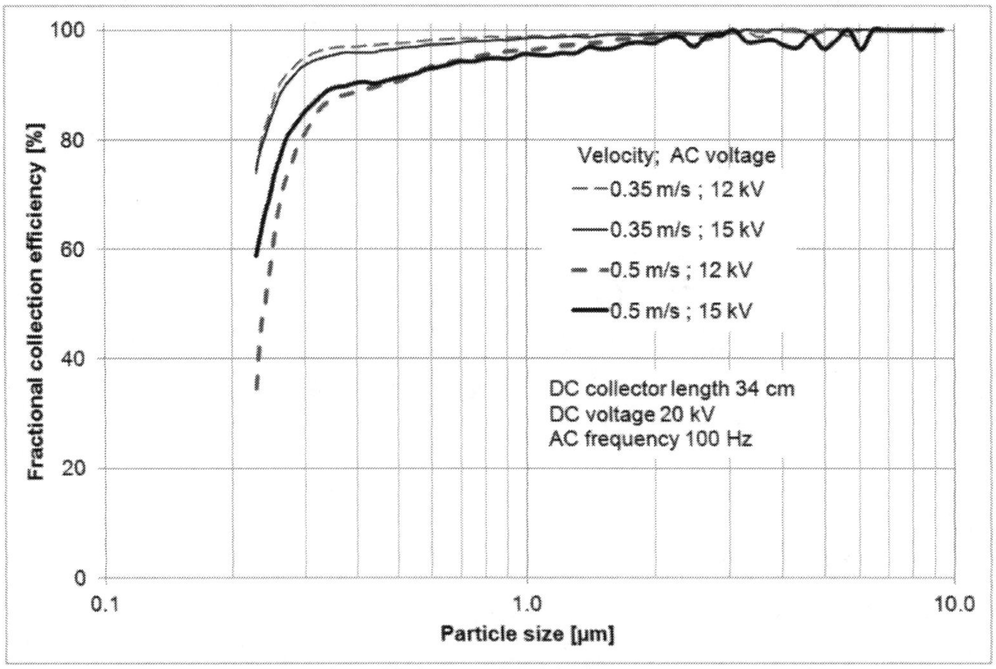

Figure 3. Fractional number collection efficiency of AC/DC electrostatic precipitator.

Fractional collection efficiency of two stage AC/DC electrostatic precipitator was determined from the measurements of size distribution of particles, from the ratio of particles concentration in each class when the voltages were switched ON n_{on} to that when the voltages were switched OFF n_{off}.

$$\eta = 1 - \frac{n_{on}}{n_{off}} \tag{1}$$

The fractional collection efficiency of AC/DC electrostatic precipitator is higher than 95% for particles larger than 1 μm, and drops below 90% for particles smaller than 300 nm. These results are higher than for conventional electrostatic precipitators. The measurements have been carried out for two gas velocities: 0.35 and 0.5 m/s, for AC voltages of 12 kV and 15 kV, and DC voltage between plate electrodes in precipitation stage of 20 kV. The total number collection efficiency was higher than 95% and mass collection efficiency >99wt.% for AC voltage of 12kV at the charger electrodes and 20 kV-DC in the collection stage. For 15 kV the collection efficiency was slightly lower (93% and 98wt.%, respectively), due probably to particle re-entrainment from the collection electrodes because of high electric field between them [7].

4. Conclusions

Fractional collection efficiency of two-stage electrostatic precipitator has been determined experimentally. Alternating electric field charger was used as a particles charging and agglomerating unit in two-stage electrostatic precipitator, which allows effective charging of particles to high electric charge. Parallel-plate collection electrodes, free of corona discharge have been used for the removal of charged particles leaving the particle charger. In this system, the number collection efficiency for particles larger than 1 μm was higher than 95%, but it dropped below 90% for particles smaller than 300 nm. Mass collection efficiency of this system was >99% for PM2.5 particles.

Acknowledgements. This work was supported by the National Centre for Research and Development (NCBiR) of Poland within the project No. 208429 "Innovative method for submicron particles emission reduction in flue and exhaust gases". The authors wish to thank Mr. Jerzy Bunikowski for his technical help in construction of experimental stand.

References

[1] Jaworek A, Krupa A, Sobczyk A T, Marchewicz A, Szudyga M, Antes T, Balachandran W, Di Natale F and Carotenuto C 2013 *J. Electrostat.* **71** 345-350
[2] White H J 1974 *J. Air Poll. Contr. Assoc.* **24**(4) 314-338
[3] Masuda S and Washizu M 1979 *J. Electrostat.* **6**(1) 57-67
[4] Masuda S and Hosokawa S 1982 *IEEE Ind. Appl. Soc. Cof. Rec.* 1094-1101
[5] Masuda S, Washizu M, Mizuno A and Akutsu K 1978 *Conf. Electrostatic Precipitation* (Leura, Australia, 21-24 Aug 1978). pp.105-118
[6] McLean K J 1988 *IEE Rev.* **135** Pt.A, No.6 347-361
[7] Jaworek A, Krupa A, Adamiak K 2003 *Inst. Phys: Conf. Ser.* **178** 343-348
[8] Lin W Y, Chang Y Y, Lien C T and Kuo C W 2011 *Aerosol Sci. Technol.* **45** 393-400
[9] Jaworek A and Krupa A 1989 *J. Electrostat.* **23** 361-370
[10] Adamiak K, Krupa A and Jaworek A 1995 *Inst. Phys: Conf. Ser.* **143** 275-278

Structure of near-electrode dissociation-recombination layers under DC stress

V A Chirkov, A M Shaposhnikov and Y K Stishkov

Physics Department, St. Petersburg State University, Ulianovskaia 1, 198504
St. Petersburg, Russia

E-mail: v.chirkov@spbu.ru

Abstract. The paper deals with processes of current passage through motionless low-conducting liquid in plane-plane electrode system at the injection and dissociation mechanisms of charge formation. The feature of the study is the joint consideration of an approximate analytical solution and the computer simulation. The latter is carried out in dimensionless form, which increases the generality of the results. The analysis of the structure of near-electrode layers in the transient regime and at various ratios between the injection and dissociation mechanisms of charge formation was performed.

1. Introduction

Near-electrode layers emerge in low-conducting liquids due to disbalance between the dissociation and recombination processes and play an important role in the electric current passage through a cell. Though the issue has a long history of investigation, it is still a topical one and is regularly discussed at the present-day conferences (e.g., [1]). The layers affect the measurements of electrophysical properties of a liquid at low voltage especially when ion mobility is determined by the analysis of polarization or depolarization currents. The corresponding currents are rather complicated, and insufficient understanding of their features can lead to incorrect interpretation of the experimental data [2]. Moreover, the presence of near-electrode layers tells on the electrohydrodynamic flow emerging in a cell under the action of a strong electric field [3]. Nevertheless, there are a lot of papers (e.g., [4, 5]) where computations or discussion of the corresponding processes are carried out disregarding the conduction processes.

2. Mathematical model and analytical description

Consider near-electrode dissociation-recombination layers in the plane-plane electrode system with an account of migration and diffusion mechanisms of ion transport and injection and dissociation mechanisms of charge formation. It should be noted that electric double layers [2] that are also located near electrode surfaces are ignored in the paper, since they demand a separate consideration and have minor effect on the processes in the bulk of a low-conducting liquid.

The current passage processes in motionless liquid dielectrics are described by coupled Poisson and Nernst–Planck equations. Consider a 1D problem statement and make the following assumptions: there are only two ion species in the liquid and they are assumed to be monovalent; diffusion coefficient and ion mobility are assumed to be the same for species of both polarities; the intensity of the dissociation charge formation is independent of the electric field strength; at last, the degree of dissociation is low. In this case, the set of equations can be written in the following dimensionless form:

$$d^2\varphi/dx^2 = -(n_+ - n_-)\, P_0/2 \tag{1}$$

$$\partial n_\pm/\partial t + \partial j_\pm/\partial x = P_0\,(1 - n_+n_-) \tag{2}$$

$$j_\pm = -\,(\pm n_\pm\, d\varphi/dx) - P_D\, \partial n_\pm/\partial x \tag{3}$$

where φ is the electric potential, n is the ion concentration, j is the density of ion flux, t is the time, x is the coordinate; P_0 and P_D are dimensionless parameters: $P_0 = \sigma_0 L^2 / bV_0\varepsilon\varepsilon_0$, $P_D = D / bV_0$, where σ_0 is the low-voltage conductivity, ε is the relative electric permittivity, ε_0 is the electric constant, L is the length of the interelectrode gap (IEG), V_0 is the voltage across IEG.

All dimensional variables were made dimensionless using their specific values—the length of IEG, the voltage across the gap, and the equilibrium ion concentration ($n_0 = \sigma_0 / 2eb$, where e is the elementary electric charge). The time span during which an ion crosses the IEG ($\tau = L^2/bV_0$) was used as the specific time scale. Chosen to be used as the specific value of the current density j_0 was its migration component: $n_0 bV_0 / L$. The following boundary conditions were utilized: $x \in [0,1]$, $\varphi(0) = 1$, $\varphi(1) = 0$, $j_+(0) = k_+$, $j_-(0) = k_-$, where $k_\pm$ is the dimensionless injection current density at the boundary ($k_\pm = j_{inj,\pm}/j_0$).

According to equation (2), P_0 parameter describes the role of dissociation-recombination processes in the bulk. If P_0 has high values, the volume charge formation plays an important role, which leads to rapid establishment of the steady-state distributions in the most part of IEG. Thus, the perturbation of the concentrations caused by the boundary conditions (e.g., by the absence of the injection charge formation at electrodes) vanishes quickly and fails to penetrate deep into the gap. This corresponds to the case of small length of non-equilibrium near-electrode layers in comparison with the length of the IEG. P_D parameter presents the ratio between the diffusion and migration current density components and appears to be low for liquid dielectrics.

The analytical solution for the set of equations (1)–(3) can be obtained under the following assumptions. Firstly, exclude from the consideration the diffusion current density and the effect of the space charge on the electric field distribution. Secondly, assume the length of the charged layer to be many times less that of the IEG. Thirdly, consider only the steady-state solution. At last, assume the absence of the injection, i.e. $k_\pm = 0$. However, the latter is not a necessary condition and the analytical solution can also be found at $k_\pm \in [0,1]$. As a result, the considered set of equations will possess the following solution for the positive ion concentration:

$$n_+ = 1 - 1 / (1 + xP_0). \tag{4}$$

Equation (4) shows the value P_0 to describe the ratio of the IEG length to that of the non-equilibrium layer. Besides, parameter P_0 designates the degree of the effect of space charge on the external electric field, according to equation (1). This all means that as soon as the volume charge formation assumes the major role the case of the strong effect of the space charge on the electric field realizes instantaneously.

3. Numerical simulation

Proceed with considering the results of numerical calculations by equations (1)–(3). The computations used software package COMSOL Multiphysics basing on the finite element method. To avoid poor numerical convergence during the transient regime, the original 1D problem was transformed into the 2D one, where the second coordinate corresponds to the time. A similar approach is of frequent use in the problem of shock wave propagation.

Examples of steady-state distributions of ion concentrations are presented in figure 1, where three cases are plotted: analytical solution and numerical ones in the simplified (i.e., with disregarding the effect of space charge density on the electric field strength) and complete approaches. The simplified case was considered in the simulation specially for comparison with the analytical solution, since they both was obtained under nearly the same assumptions. Both plots in figure 1 present only the left half of the IEG. The analytical and simplified solutions are close to each other and differ from the results of the exact computation. The main cause of the corresponding discrepancy is the ignoring the change of the electric field distribution under the effect of space charge density localized in the near-electrode layers. However, the specific length of the layers is described in a correct way in all above cases.

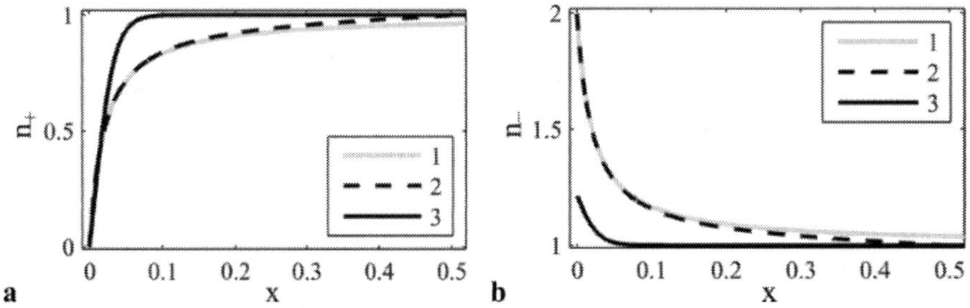

Figure 1. Distributions of dimensionless concentrations of positive (a) and negative (b) ions at $P_0 = 50$: 1—the analytical solution, 2—the simplified and 3—exact solutions.

Briefly analyze the numerical solution. The migration ionic flow provided by the conductivity emerges in the cell when the voltage is applied. However, both the concentration and current density of positive ions is zero at the surface of the positive electrode, since the latter presents itself a "wall" where there is no surface charge forms. Thus, a region of ion deficit arises near the electrode, which is clearly seen in figure 1a. Inside the deficit region, the ion concentration grows from zero level up to its equilibrium value owing to dissociation charge formation. The specific length of the layer is determined by the ratio between the intensity of dissociation charge formation and the migration current density or, according to the analytical solution, by dimensionless parameter P_0.

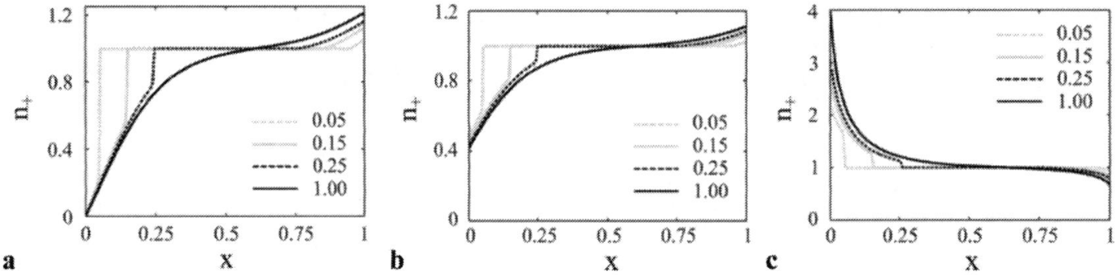

Figure 2. Distributions of dimensionless concentrations of positive ions during the transient regime (with the dimensionless time being specified in the legend) at $P_0 = 5$ for different injection currents: $k_\pm = 0$ (a), 0.5 (b) and 2 (c).

Basing on the computation results, consider how the injection charge formation affects the structure of the near-electrode layers (figure 2). When $0 < k_\pm < 1$ (figure 2b), the injected ions cause a rise in ion concentration at the boundary comparing to the case without injection (figure 2a); however, the value of the total space charge remains below zero. Emergence of a homocharge takes place only if the injection current prevails over the migration one (i.e., at $k_\pm > 1$); however, ion concentration quickly decreases when ions propagate inside the IEG due to strong recombination (figure 2c), with the length of the layer being nearly the same as that in the case without the injection.

Ion distributions presented above were calculated at $P_0 = 5$, which corresponds to the case of the strong effect of the space charge on the electric field strength (figure 3a). Thus, the strength in near-electrode area increased by 40% due to the presence of heterocharge layer at $k_\pm = 0$, while reduced by near 50% under the effect of homocharge at $k_\pm = 2$.

Lastly, consider the charging current or the so-called current-time characteristic of the transient regime taking place after the pulsed application of the voltage. The computation technique of charging current in the unsteady-state regime is described in [6]. The current starts at the same dimensionless value—2, which corresponds the classical conduction current provided by initial (equilibrium) ion concentration. Then, however, the current either decreases due to the formation of areas with ion deficit or increases due to the emergence of additional (injected) charge. Besides, it is worth noting that despite quite high injection current density (that is twice the conduction one), the total current increases only by 15%. The latter can be explained by important role of the recombination processes.

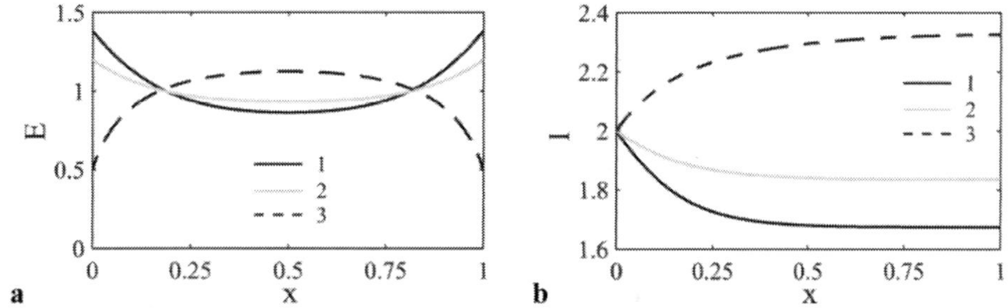

Figure 3. Steady-state distributions of the electric field strength (a) and charging currents (b) at $P_0 = 5$ for different injection currents: 1—$k_\pm = 0$, 2—$k_\pm = 0.5$ and 3—$k_\pm = 2$.

4. Conclusions

The joint consideration of the analytical solution along with the conduction of numerical calculations in dimensionless form allows formulating the following conclusions.

First of all, the degree of the electric field screening by the space charge emerged in the bulk is determined by the same dimensionless parameter P_0 as the degree of the effect of dissociation-recombination processes on the current passage in a low-conducting liquid.

Secondly, the injected ions fail to penetrate inside the bulk when parameter $k_\pm < 1$; thus, the homocharge layer is absent in this case. When $k_\pm > 1$, the homocharge penetrates into the bulk in the efficient way only at low values of parameter P_0; otherwise (at $P_0 > 1$), the injected charge rapidly recombines during its propagation deep into the gap, with the specific length of penetration being approximately equal to $1/P_0$. This all, in turn, signifies that the consideration of the ion transport and, in particular, electrohydrodynamic flows disregarding the dissociation-recombination processes in the corresponding range of liquids and voltage values (when $P_0 > 1$) is highly ungrounded. Thus, for example, the pure injection model of charge formation fails to describe the structure of electrohydrodynamic flows for liquids with enhanced conductivity or lessened ion mobility.

Thirdly, layers with reduced effective conductivity form near surfaces of both electrodes when the injection is absent, which expresses in the decay of the integral electric current passing through the cell. Accordingly, to obtain correct results, measurements of the low-voltage conductivity of liquid dielectrics are to be made only when the experimental conditions satisfy the following: $P_0 >> 1$.

Acknowledgments

The authors acknowledge St. Petersburg State University for a research grant 11.0.65.2010. Research was carried out using computer resources provided by Resource Center "Computer Center of SPbU".

References

[1] Vahidi F, Tenbohlen S, Rösner M, Perrier C and Fink H 2014 Space charge formation in insulating liquids under DC stresses *Proc. Int. Conf. on Dielectric Liquids ICDL (Bled, Slovenia, 2014)* 1–5
[2] Zhakin A I 2006 *Phys. Usp.* **49** 275–295
[3] Stishkov Y K and Chirkov V A 2013 *Tech. Phys.* **58** 1822–30
[4] Traoré Ph Daaboul M and Louste C 2010 *J. Phys.* D: *Appl. Phys.* **43** 1–8
[5] Shrimpton J 2009 *Charge injection systems: physical principles, experimental and theoretical work* (Berlin: Springer)
[6] Chirkov V A and Stishkov Y K 2013 *J. Electrostatics* **71** 484–488

Comparative analysis of numerical simulation and PIV experimental results for a flow caused by field-enhanced dissociation

V A Chirkov, D K Komarov, Y K Stishkov and S A Vasilkov

Physics Department, St. Petersburg State University, Ulianovskaia 1, 198504 St. Petersburg, Russia

E-mail: sergeyvasilkov@yandex.ru

Abstract. The paper studies a particular electrode system, two flat parallel electrodes with a dielectric plate having a small circular hole between them. Its main feature is that the region of the strong electric field is located far from metal electrode surfaces, which permits one to preclude the injection charge formation and to observe field-enhanced dissociation (the Wien effect) leading to the emergence of electrohydrodynamic (EHD) flow. The described electrode system was studied by way of both computer simulation and experiment. The latter was conducted with the help of the particle image velocimetry (or PIV) technique. The numerical research used trusted software package COMSOL Multiphysics, which allows solving the complete set of EHD equations and obtaining the EHD flow structure. Basing on the computer simulation and the comparison with experimental investigation results, it was concluded that the Wien effect is capable of causing intense (several centimeters per second) EHD flows in low-conducting liquids and has to be taken into account when dealing with EHD devices.

1. Introduction

Specific mechanisms of high-voltage charge formation, i.e. the surface charge injection and field-enhanced dissociation (or the Wien effect), take place in low-conducting liquids under the influence of a strong electric field. Apart from their effect on the total value of the electric current passing through the cell, they can ensure the formation of space charge density inside the bulk in the case of non-uniform electric field and lead to the emergence of electrohydrodynamic (EHD) flows [1]. The field-enhanced dissociation is frequently termed as a purely theoretical phenomenon and its significance in the EHD flow formation is usually disregarded. The main problem in the context of revealing the role of field-enhanced dissociation in EHD flow formation is caused by similarities in the corresponding flow structure with that taking place in the injection model. Thus, the EHD flow formation is frequently ascribed only to the injection even if two mechanisms of charge formation are active [2, 3].

Electric field intensity can exceed 10^7 V/m, and dissociation rate, according to [4], increases by a factor of tens. But the presence of injection from metal electrodes creates difficult conditions for experimental study of EHD flows, caused by Wien effect. Thus, to investigate EHD flows with field-enhanced dissociation charge formation mechanism, one has to exclude injection from electrode surfaces. To do this, one can use relatively weak electric fields on the order of 10^6 V/m [5], when no injection was observed (dissociation rate increases less than twice in the case), or create a region of a strong electric field far from electrode surfaces. In the present work, the second way was chosen.

To resolve the issue, a particular setup was used: flat parallel electrodes and a dielectric plate (a barrier) having a small circular hole (figure 1) between them. The barrier surface accumulates ions, so all electric field lines pass through the hole and the intensity attains the maximum here, i.e., far from any electrode, which prevents the injection charge from forming inside the enclosure and provides favorable conditions for the Wien effect emergence. As the electric field intensity distribution and that of the increase in dissociation rate are non-uniform (figure 1a), they cause the appearance of space charge and hence EHD flow, which was studied numerically and experimentally.

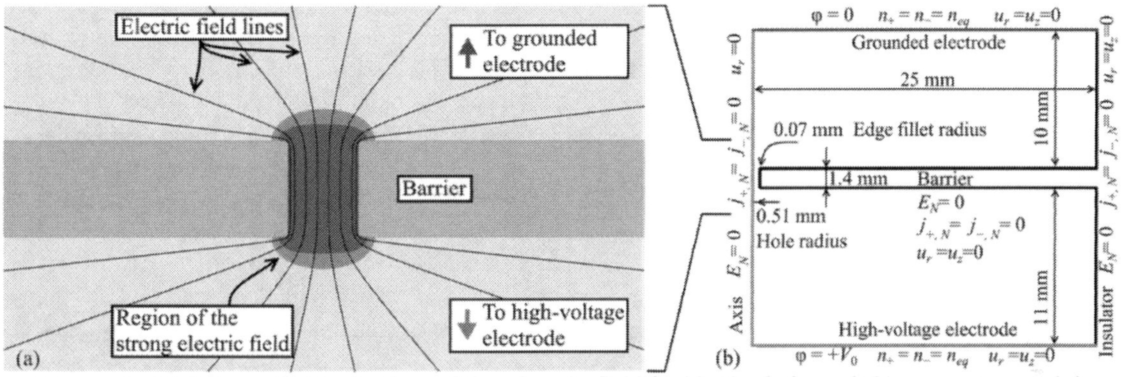

Figure 1. (a) Schematic illustration of the electric field inside the hole and (b) computer model with marked dimensions and boundary conditions.

2. Simulation and experimental techniques

The present work includes computer simulation of an EHD flow with the corresponding technique being described in [6]. The computations were carried out using software package COMSOL Multiphysics® 4.2 based on the finite element method. The complete set of equations (as in [6]) was solved. The dissociation intensity, the part of the source function for transport equations, is $W_0F(p)$ where W_0 is that in the absence of electric field and F is the relative increase in the dissociation rate:

$$F(p) = I_1(4p)/(2p), \quad p = e^2/(2k_BT)\sqrt{(E/(4\pi\varepsilon\varepsilon_0 e))} \tag{1}$$

Here I_1 is the modified Bessel function of the first kind, e is the elementary electric charge, k_B is the Boltzmann constant, T is the temperature, E is the electric field strength, ε is the relative electric permittivity, ε_0 is the electric constant.

Two-dimensional model with the axial symmetry was built (figure 1b). Dimensions used in the model correspond to those of experimental setup. However, there is a difference in barrier shape: it is a circle in the case of the computer simulation and a rectangle in the case of the experiment. Nevertheless, a change in shapes of boundaries that are far from the region of interest (the hole) does not affect EHD flow near it. It is difficult to measure the fillet radius of the hole edges but the computer simulation shows that if its value is relatively small, it has little influence on velocity distribution. Figure 1b also lists boundary conditions, where φ is the electric potential, n is the ion concentration, $\boldsymbol{j}$ is the ion flux density, $\boldsymbol{u}$ is the fluid velocity; $\boldsymbol{N}$ is the outward normal, r and z are coordinates; n_{eq} is the equilibrium concentration. As the electrodes are far from the hole, approximate conditions for transport equations can be used. The condition E_N=0 was used on each dielectric surface, i.e. the total screening of electric field by accumulated surface charge was assumed. In this model, electro-osmotic effects are disregarded.

To obtain the velocity distribution, the particle image velocimetry (or PIV) technique was used. This requires seeding the flow with small tracer particles that are illuminated during the measurement by pulsed laser sheet (for more details, see [7]). Two consecutive recorded frames are used by the DaVis software to compute the velocity distribution. Working liquid was chosen to be transformer oil mixed with 5% of butanol. Its conductivity was $3.3\cdot10^{-10}$ S/m, with other properties being chosen typical for the transformer oil [1].

3. Results and discussion

Consider results for 15 kV. Figure 2a shows all electric field lines to pass through the hole. So, the electric field intensity increases inside the hole and has the values from $0.6 \cdot 10^7$ to $1.1 \cdot 10^7$ V/m (with the maximum observed on the hole edges). At the same time, the electric field intensity has the value of order 10^4 V/m near the electrode surfaces. According to equation (1), at the hole center, the dissociation rate increases by the factor of eight, while near the edges — more than by the factor of twenty (figure 2b). The non-uniform dissociation rate distribution results in space charge emergence. The distributions of the space charge (figure 2c) and the velocity field (figure 2d) are highly interrelated: the former governs the latter via the Coulomb force, and the latter deforms the former. The velocity magnitude is of the order of several centimeters per second, and the flow has following structure: liquid comes to the hole from the bulk and then spreads radially along the barrier.

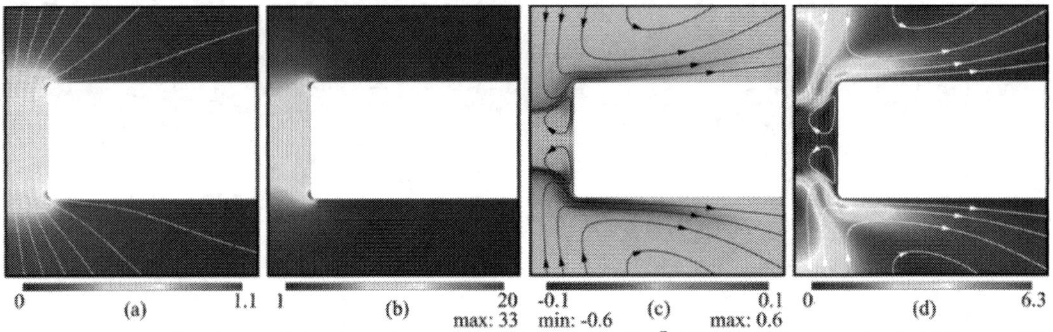

Figure 2. Distributions of (a) the electric field intensity (10^7 V/m) with electric field lines, (b) the relative increase in the dissociation rate (1), (c) space charge density (C/m^3) with streamlines, (d) velocity magnitude (cm/s) with streamlines.

The PIV technique was applied to study the velocity distribution below the barrier. Figure 3 shows the comparison of computer simulation and experiment results. It is noteworthy here that the experimental results have no symmetry in contrast to simulation, but the flow structure is quite similar: liquid spreads (but non-uniformly) from the gap center to the edges (figure 3b). Besides, it is worth noting that barrier perimeter was pressurized during the enclosure assembly and zero mass flux thought the hole was guaranteed.

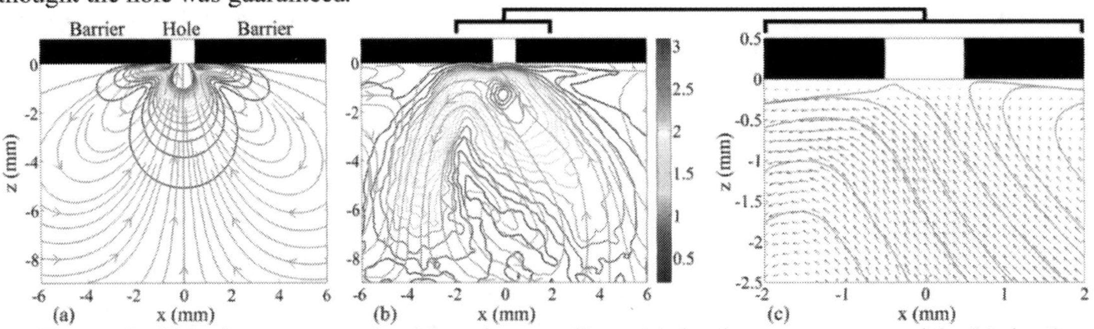

Figure 3. Velocity contours (cm/s) and streamlines (a) in the computer model, (b) in the experiment, and (c) velocity vectors and streamlines near the hole.

At distances closer to the barrier than 0.5 mm, the velocity values are underestimated (and are incorrect) due to the data processing specifics. As far as 1 mm from the hole, the velocity in both computer simulation and experimental study has values on order of 3 cm/s, but the values in the latter case decrease slower with growing distance from the barrier. This happens because a stream coming to the hole remains relatively thin far from the barrier surface compared to simulation. Also, after passing the hole most liquid moves to the left, but there also exists a flow directed to the right, with its size agreeing to that from simulation.

As it was noted above, the computer model disregards electro-osmotic effects. In the case of pure electro-osmosis, the flow should be directed in the opposite direction [8]. But indeed, such flow was observed at 5kV, when the field-enhanced dissociation effect is insignificant.

In addition, the local heating of the liquid inside the hole can theoretically lead to the observed flow structure (the temperature gradient is frequently used to create the space charge inside induction pumps [9]). But the computer simulation shows that in the case of considered liquid, the overheat is less than 0.01 K inside the hole so that the temperature gradient is negligible.

This means that the observed flow structure can be explained only by the Wien effect. The absence of symmetry of experimental data precludes the accurate quantitative comparison (but the flow proves to be more intensive far from the hole).

4. Conclusions

A particular experimental setup was arranged to create a strong electric field far from the electrode surfaces and to study EHD flows caused by the Wien effect. The system was investigated by means of computer simulation: the velocity values of order of several centimeters per second were obtained, the flow structure was described. It was demonstrated by the experiment that the Wien effect actually causes such flow, since other charge formation mechanisms are excluded or negligible. This means that the Wien effect should be taken into account when EHD flows are investigated as its role in EHD flow formation can be comparable with that of injection and it can affect dissociation-recombination layers (and hence conduction pump performance [10]).

Acknowledgments

The authors acknowledge St. Petersburg State University for a research grant 11.0.65.2010. Research was carried out using computer resources provided by Resource Center "Computer Center of SPbU" (http://cc.spbu.ru) and Resource Center "Geomodel".

References

[1] Stishkov Y and Chirkov V 2011 Dependence of the electrohydrodynamic flows structure in very non-uniform electric field on the charge formation mechanism *Proc. Int. Conf. on Dielectric Liquids, ICDL (Trondheim, Norway, 2014)*

[2] Atten P and Seyed-Yagoobi J 2003 Electrohydrodynamically Induced Dielectric Liquid Flow Through Pure Conduction in Point/Plane Geometry *IEEE Trans. Dielectr. Electr. Insul.* **10** 27–36

[3] Daaboul M, Louste C and Romat H 2009 PIV measurements on charged plumes – influence of SiO2 seeding particles on the electrical behavior *IEEE Trans. Dielectr. Electr. Insul., special issue ''EHD & Flow Electrification''.*

[4] Onsager L 1934 Deviations from Ohm's law in weak electrolytes *J. Chem. Phys.* **2** 599–615.

[5] Suh Y K, Baek K H and Cho D S 2013 Asymptotic and numerical analysis of electrohydrodynamic flows of dielectric liquid *Phys. Rev.*E **88** 023003.

[6] Chirkov V A, Stishkov Yu K 2013 Current-time characteristic of the transient regime of electrohydrodynamic flow formation *J. Electrostat.* **71** 484–488.

[7] Willert C E , Gharib M 1991 Digital particle image velocimetry *Exps. Fluids* **10** 181–193.

[8] Squires T M and Bazant M Z 2004 Induced-charge electro-osmosis *J Fluid Mech* **509** 217–252.

[9] Seyed-Yagoobi J 2005 Electrohydrodynamic pumping of dielectric liquids *J. Electrostat.* **63** 861–869.

[10] Stishkov Y, Chirkov V and Vasilkov S 2015 Characteristics of electrohydrodynamic pump of the dissociation type: low- and high-voltage ranges *IEEE Trans. Dielectr. Electr. Insul., in press.*

Time resolved strain dependent morphological study of electrically conducting nanocomposites

Imran Khan[1], Geoffrey Mitchell[1], Artur Mateus [1] and Christina S Kamma-Lorger[2]
[1] Centre for Rapid and Sustainable Product Development, Institute Polytechnic Leiria, Marinha Grande, Portugal
[2] NCD beamline, Alba Synchrotron Light Source, Cerdanyola del Vallès, Barcelona, Spain

Corresponding author email: geoffrey.mitchell@ipleiria.pt

Abstract. An efficient and reliable method is introduced to understand the network behaviour of nano-fillers in a polymeric matrix under uniaxial strain coupled with small angle x-ray scattering measurements. The nanoparticles (carbon nanotubes) are conductive and the particles form a percolating network that becomes apparent source of electrical conduction and consequently the samples behave as a bulk conductor. Polyurethane based nanocomposites containing 2% w/w multiwall carbon nanotubes are studied. The electrical conductivity of the nanocomposite was ($3.28x10^{-5}$s/m).The sample was able to be extended to an extension ratio of 1.7 before fracture. A slight variation in the electrical conductivity is observed under uniaxial strain which we attribute to the disturbance of conductive pathways. Further, this work is coupled with in-situ time resolved small angle x-ray scattering measurements using a synchrotron beam line to enable its measurements to be made during the deformation cycle. We use a multi-scale structure to model the small angle x-ray data. The results of the analysis are interpreted as the presence of aggregates which would also go some way towards understanding why there is no alignment of the carbon nanotubes.

1. Introduction

The synthesis and fabrication of nanometer-sized particles has reached a mature state and they can be made with controlled composition and size [1]. Nanoparticles offer high surface area to interact with polymer phase due to which a new interphase develops. Such interphases play a key role in developing the microstructure of the nanocomposite material. The microstructure changes which take place in the nanocomposite, under different conditions ranging from temperature, pressure and mechanical deformation have not been properly investigated. Moreover, nanoparticles in a polymer matrix represent a more dynamic system that has greater freedom for reversible nanoscale restructuring, which is essential for stretchability [2]. A disturbance in a multi wall carbon nanotube (MWCNT) network can also disturb the microstructure of the nanocomposite material and consequently the physical properties are altered [3]. Miao and his co-workers examined the piezo resistive response of CNT–polymer nanocomposites and reported a decrease in electrical conductivity [4]. Highly stretchable conductive polymer composites with MWCNT and carbon nanospheres subjected to mechanical strain using polyurethane as the matrix material exhibited an increase of the electrical conductivity of the nanocomposite by 20% under strain, followed by a decrease in conductivity upon further elongation [5]. Such studies suggested that there is a significant effect on electrical conductivity during deformation. These may well reveal structural information about the percolating network. Small angle x-ray scattering (SAXS) is a versatile technique to probe the structure, orientation and network of nano scale particles within a matrix material.In this work, we present small angle x-ray scattering (SAXS) coupled with four probe method to simultaneously measure electrical conductivity to understand the MWCNT network and its behaviour under uniaxial deformation.

2. Materials & Methods:

The carbon nanotubes (MWCNT) purchased from Sigma-Aldrich, Portugal. Commercially available Diphenylmethane diisocyanate, polyol and Dimethyl sulphoxide (DMSO) were used without further treatment. Thermoset polyurethane samples were prepared with 1:1 ratios. figure 1 gives a schematic of the steps in the synthesis of the nanocomposites. The electrical conductivity is measured using a four probe method coupled with tensiometer to enable measurements during deformation as shown in figure 2. The SAXS detector was mounted at a sample-to-detector distance of 6.3m which provides a q range of 0.001 - 0.020 $Å^{-1}$, using a beam of wavelength 1Å.

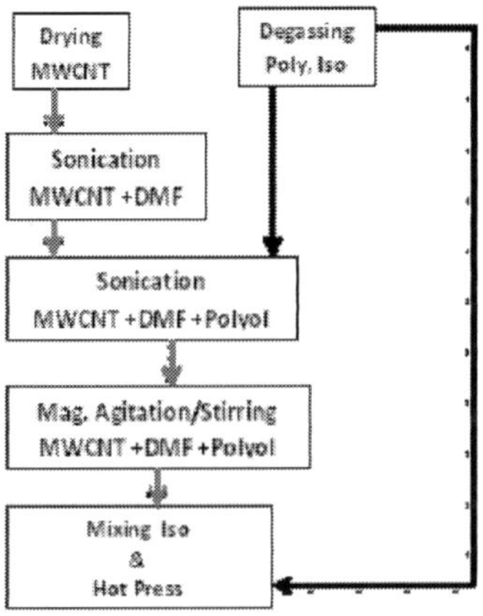

Figure 1. The nanocomposite synthesis mechanism

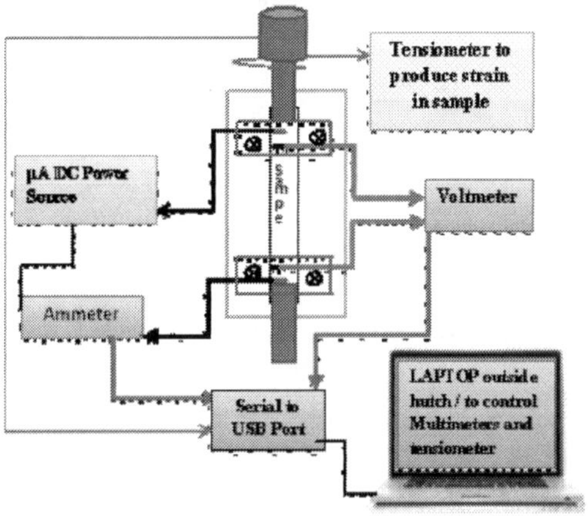

Figure 2. The four probe set up coupled with the tensiometer used in the time resolved SAXS experiment described in the text.

3. Results and Discussion:

3.1 Electrical Conductivity under uniaxial strain:

The measured electrical conductivity (σ) with extension ratio λ is plotted in figure 3. The overall electrical conductivity change with extension ratio is slight and shows a decreasing behavior from $3.07\text{x}10^{-5}$ s/m to $2.70\text{x}10^{-5}$ s/m. The SAXS data was used to evaluate the level of preferred orientation $\langle P_2 \rangle$of the MWCNTs which at its maximum was found to be 0.02. (1.0 corresponds to perfect common alignment and 0.0 to random alignment). The fluctuations in electrical conductivity under uniaxial strain lead to slight decrease in magnitude. The same kind of behavior was reported by Shang and his coworkers in which a nanocomposite was stretched up to 100% without significant variation in electrical conductivity [6]. The electrical conductivity initially increases from $3.07\text{x}10^{-5}$ s/m to $3.28\ \text{x}10^{-5}$s/m for an extension ratio (λ) of 1.04 and then decreases to $2.80\text{x}10^{-5}$s/m. We attribute the decrease in electrical conductivity to be caused by a strained network leading to breaking of certain MCNT contacts. An increase and decrease in electrical conductivity at 20.0% strain was also reported in literature [5]. This is also most likely caused by MWCNT reorganization and stretching under uniaxial strain causes in breaking and relaxation of certain networks to increase and decrease electrical conductivity. The fluctuations in electrical conductivity continue until λ=1.37. A further decrease in electrical conductivity results in breaking of more number of contacts between MWCNT while an increase in electrical conductivity is due to new contact formation. Since, the polyurethane matrix is elastic and can be stretched while MWCNT are rigid and retain their position. These contacts breaking and formation is locally, but when they are significant in number, they will have a global effect.

Electrical conductivity continuously decreases for λ=1.37-1.41 and attains a value2.78x10^{-5} s/m. When λ exceeds a value of 1.41 to 1.45, the electrical conductivity remains constant. An equal number of contact breaking and formation lead towards a minute change in electrical conductivity or either no change at all. An increasing behavior in electrical conductivity is observed for value of λ between 1.45 and1.47 and σ reaches 2.99x10^{-5} s/m and after that till the value of λ=1.55, σ remains constant. The initial increase-decrease behavior in σ again comes into play with a final value 2.70x10^{-5} s/m till $\lambda_{max.}$ =1.70. The sample is stretched up to 70% of original length. The maximum electrical conductivity (3.30x10^{-5} s/m) is observed for λ=1.22 i.e when sample was strained 22% of original length while minimum electrical conductivity was 2.66x10^{-5} s/m for λ=1.63. In figure 3 the arrows are used to indicate fluctuations in electrical conductivity with respect to the extension ratio while the trend line indicates an overall deceasing trend in electrical conductivity. The decrease in electrical conductivity is consistent with theoretical modelling [7] as well as other reported experimental results under strain [4].

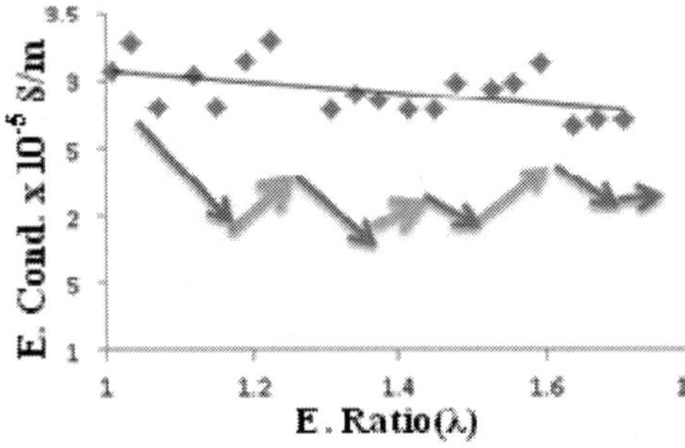

Figure 3. Variation in electrical conductivity vs the extension ratio λ. The red arrows indicate a decrease and green arrows an increase in electrical conductivity while length indicates the magnitude of change

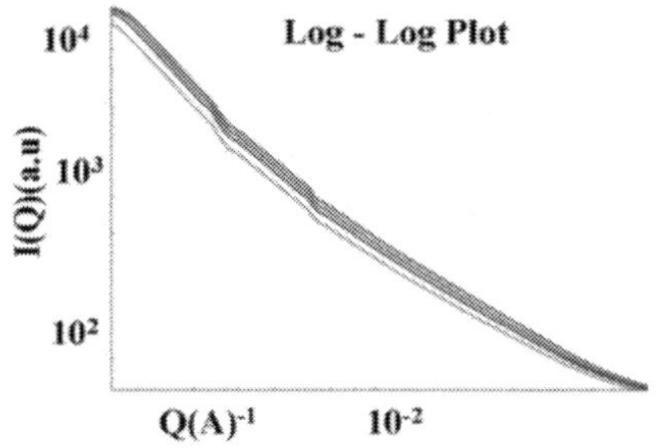

Figure 4. The SAXS curves I(q) plotted on a log-log scale as function of q for samples held at the extension ratios λ=1.45 (green), 1.55 (red), 1.59 (cyan), 1.70 (magenta).

3.2 Morphology of nanocomposite under uniaxial strain:

SAXS 1-D patterns were obtained by integrating the 2D SAXS images over the azimuthual angle and these are shown in figure 4. The data comes close to a power law curve with an exponent of between 2 and 3. We interpret this as the scattering arising from aggregates or mass fractals lying between disks and spheres. The variation in the value of the slope is associated with a variation in microstructure caused by the disturbance produced by stretching of network of MWCNT which leads to a decrease in electrical conductivity. It is clear that the curve cannot be described using a single structural scale and so we have employed a multiple scale model with aggregates clustered around different dimensions. This is described in the next section.

4. Model:

An empirical unified Power law and R_g Model with level 3, developed by G. Beaucage [8] is used to fit the integrated SAXS data (figure 5). The empirical expressions are used to approximate the scattering from different type of fractal structures. The empirical fit function is

$$I(q) = Bkgd + \sum_{i=1}^{N} G_i \exp\left(-q^2 R_{g,i}^2/3\right) + \frac{B_i\left[erf(qR_{g,i}/\sqrt{6})\right]^{3P_i}}{q^{P_i}} \quad \ldots\ldots\ldots\ldots (1)$$

Table 1

Data	λ	Rg1	Rg2	G1	G2	B1	B2	P1	P2
S_TB3c_0015_out.txt	1.15	20118	757.1	5.44E+05	3.41E+03	2.87E-01	9.26E-05	1.43658	2.641311
S_TB3c_0023_out.txt	1.450058	15783	819.6	4.37E+06	4.93E+03	5.86E-03	2.89E-05	1.931631	2.850562
S_TB3c_0026_out.txt	1.555911	18987	805.4	1.39E+06	5.46E+03	1.37E-01	5.31E-05	0.432537	2.74713
S_TB3c_0030_out.txt	1.7	12904	760.6	4.59E+06	2.64E+03	4.64E-03	3.05E-05	2.139953	2.817482

The parameters, in Table1, are obtained using the empirical model (eq.1) to fit the SAXS experimental data. All parameters from Table1 are directly or indirectly related to microstructure of nanocomposite. In figure 6, we show the radius of gyration (Rg_1) of the largest aggregates which fluctuates and reduces slightly with uniaxial strain as does the smaller scale structures with Rg_2.

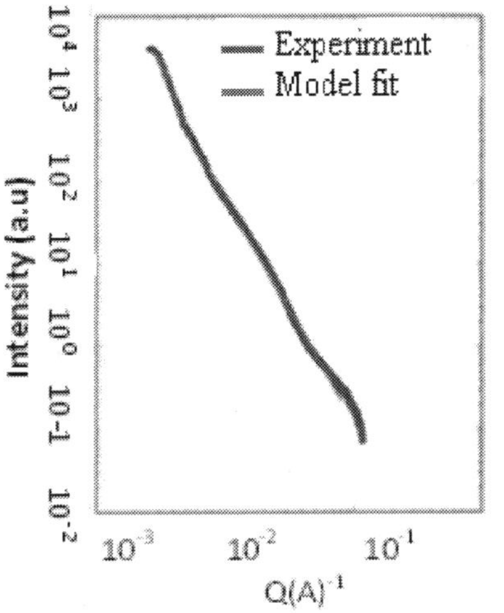

Figure 5. A Comparison of the experimental SAXS data and the Unified model fit (equation 1)

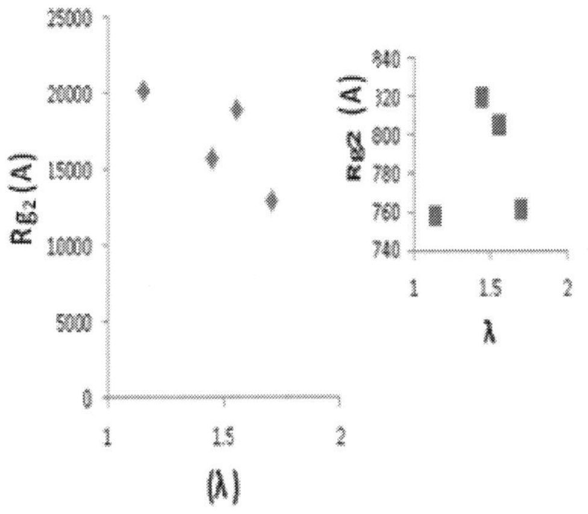

Figure 6. Variation in RG_1 and RG_2 (equation 1) with respect to the extension ratio, λ

5. Conclusion:

Electrically conductive Polyurethane nanocomposites with the 2% w/w MWCNT exhibited a conductivity of 3.28 x 10^{-5} s/m. The effect of uniaxial strain on the electrically conductive nanocomposite was studied. Remarkably, despite an extension ratio of 1.7 the overall electrical conductivity decreased slightly although fluctuations were observed under uniaxial strain. At the same time the measured SAXS data showed no alignment of the MWCNTs as might be expected from their aspect ratio. We interpret this as a consequence of the percolating network being formed from aggregates with some distributions in the size of those aggregates. The combination of strain coupled with in-situ SAXS measurements and accompanying electrical conductivity measurements has proved to be an effective tool to develop an understanding of these complex materials.

Acknowledgements: The work at IPL was supported by FCT (Portugal). These experiments were performed at the NCD beamline at ALBA Synchrotron with the collaboration of ALBA staff.

References

[1] Cao Y, Irwin P C, Younsi K 2004 *IEEE Trans. Dielectr. Electr. Insul.* **11** 797-807
[2] Kim Y, Zhu J, Yeom B, Prima M D, Su X, Kim J G, Yoo S J, Uher C and Kotov N A 2013 *Nature* **59** 500
[3] Bikiaris D 2010 *Materials* **3** 2884-2946
[4] Miao Y, Chen L, Lin Y Z, Sammynaiken R and Zhang W J 2011 *J. Intell. Mater. Syst. Struct.* **22** 2155–2159
[5] Shang S M, Zeng W and Tao X 2010 *Adv. Mater. Res.* **123-125** 109 - 112
[6] Shang S, Zeng W and Tao X. 2011 *J. Mater. Chem.* **21** 7274-7280
[7] Feng C and Jiang L Y 2014 *J. Phys. D: Appl. Phys.* **47** 405103
[8] Beaucage G and Schaefer D W 1994 *J. Non-Cryst. Solids* **172-174** 797 - 805

Measurement of the electric potential at the surface of non-uniformly charged polypropylene nonwoven media

Ali Fatihou[1], Noureddine Zouzou[1], Gheorghe Iuga[2] and Lucian Dascalescu[2]

Institut PPRIME, UPR 3346, CNRS - Université de Poitiers - ENSMA,
[1] SP2MI, BP 30179, 86962 Futuroscope Chasseneuil Cedex, France
[2] Institut Universitaire de Technologie, 16021 Angoulême, France

E-mail: lucian.dascalescu@univ-poitiers.fr

Abstract. The aim of this paper is to establish the conditions in which the vibrating capacitive probe of an electrostatic voltmeter could be employed for mapping the electric potential at the surface of non-uniformly charged insulating bodies. A first set of experiments are performed on polypropylene non-woven media (thickness: 0.4 mm; fiber diameter: 20 μm) in ambient air. In a second set of experiments the non-uniformity of charge is simulated using five copper strips (width: 2 mm or 3 mm; distance between strips: 2 mm). All the strips are connected to a high-voltage supply (V_s = 1000 V). The sample carrier is attached to a computer-controlled positioning system that transfers it under the capacitive probe (TREK, model 3451) of an electrostatic voltmeter (TREK, model 1341B). The measurements are performed at various relative speeds V_b between the sample and the probe, and for various sample rates F_e. A first set of experiments point out that the electric potential displayed by the electrostatic voltmeter depends on the spacing h between the sample and the probe. The diameter D of the spot "seen" by the probe is approximately $D \approx 8h/3$. From the second set of experiments performed with the test plate, it can be concluded that the surface potential can be measured with the media in motion, but the accuracy is limited by the spatial resolution defined by $k = V_b/F_e$.

1. Introduction

Electrostatic charging enhances the collection efficiency of nonwoven fibrous media employed as air filters [1]. Surface potential decay measurement [2] - [4] is a standard technique for the characterization of such media with respect to their ability to preserve the electric charge [5], [6]. Electric potential measurements have also been used for characterizing the distribution of the electric charge at the surface of non-woven media for air filtration [7] - [9]. Some of the peculiarities of non-contact techniques employed for the measurement of the electric potential at the surface of insulating bodies have been pointed out in [10] and [11].

In a recent paper [12], the authors have analyzed the correlation between the value displayed by the instrument and the distribution of the potential across the surface examined by the probe. In order to reduce the errors due to potential decay in time, all measurements need to be done as fast as possible, ideally with the sample moving at constant speed in front of the probe. The question is: "To what extent the relative speed between the sample and the probe affects the accuracy of the measurement?" The aim of the present paper is to answer this question, taking also into consideration the effect of the sample-rate at which the acquisition of the date is performed.

2. Experimental set-up and material

In the corona-charging experiments, carried out under stable ambient conditions (temperature 20 – 25°C; relative humidity 30 – 48%), a triode-type electrode system (figure 1) is used to corona-charge PP nonwoven samples (surface: 95 mm x 95 mm; thickness: 0.4 mm; fiber diameter: 20 µm), the aspect of which can be examined in figure 2. The ionizing electrode consists in a tungsten wire (diameter 0.2 mm) supported by a metallic cylinder (diameter 26 mm) and distanced at 34 mm from its axis. The wire and the cylinder are energized from the same high-voltage amplifier 30 kV, 20 mA (model 30/20A, Trek Inc., Medina, NY). The amplitude V_s of the high-voltage is adjusted using a synthesized function generator (model FG300, Yokogawa, Japan).

The grid of the triode-type electrode system is connected to the ground through a resistance R. For an intensity I, a potential $V_g = RI$ is imposed to the grid. Part of the charge carriers generated by the ionizing electrode pass through the grid and are driven by this potential to the surface of the sample laid on the grounded plate electrode. The deposition of the charge stops when the electric potential at the surface of the sample is equal to that of the grid. In all the experiments the samples are exposed for 10 s to the corona generated at an applied high voltage $V_s = 11$ kV and the grid potential $V_g = 1$ kV.

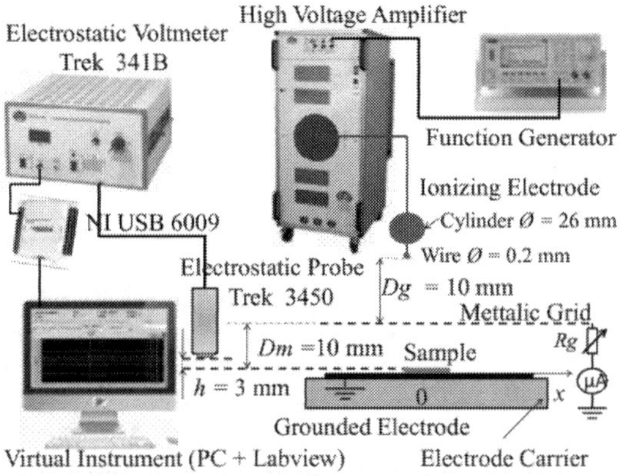

Figure 1. Experimental setup for the corona-charging of PP non-woven fabrics and the measurement of the electric surface potential.

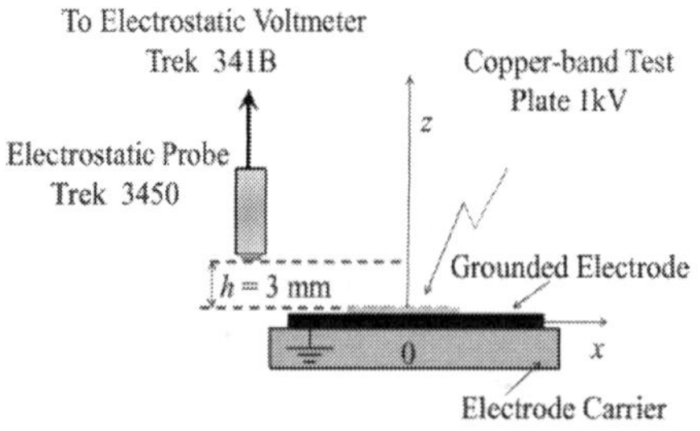

Figure 2. Experimental set-up for the surface-potential-probe-characterization experiments.

The probe-characterization experiments (figure 2) are performed using two test plates fabricated of a PCB. One plate has a central strip 20-mm-wide, 80-mm-long, connected to high potential (1 kV or 0.5 kV), the rest of its surface being grounded (figure 3, a). The other plate has five strips 80 mm - wide, distanced at $d = 2$ mm (figure 3, b). Each strip can be connected at a different electric potential, to simulate the non-uniformity of charge distribution at the surface of a sample. The four 2-mm-wide and one 3-mm-wide strips are separated from the rest of the copper layer at the surface of the PCB by 0.1 mm - wide insulating strips. In these experiments all the copper strips are connected to a high potential $V_s = 1$ kV, and the rest of the plate is grounded.

The sample carrier is attached to a computer-controlled positioning system (minimum resolution: 10 µm; minimum speed: 1.5 mm.s^{-1}; maximum speed: 30 mm.s^{-1}). As soon as the corona discharge is turned off in the corona-charging experiments or the bands are energized from an appropriate power supply in the probe-characterization experiments, the sample carrier is moved under the electrostatic probe (model 3450, Trek Inc., Medina, NY), of an electrostatic voltmeter (± 20 kV; model 341B, Trek Inc., Medina, NY). The translational speed of the sample-carrier relative to the probe can be adjusted at various values. The measured values of the electric potential are transferred to a computer via an acquisition card (model NI USB 6009), and processed by an ad-hoc virtual instrument, developed in LabView. The sample rate is limited to 5 kHz. Different sample rates (1 to 64Hz) and various scanning speeds (1.5 to 24 mm.s^{-1}) are compared in the present study.

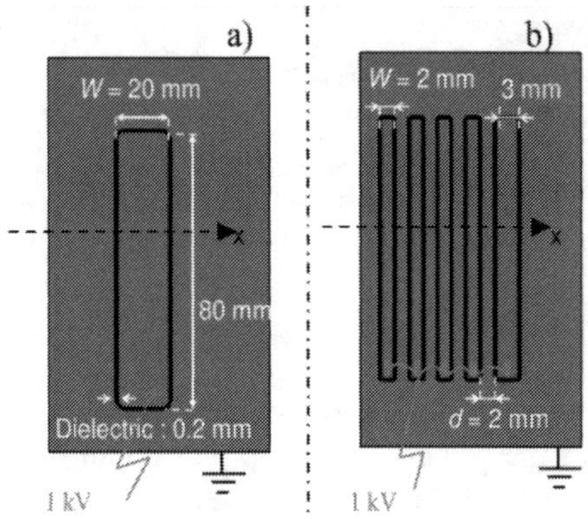

Figure 3. Schematic representation of the test plates for spot diameter study (a) and for non-uniformity simulation (b).

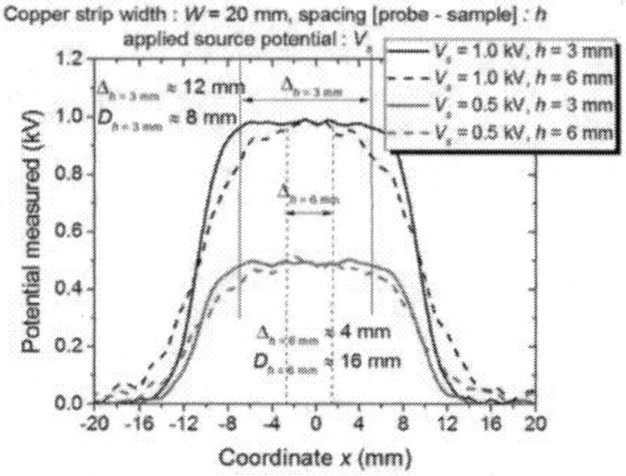

Figure 4. Repartition of the potential along the x-axis in function of relative position between the probe and the copper strip center, for two values of spacing h and source potential V_s

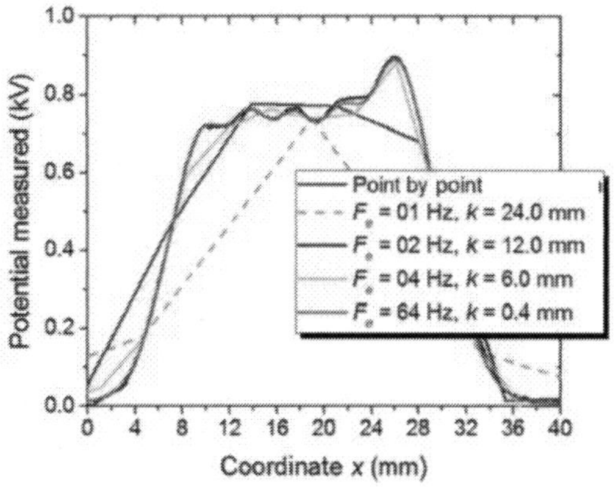

Figure 5. Repartition of the potential along the x-axis of the five-strip test plate, for different sample rates (F_e = 1 to 64 Hz); V_b = 24 mm.s^{-1}.

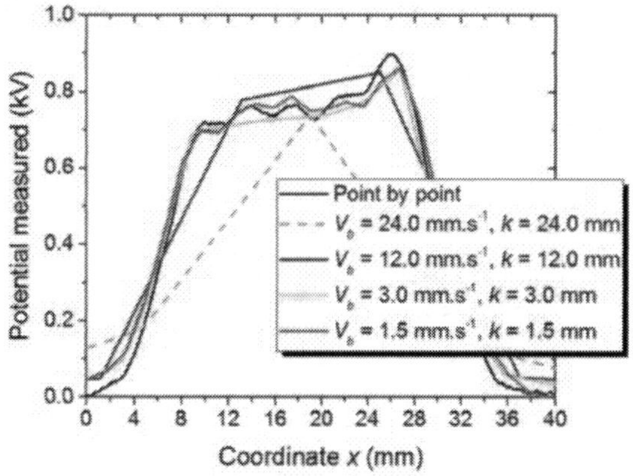

Figure 6. Repartition of the potential along the x-axis of the five-strip test plate, for different speeds (V_b = 1.5 to 24 mm.s^{-1}); F_e = 1 Hz.

3. Results and discussion

The electric potential values displayed in figure 4 are obtained as averages of 10 measurements, at constant sample rate F_e = 10 Hz in "static mode" (i.e., every 2 s, the electrode carrier is moved 0.1 mm along the x axis and the data are recorded point by point), for different values of h and V_s. The probe "sees" the potential in a circular area (spot) of diameter D on the surface of the test plate facing it. The spot diameter D can be estimated from the curves in figure 5 as the difference between the first position x_{max} where the voltmeter displays the potential $V = V_s$ and the last position x_0 where $V = 0$. The diameter D does not depend on V_s, but is affected by the probe-sample spacing h: $D \approx 8h/3$.

In figure 5, the data recorded point-by-point for the five-strips test plate are compared to the results recorded in the "dynamic mode" for different values of the sample rate F_e = 1 to 64 Hz, at constant scanning speed: V_b = 24 mm.s^{-1}. Similarly, in figure 6, the "static mode" data are compared to the results of the measurements performed at various scanning speeds V_b = 1.5 to 24 mm.s^{-1}, and constant sample rate: F_e = 1 Hz. In neither modes, the 0 V zones between two adjacent strips cannot be detected, as their width (2 mm) is smaller than the spot diameter (D = 8 mm, for h= 3mm).

The spatial resolution is better for the lower values of $k = V_b/F_e$. For $k < D/20$, the values of electric potential measured in the "dynamic mode" are close to those obtained in "static mode".

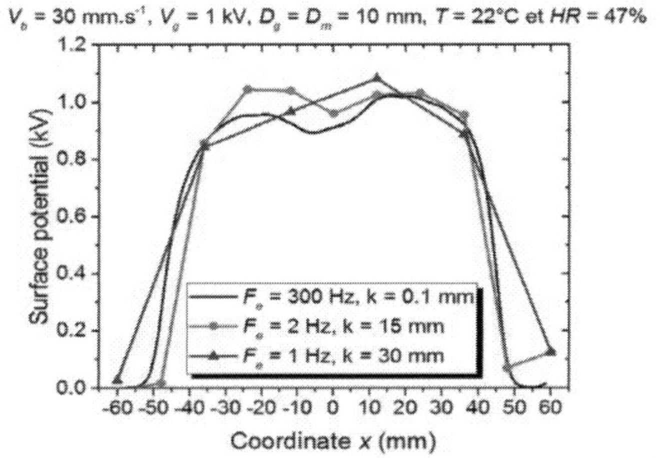

Figure 7. Repartition of the electric potential along the x-axis of a corona-charged non-woven PP sample, for various values of the sample rate F_e. A satisfactory accuracy is obtained for V_b = 30 mm.s^{-1} and F_e = 300 Hz, as $k = V_b / F_e = 0.1$ mm $< D/20 = 8/20 = 0.4$ mm.

The results of an experiment performed with a corona-charged non-woven PP sample are represented on figure 7. They have been obtained with the maximum scanning speed of the sample-positioning system (V_b = 30 mm.s^{-1}). The comparison between the curves corresponding to three different values of F_e show that the capacitive probe is able to accurately detect the variations of the potential at the surface of a sample in motion, provided that $k < D/20$.

4. Conclusion

(i) In the case on non-uniformly charged media, the values recorded by an electrostatic voltmeter can follow the rapid variations of the actual potential at the surface of a moving sample, if the spatial resolution is $k < D/20$, where D the diameter of the circular area "seen" by the capacitive probe.
(ii) At high scanning speeds, the increase of the sample rate is able to significantly improve the accuracy of the electric potential measured by the electrostatic voltmeter, but the recorded values must be filtered to reduce the noise.

References

[1] Łowkis B and Motyl E 2001 *J. Electrostat.* 51(52) 232-238
[2] Taylor D M and Secker P E 1994 *Industrial Electrostatics: Fundamentals and measurements* (New York: Research Studies Press, John Wiley)
[3] Molinié P and Llovera P 2004 *IEEE Trans. Diel. Electr. Insul.* 11 1049-1056
[4] Antoniu A, Dascalescu L, Vacar I V, Plopeanu M C, Tabti B and Teodorescu H N 2011 *IEEE Trans. Ind. Appl.* 47 1118 – 1125
[5] Tabti B, Mekideche M, Plopeanu M, Dumitran L M, Herous L and Dascalescu L 2010 *IEEE Trans. Ind. Appl.* 46 634 – 640
[6] Tabti B, Mekideche M, Plopeanu M, Dumitran L M, Antoniu A and Dascalescu L 2010 *IEEE Trans. Ind. Appl.* 46 1586-1592
[7] Plopeanu M C, Dascalescu L, Yahiaoui B, Antoniu A, Hulea M and Notingher P V 2012 *IEEE Trans. Ind. Appl.* 48 851-856
[8] Yahiaoui B, Megherbi M, Samuili A, Antoniu A, Tabti B and Dascalescu L 2013 *IEEE Trans. Ind. Appl.* 49 1758-1766
[9] Plopeanu M C, Dascalescu L, Neagoe B, Bendaoud A and Notingher P V 2013 *J. Electrostat.* 71 517-523
[10] Noras M A and Pandey A 2010 *IEEE Ind. Appl. Mag.* 16 41-47
[11] Kachi M, Nemamcha M, Tabti B and Dascalescu L 2011 *J. Electrostat.* 69 394-400
[12] Fatihou A, Dascalescu L, Zouzou N and Dumitran L M 2014 *Conf. Rec. IEEE IAS Ann. Meet.* DOI: 10.1109/IAS.2014.6978354

Techniques for Investigating the Effects of ESD on Electronic Equipment

D C Smith

University of Oxford, Department for Continuing Education, Course Tutor
Rewley House, 1 Wellington Square, Oxford, Oxfordshire, OX1 2JA, UK
US address: 1305 Arizona St. #101, Boulder City, NV, USA 89005

Email: doug@dsmith.org

Abstract. During of course of determining the mechanism by which electrostatic discharge affects electronic equipment, the process is complicated by the fact that the energy of the ESD event is dispersed across the whole of the electronic system and its connections. A collection of methods is presented whereby the effects of ESD on electronic equipment and potentially other objects can be determined. These methods narrow down the parts of the system affected by applying E and H field stresses locally in the system. Although the stresses are not necessarily ESD, they model the local effects of ESD within the system and allow determination of which parts of the system are affected by ESD so the specific design fixes can be applied to the appropriate parts of the system.

1. Introduction

When trying to determine the mechanisms whereby ESD affects electronic equipment, the results of the ESD test itself may not be useful. This is because the ESD illuminates the whole system, making it difficult to determine the mechanism(s) by which the ESD affects the system. One would like to determine the specific way ESD gets into the circuitry. There are several quick techniques that can be used right on the lab bench to illuminate specific parts of the system in order to determine how ESD is causing problems. One of these techniques uses capacitive coupling.

2. Capacitive Coupling

Figure 1 shows an ESD simulator fitted with an air discharge tip that is insulated to insure that no direct discharge is made to the PCB. A contact discharge is made and capacitance between the insulated tip and PCB couples high frequency energy to the area of the PCB closest to the simulator tip. The tip is then moved around the board and additional contact discharges are made to test areas of the PCB for sensitivity to the electric fields produced by the simulator tip.

Generally, contact discharge levels up to about two to four kilovolts are used. The charge on the tip must be bled off after each firing of the ESD simulator. This can be done using a one megohm resistor between the tip of the simulator and its ground lead. The simulator shown provides a discharge path back into the simulator, but many simulators do not.

Figure 2 shows an ESD simulator with an air discharge tip that is covered with copper tape applied over insulation to prevent a direct discharge from the tip to the copper tape.[1] A contact discharge is applied and the capacitance between the tip and the copper tape forms a high pass filter that only allows the high frequency portions of the ESD to be injected into a system at the point of contact with the copper tape. As before, one must insure the tip of the ESD simulator is allowed to bleed off charge before the next simulator discharge.

Figure 1. Coupling ESD energy to a PCB.

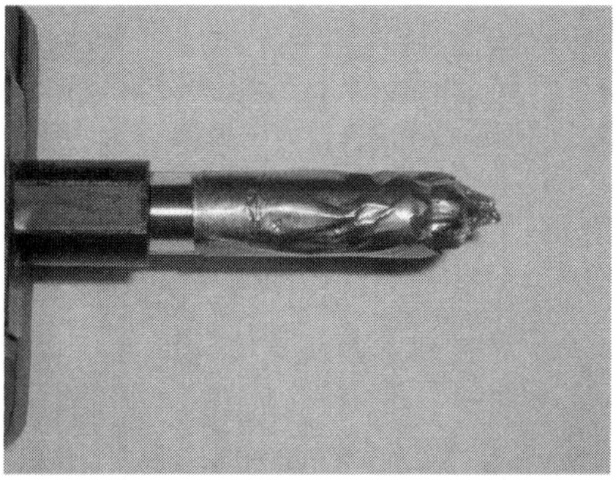

Figure 2. High pass filter on ESD simulator.

Figure 3 shows the discharge current into a ground plane measured with a current probe. The resulting waveform is displayed in Figure 4. Only the high frequency components of ESD are able to pass though the copper tape high pass filter and appear as a single pulse about two nanoseconds wide. If a system responds to this, when the peak is adjusted to be the same amplitude as the peak of a normal contact discharge at the desired voltage, then the system is responding to the high frequency components of ESD. If the system responds to a normal contact discharge but not after the high pass filter is added, then the system is responding to the lower frequency components of the applied ESD. The circuit mechanisms are quite different for the two cases and knowing whether high or low frequency components are dominating the system response is valuable information.

Figure 3. Measuring the filtered discharge current.

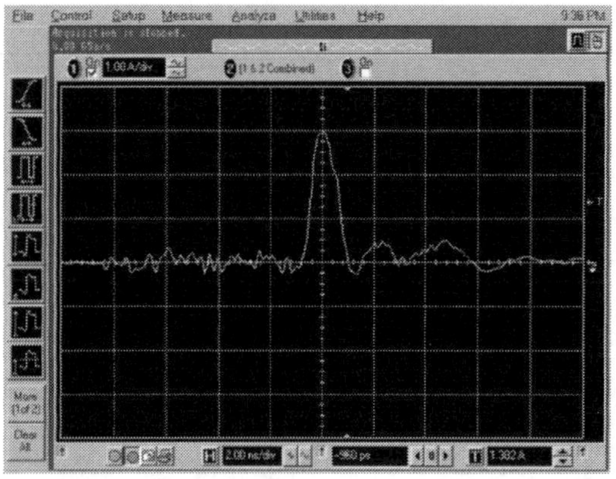

Figure 4. Measured discharge current. Scale factors are 1 A/div and 2 ns/div.

In a system with many cables, one question arises in cases of problems caused by ESD. Which cable or cables are involved? Figure 5 shows one technique for injecting ESD "noise" into a system cable.[2] A contact discharge is delivered to a length of foil wrapped around the cable in question. About 10 cm is an adequate length of foil to get most of the high frequency energy of the ESD into the cable and some of the lower frequency components as well. It is a good idea to monitor the injected current with a current probe on the cable as it enters the system.

Capacitive coupling, as in Figure 5, results in a current injection and this current splits in two directions, entering the system and traveling in the opposite direction away from the system. Both to force more of the current into the system and make that current less dependent of external cable arrangements, a ferrite core can be added as shown in Figure 5 to increase the cable common mode impedance for currents traveling away from the system.

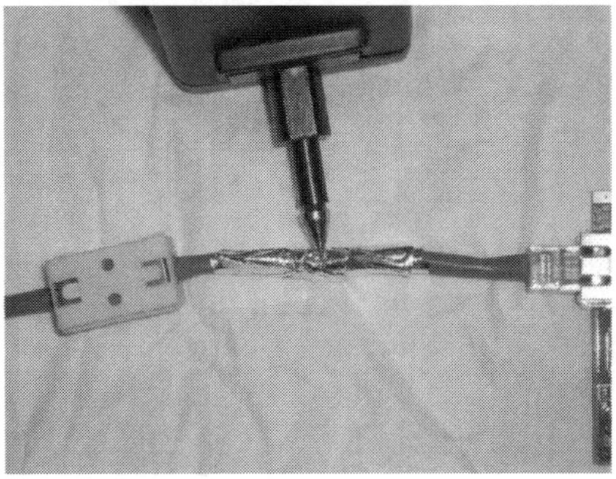

Figure 5. Injecting ESD energy into a cable.

Be sure to discharge the foil between contact discharges if the ESD simulator does not do this. As before, a one megohm resistor can be used, connected between the foil and ground. Or one can simply touch the foil with a finger while wearing a wrist strap. The foil will not store enough charge to be a problem.

The ESD simulator should be set at a low enough voltage so as not to punch through the cable insulation, 2 kV is generally safe. ESD current gets distributed throughout a system, its cables, and through displacement current to other objects, so only a fraction of the energy ends up on a single cable. Thus applying 2 kV on one cable this way is a reasonable stress.

3. Inductive Coupling

Inductive coupling can also be used to find ESD sensitivities in a system and can be used with much more spatial resolution than capacitive coupling. In this case, the stress is not derived from an ESD simulator but from an Electrical Fast Transient, EFT, simulator (IEC 61000-4-4 generator or any pulse generator that can furnish a high open circuit voltage, high current pulse with a fast rise time and slow fall time).

Figure 6 shows a current probe (Fischer Custom Communications F-65 with a flat frequency response from 1 MHz to 1000 MHz) used to measure the current flowing in a square 2.5 cm loop being fed an EFT like pulse. The current waveform is shown in Figure 7 along with its time derivative as calculated by the oscilloscope.

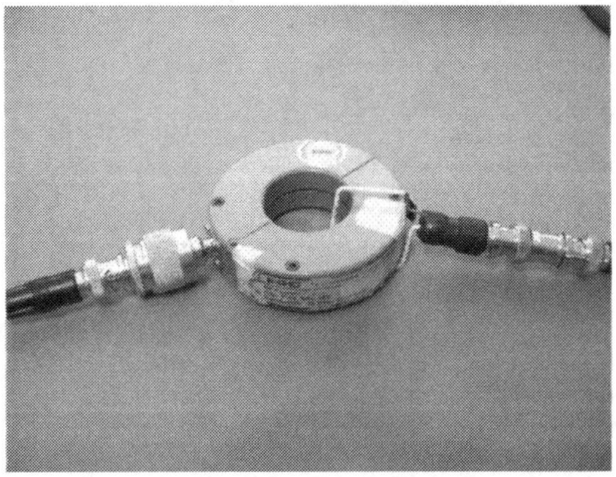

Figure 6. Measuring loop probe current.

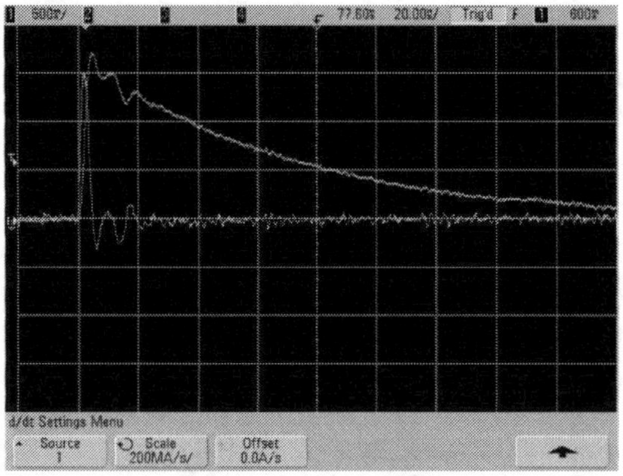

Figure 7. Loop probe current and its di/dt. Scale Factors are 0.5 Amp (Current), 0.2 A/ns (di/dt), and 20 ns/div.

Note the current rises to 1.5 Amperes in a few nanoseconds and falls over a period of over 100 ns. This results in a di/dt consisting of a single positive pulse as the current falls so slowly that it cannot be seen in the di/dt plot. Since the induced voltage in a circuit adjacent to the loop is Mdi/dt, where M is the mutual inductance between the circuit and the loop, the induced voltage will have the same wave shape as the di/dt plot in Figure 7, different only by a constant. If this loop is then held up to a path on a PCB or a wire in the system, a controlled positive or negative (by reversing the loop) pulse can be injected into the circuit.

Figure 8 shows the loop positioned over a PCB trace and the resulting induced voltage is shown in Figure 9. The induced voltage is very similar in shape to the di/dt waveform in Figure 7 as expected.

Figure 8. Injecting pulses into a PCB trace.

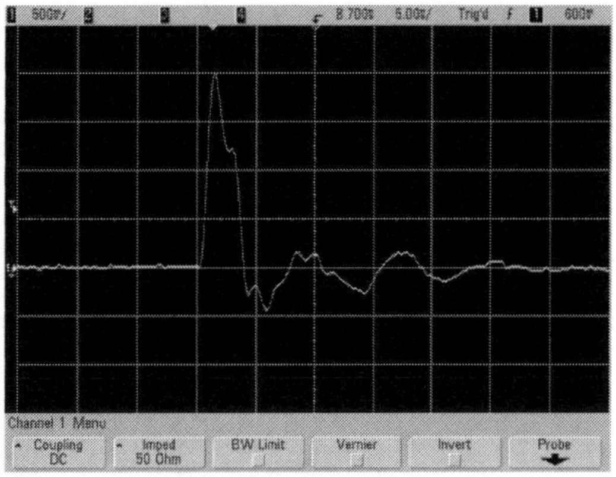

Figure 9. PCB induced voltage. Scale factors are 0.5 Volt and 5 ns/div.

Applying this technique to a PCB is demonstrated in a YouTube video by the author.[3] Another treatment of this inductively coupled method is given in a paper published by the author in 2009.[4] Generally the amplitude of the injected pulse is slowly increased as the board is scanned by the loop. If the first symptom to appear is the same as an observed symptom in response to ESD, that part of the circuit is likely involved. In the above figures, a 100 Volt open circuit pulse was applied to the loop from an EFT pulse source.

4. Summary and Conclusions

By applying stresses locally within a system, ESD induced problems can generally be tracked to their cause in the system design. Both capacitively and inductively coupled stresses are useful. Inductive coupling has the better spatial resolution, but electric field sensitivities are also important in system design.

Application of the methods described here to otherwise intractable ESD problems have found the root causes of those problems in hours after weeks and months of traditional ESD troubleshooting methods, basically trial and error methods in an ESD test area, had failed.

References

[1] "Filtering Contact Discharges - Part 2, High Pass Filtering," Douglas C. Smith, http://www.emcesd.com/tt2006/tt100106.htm

[2] "Injection of ESD Current Into Cables Using Capacitance," Douglas C. Smith, http://www.emcesd.com/tt2009/tt010109.htm

[3] Demonstration video at: http://youtu.be/CdJuaNlUTqE

[4] "Noise Injection for Design Analysis and Debugging, Douglas C. Smith, 2009 Design Con paper, available at: http://emcesd.com/pdf/DC09_DCSmith.pdf

Uncertainty of potential measurements of Q-constant objects

R Kacprzyk

Department of Electrical Engineering Fundamentals, Wroclaw University of Technology, 50-370 Wroclaw, Wyb. Wyspianskiego 27, Poland

E-mail: ryszard.kacprzyk@pwr.edu.pl

Abstract. Direct and compensating methods of non-contact potential measurements on constant charge objects are distinguished and discussed in the paper. It is well known that in case of potential measurements using direct method the measured voltage value is lower in comparison to that in a non disturbed state. However in case of application of a compensating method, values of measured potential are higher. It is shown that in both of discussed cases approaching of the conducting object by the particular voltmeter (probe) leads to changes of the total capacitance of the object. Approximated relations for estimation of relative error of potential measurements for both of distinguished groups of methods were also given.

1. Introduction

Description of a particular object or an environment from the point of view of electrical field distribution may require field or potential measurements of the particular charged objects which creats the field. Contactless potential probes (voltmeters) are usually applied for potential measurements of extremely high resistance objects [1-7]. All of the methods applied for non-contact potential measurements can be divided into two basic groups, namely "direct" and "compensation" methods. In the case of direct methods the measured quantity (potential or field intensity) is derived and determined directly from the values of "other quantities" unequivocally related to the field intensity in the vicinity of the investigated object. "Other quantities" could be considered as: force (electrostatic voltmeters), current (radio-active or ion-path field-meters); electric charge induced on the measuring electrode (induction methods) or changes in the refractive index (Kerr or Pockel's electro-optic phenomena). Each of the mentioned variants of the "direct" methods may be used in a compensation system – with an automatic or hand-driven setting of the compensation state [5]. It is generally established that approaching of an investigated object with any earthed meter may change its potential trough its additional capacitive loading (increase the effective capacitance) [2, 3, 6]. This phenomenon leads to reduction of the potential of the investigated object (when conducting) or to a local potential reduction in case of weakly conducting or insulating objects [e.g. 6] in comparison to the values expected in a non disturbed state.

Compensating voltmeters also change the field distribution in the vicinity of an investigated object and, finally, may influence both, its effective capacitance and potential. The last may be of special importance in case of low capacitance objects as well as during potential measurements on 3D dielectric objects [5].

2. Potential measurements using direct method

An insulated conducting object with constant charge Q, total own capacitance to the earthed surrounding C_0 and potential U will be considered. In case of potential measurements with a non-contact method, e.g.

using an earthed field meter FM, the electric field E is disturbed mainly in the space between the field-meter and the investigated object. The space is limited by a surface S_F as shown in Fig. 1. Due to additional capacitance C_M, introduced by the field-meter, the measured potential of the object will be decreased to U_D value given by:

$$U_D = U \frac{C_O}{C_O + C_M} \tag{1}$$

The equivalent capacitance of the object C_{ED} will be determined by the formula:

$$C_{ED} = C_O + C_M \,. \tag{2}$$

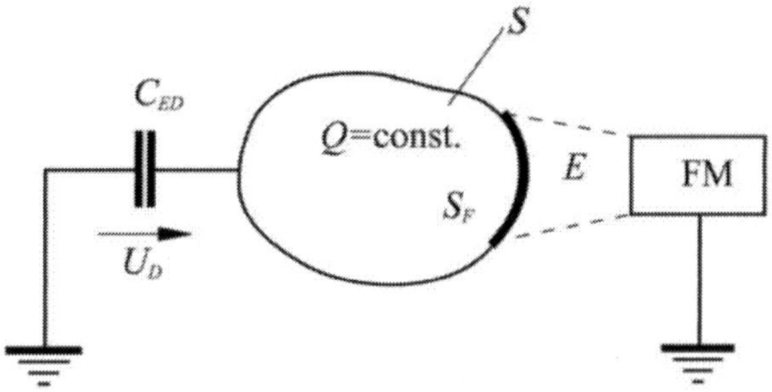

Figure 1. Potential measurements using non-contact direct method. FM – field meter.

Error ΔU_D and relative error δU_D values of potential measurement for direct method are determined by equations, respectively:

$$\Delta U_D = U_D - U \tag{3}$$

$$\delta U_D = -\frac{\Delta U_D}{U} = -\frac{C_M}{C_O + C_M} \tag{4}$$

Assuming, that the capacitance C_M can be approximated by a mutual meter-object capacitance, its value can be determined from the approximated relation [1]:

$$C_M \approx \frac{\varepsilon_0 S_F}{l} \tag{5}$$

where ε_0 - electrical permittivity of a free space, S_F - surface determining space of the field distortion, assumed to be equal to the "sensitive surface" (sensitive aperture) of the meter (probe), l – object-meter distance average value (in the space limited by a surface S_F). Generally the range of C_M variability is from 0 to infinity. C_M capacitance is equal to 0 when the electrical field around the object is not disturbed by any meter. Capacitance C_M increases to infinity for infinitely small distance between the meter and the investigated object. Finally, the undisturbed potential U and relative error δU_D values can be determined from the approximated relations, respectively

$$U \approx U_D \left(1 + \frac{\varepsilon_0 S_F}{C_O l} \right) \tag{6}$$

$$\delta_D U = \frac{\Delta U_D}{U} \approx -\frac{\varepsilon_0 S_F}{C_O l + \varepsilon_0 S_F} \tag{7}$$

The preceding discussion indicates that the measured potential may be strongly influenced by the earthed meter even in case of contactless measurements. In practical situations the effect of potential reduction is easy to observe in case of potential measurements of low capacitance objects. It should be emphasized, that potential measurement by an earthed meter gives (always) lowered voltage values in comparison to the non-disturbed state.

3. Potential measurements using compensation method

Application of a compensation method requires zeroing of the average field value in the space between the probe and the investigated object (in the air gap over the surface S_F), see Figure 2. Zeroing of the electric field in the region S_F means zeroing of the surface charge density in the same area. Charge loss

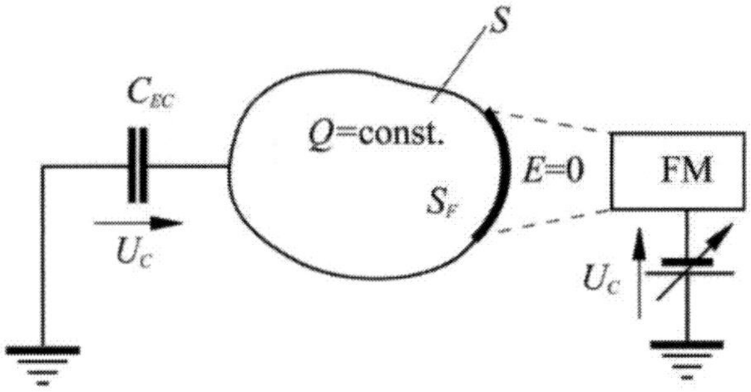

Figure 2. Potential measurements using compensation method

on the surface S_F, implicates an increase in surface charge density on the remaining part of the object (remaining part of its total surface S). Thus, the electric field and potential of the object must increase to the value U_C given by [5]:

$$U_C = U \frac{C_O}{C_O - C_C}, \tag{8}$$

where C_C is the capacitance related to the effect of "cutting out" of the surface S_F from the total surface of the object S. Consequently, the equivalent capacitance of the object will be now given by equation:

$$C_{EC} = C_O - C_C. \tag{9}$$

Capacitance C_C is equal to 0 when the electrical field around the object is not disturbed by any meter or when the S_F surface (more directly the S_F/S ratio) approaches zero. Capacitance C_C increases to C_O when the meter (compensating electrode) completely encloses the investigated object.

In case of a compensation method values of the error ΔU_C and the relative error δU_C of potential measurement are given by equations, respectively [7]:

$$\Delta U_C = U_C - U \tag{10}$$

$$\delta U_C = \frac{\Delta U_C}{U} = \frac{C_C}{C_O - C_C} \tag{11}$$

Assuming field intensity equal to zero in the S_F (compensation) region and uniform distribution of the surface charge on the remaining part of surface S of the investigated object, C_C capacitance can be determined from the approximated equation [7]:

$$C_C \approx C_O \frac{S_F}{S} \tag{12}$$

Thus, the undisturbed potential U and relative error δU_C values can be determined from the equations, respectively:

$$U \approx U_C\left(1 - \frac{S_F}{S}\right), \tag{13}$$

$$\delta_C(U) = \frac{\Delta U_C}{U} \approx \frac{S_F}{S - S_F}. \tag{14}$$

It should be noted, that the results of potential measurements obtained by a compensating voltmeter will give (always) higher voltage values in comparison to that in a non-disturbed state. Equations (7) and (14) allow determine the relative uncertainty of potential measurements (uncertainty of the applied method) without consideration of accuracy of applied measuring devices. Both of the mentioned equations confirm, that minimal error of potential measurements of conducting objects require application of voltmeters (probes) with minimal "sensitive surface" (aperture) S_F.

Equations (6) and (13) allow to determine the potential value of the object in a non disturbed state and may explain some incompatibilities which can arise between results of measurements made by different meters or by application of different methods.

4. Conclusions

Consideration of the influence of non-contact measuring method on the results of potential measurements of Q-constant and conducting objects, leads to the following conclusions:

- potential and equivalent capacitance of Q-constant objects are changed by a measuring system operating in both, direct and compensation circuitry;
- potential values obtained by application of a direct method are (always) lower than those in a non-disturbed state and the value of the relative error δU_D depends on the distance l between the object and the probe (equation 7);
- potential values obtained by application of a compensation method are (always) higher than that in a non-disturbed state and the value of the relative error δU_C does not depend on the distance between the object and the probe (roughly, see equation 14);
- proper determination of the relative error value requires knowledge of the particular method applied for potential measurement;
- for non-uniform distribution of surface charge density on the investigated object, results of potential measurements may depend on the place of measurement.

References

[1] BS 7506:1996 *Methods for Measurements in Electrostatics*. Part 2. Test methods
[2] Taylor M, Secker P A 1994 *Industrial Electrostatics, Fundamentals and Measurements* (J. Wiley & Sons Inc., New York, Research Studies Press Ltd., Taunton, Somerset, England)
[3] Cross J 1987 *Electrostatics: Principles, Problems and Applications* (A Hilger, IOP Publ. Ltd. Bristol, England)
[4] Chubb J *An introduction to electrostatic measurements* http://www.jci.co.uk
[5] Kacprzyk R 2012 IEEE TDEI **19** 134
[6] Horenstein M N 1995 J. Electrostat. **35** 203
[7] Kacprzyk R 2013 *Measurement Methods in Electrostatics* (Wroclaw: OWPWr, in Polish)

Simultaneous measurement of particle charge, distance and size using coaxial induction probe

M Murtomaa and J Salonen
Department of Physics and Astronomy, University of Turku, 20014 Turku, Finland

E-mail: matti.murtomaa@utu.fi

Abstract. A novel probe for charge measurements was simulated, designed and tested. It consisted of two coaxial induction probes which were grounded via a current sensing circuit. As a charged spherical object passed the probe, data was collected and analyzed using a PC running a program compiled with LabVIEW. Two simultaneous current signals were integrated to obtain induced charges as a function of time. Then, Gaussian curves were fitted to the data and widths and amplitudes were collected for further processing. According to simulations performed with Comsol MultiPhysics, object charge, distance from the probe and size could be calculated using the above mentioned Gaussian peak parameters. The probe was calibrated and tested using induction charged water droplets and frictionally charged insulating spheres.

1. Introduction.

There is a variety of different methods for measurement of particle charge [1-3]. For single objects or for an amount of powder, traditional Faraday pail is a good and reliable option. However, using Faraday pail requires sampling and in some cases sampling may cause additional charging or it may be inapplicable for other reasons. For online charge measurement different methods based on induction are widely used. In the most simple induction measurement, a probe, which can be a conducting rod or just a piece of wire, is placed near a charge to be measured. The probe can be set to measure either potential or current. If potential is measured, the potential depends on the capacitance of the probe and electronics. If, on the other hand, current is measured, the probe potential is maintained at zero and a bipolar signal is recorded as a charged object passes the probe.

If we consider a charged object passing the probe, signal obtained using a single induction probe is a function of object's charge, distance from the probe, size, speed and shape. To get an accurate value of some of the parameters listed above, some other parameters need to be fixed. As an example, an open-ended Faraday pail can be considered as a single induction probe. When a charged object passes through the probe, an accurate value of charge (and speed to some extend) can be obtained. However, this setup does not give information about the location or size of the object. If another probe with different geometry is added, more data can be obtained and more object parameters can be calculated. As far as authors are aware, a system which would allow a measurement of charge, location and size simultaneously has not been presented.

In the present work, induction method in current mode was used. Instead of one induction probe, two concentric probes were used. Due to current mode, both probes were kept in zero potential which means that an induced charge on one of the probes does not interfere with the other. Recently, somewhat different two probe system in a two dimensional fluidization column has been proposed by

He et al. [4]. A concentric geometry was chosen because that enables quite easy selection of right induction signal peaks from possibly noisy data and many passing charges. In other words, the recorded peaks from both probes are exactly simultaneous for same passing object. The probe was manufactured and assembled perpendicular to a grounded steel pipe so that the probe surface was in the same plane as the inner surface of the pipe.

The system was modeled and simulated using COMSOL multiphysics. It was observed that particle size, location and charge affect the signal properties. Using the simulated results, it was possible to write equations into measurement program which gives particle distance, size and charge as a result.

2. Simulations

The probe and the pipe assembly were modeled using COMSOL Multiphysics 4.3b and the model was identical to the actual probe. In the simulations, spherical particles were used. Radius of the particles was varied from 1 mm to 7 mm. Distance between the probe and the particle was varied between 1 mm and 14 mm. Total charge which was spread uniformly either over surface or within volume, was varied from 0 nC to 1 nC.

Inner probe had a radius of 1 mm. It was separated from outer probe by 1 mm thick insulator. Outer electrode was cylindrical and its thickness was 1 mm. Outer probe was again surrounded with an insulator. The probe was attached to a grounded steel pipe (inner diameter 35 mm). Figure 1 presents a screenshot from COMSOL Multiphysics. In the figure, a positive particle is shown in front of the probe, a few millimeters below the probe axis. Grayscale image shows the surface charge density on the conducting surfaces. Insulating rings between the probes have been set transparent so that surface charge density in these probe regions can be seen.

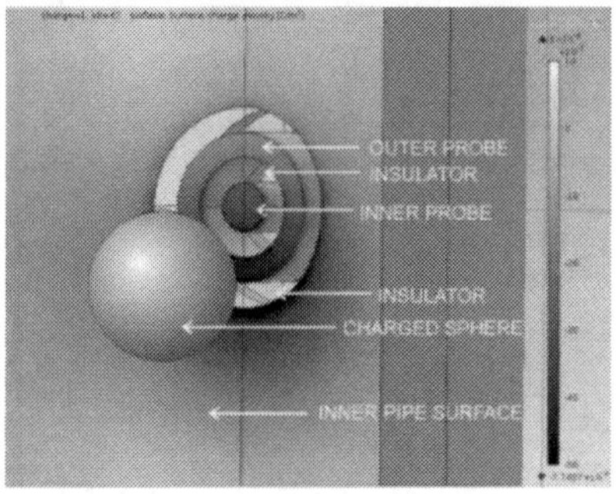

Figure 1. Snapshot from the simulation showing the geometry of the probe and surface charge density on the probe surfaces.

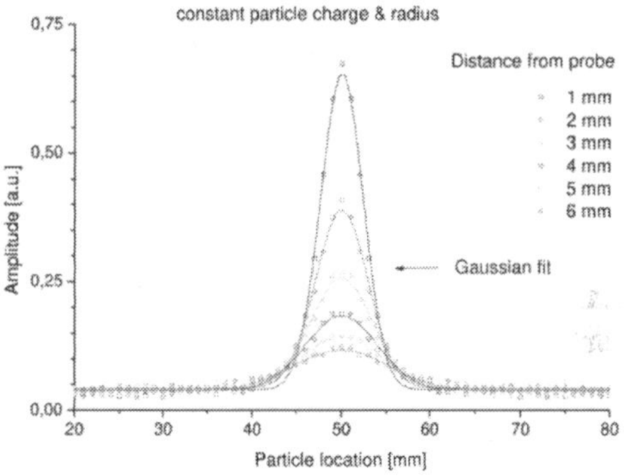

Figure 2. Induced charge on inner probe as a function of height of the particle (50 mm = probe height).

In the simulations, particle was set to move vertically along the pipe at 1 mm steps. Induced charge was calculated by integrating surface charge density over probe surfaces. The shape of the induced charge curve as a function of particle height was of interest and is shown in Figure 2. It was noticed that a Gaussian function fits to the datapoints very well, and it was used in calculations.

In the simulations, particle parameters: size, distance from the probe and charge were varied as mentioned above. Gaussian peaks were fitted to each data and peak parameters: amplitude and width were collected. For further analysis, also amplitude ratio (A_{outer} / A_{inner}) and width ratio (W_{outer} / W_{inner}) were calculated. Based on the computer simulations, above mentioned peak parameters were mathematically expressed as a function of particle parameters. The constants in these equations were then refined using experimental data.

3. Probe structure and experimental methods

Dimensions of the probe were same as in the computer model. The induction probes were made of brass and the insulator was non-conducting epoxy. Current from the probes was measured using LMC662 CMOS dual operational amplifiers (National Semiconductor) with 1 GΩ feedback resistors.

The circuit was powered with 9 V battery and was shielded inside a grounded case. Amplified signals from both probes were recorded using NI USB-6008 data acquisition system (National Instruments) at sampling rate of 1 ms.

At the present, charged objects which were dropped through the vertical pipe were either induction charged water droplets (diameter 3.5 mm) or frictionally charged insulating spheres (diameters 2, 6, 10 and 13 mm). The charge on water droplets could be continuously adjusted. It is important to note that at the present the system is not capable to analyze particles with different velocity. Higher velocity would narrow the peaks but would not change the amount of induced charge on the probe surfaces. Thus, all particles were dropped under gravity from the height of 30 mm with zero initial velocity. Particles were collected directly in a Faraday pail connected to an electrometer (Keithley 6517A, Keithley Instruments).

Data was collected and analyzed using a virtual instrument (VI) compiled in LabVIEW 2009. The VI scanned both signals and when concurrent peaks occurred, signals were integrated. Next, Gaussian curves were fitted to the signals, and amplitudes and widths were taken to further processing.

4. Calculation of particle charge, distance and size

4.1. Induction charged water droplets

Expressions obtained using simulations could be verified using induction charged water droplets. Figure 3 presents Gaussian amplitudes measured from outer and inner probes (A_o and A_i respectively) and amplitude ratio as a function of distance from the probe. In this experiment, charge was kept constant (169 pC) but the distance from the probe was varied.

Figure 4 proves that amplitude increased linearly as a function of water droplet's charge which was measured using a Faraday pail and an electrometer. In Figure 4, the distance was kept constant at 6 mm from the probe. From the same figure, it can also be noticed that the widths of Gaussian peaks did not vary with charge. All datapoints in Figs 3 and 4 are mean values of 10 measurements. Standard deviations are too small to notice except in Figure 4 (widths), where SD's are significant especially, when the charge was small and signal quite noisy.

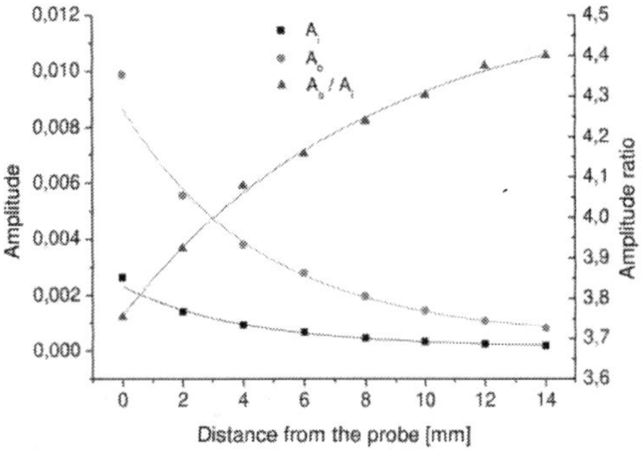

Figure 3. Gaussian peak amplitudes fitted to outer and inner probe signals (A_o and A_i respectively) and amplitude ratio (A_o / A_i) as a function of distance from the probe.

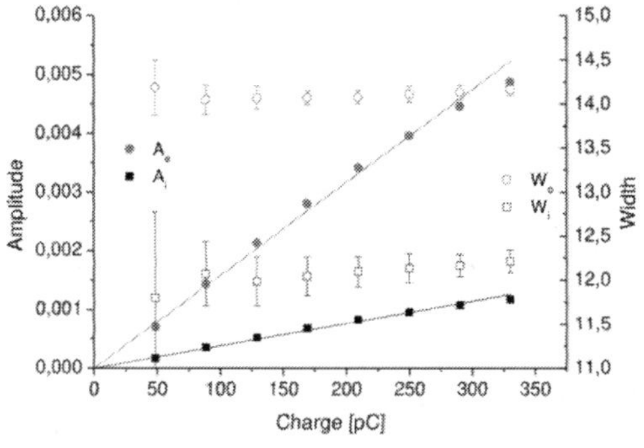

Figure 4. Gaussian peak amplitudes fitted to outer and inner probe signals (A_o and A_i respectively) and fitted widths (W_o and W_i) as a function of droplet charge.

Situation was quite easy with induction charged water droplets which had constant diameter and velocity. The distance from the probe could be calculated from the amplitude ratio and when the distance was known, charge of the droplet could be calculated from either amplitude. So, in case of water droplets, only amplitude information was needed to evaluate both charge and distance.

4.2. Frictionally charged insulating particles

Variable size of the particles brought another parameter to be solved from the signals. Additionally, charge distribution along the sphere surface was probably not uniform. Thus, more measurements were required to get some statistics. Different sized particles were frictionally charged, passed by the probe at different distances, and their charges were measured using the Faraday pail.

Figure 5 shows a good correlation between simulated and measured data. Both amplitude ratio and width were functions of particle distance d and size r. These two equations (Equations 1 and 2) were defined using experimental data and written into the VI. VI measured the values of (A_0 / A_i) and W_o and solved these equations simultaneously to get particle radius and distance. Once these were solved, charge q could also be solved using A_o (Equation 3). Amplitude A_o was a linear function of particle charge but the slope was actually a function of particle size and distance. The (exponential) dependence of the slope on the particle radius and distance was verified experimentally and the resulting equation was written into the VI.

$$\frac{A_o}{A_i}(r,d) = 4.7 + (0.194 \cdot r - 2.012) \cdot e^{(-d/(0.2122 \cdot r + 6.333))} \quad (1)$$

$$W_o(r,d) = 0.61655 \cdot d + 0.64564 \cdot r + 4.39215 \quad (2)$$

$$q = \frac{A_o}{5.6 \cdot 10^{-4} \cdot e^{-r/1.4} + 5.8 \cdot 10^{-5} \cdot e^{-d/(0.7 \cdot r + 3.1)}} \quad (3)$$

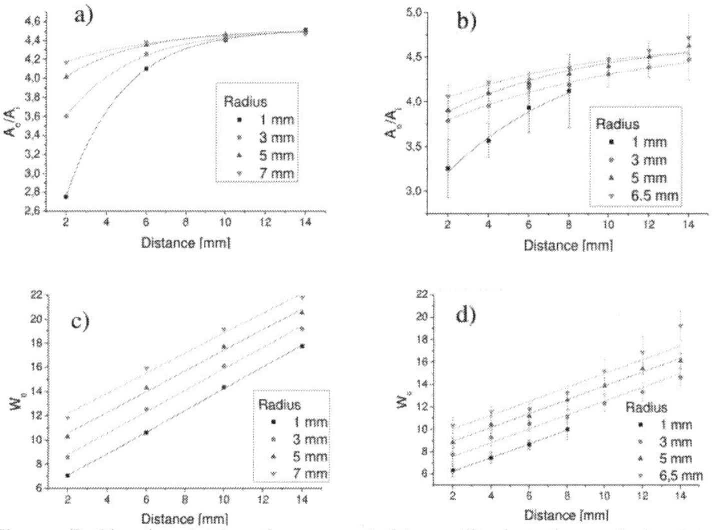

Figure 5. Simulated (a) and measured (b) amplitude ratios and simulated (c) and measured (d) outer probe widths for different sized particles as a function of the distance from the probe.

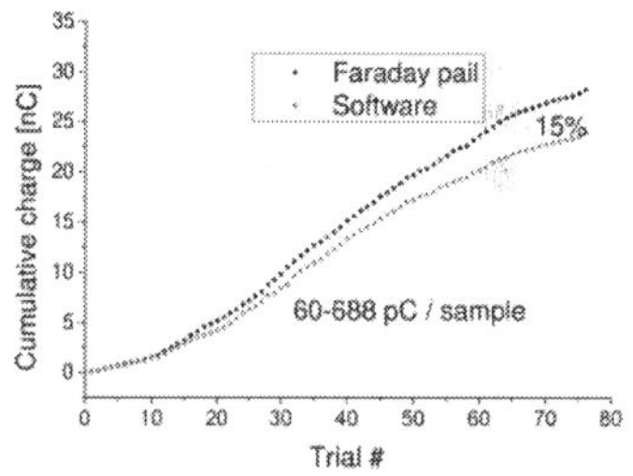

Figure 6. Cumulative charge of 3 mm radius spheres measured with Faraday pail and by coaxial probe. Particle's charge and the distance from the probe were randomly varied from 60 to 688 pC and 1 to 10 mm respectively. Difference between measured and calculated charge was 15%. Average calculated radius was 3.7 mm.

5. Results and conclusions

Figure 6 shows example data measured using spheres of 3 mm radius and random charge and distance. Although a single measurement might have some error, cumulative charge graph shows that on average, the system was able to measure charge reasonably well. Measurements performed with the coaxial probe agree well with computer simulations. By measuring current signals simultaneously from both probes and performing necessary signal analysis, it was possible to determine particle size, charge and distance from the probe when the particle velocity remained constant. Ongoing work includes gathering more calibration data to improve the calculation accuracy. Also, probe geometry is being changed in order to get the velocity data as well.

6. References

[1] Taylor D M 2001 *J. Electrostat.* **51-52** 502–8
[2] Cross J A 1987 *Electrostatics: principles, problems and applications* (Bristol: IOP Publishing)
[3] Mazumder M K, Ware R E, Yokoyama T, Rubin B J and Kamp D 1991 *IEEE T. Ind. Appl.* **27** 611–9
[4] He C, Bi X T and Grace J R 2015 *Chem. Eng. Sci.* **123** 11–21

Importance of the substrate nature to preserve microorganisms' cultivability in electrostatic air samplers

Jean-Maxime Roux, Anaëlle Rongier and Dorothée Jary

CEA, 17 rue des Martyrs, 38054 Grenoble - France

E-mail: jean-maxime.roux@cea.fr

Abstract. Recent research shows that electrostatic precipitation is a gentle method to collect airborne microorganisms and preserve their cultivability. However, the corona discharge used to charge the particles and the high electric field used to capture them are known to have a germicidal effect. The present paper investigates this paradoxical situation. Vegetative cells of *E. coli* and *B. subtilis* and spores of *A. fumigatus* and *B. subtilis* were deposited on different media and subjected to electrostatic fields of different strengths and polarities for controlled time periods. Vegetative cells are inactivated on cultivation agar plates, but remain cultivable when exposed on a stainless steel electrode and transferred afterwards onto agar plates. For the investigated conditions, spores were not affected by the corona discharge. Further experiments with a pH indicator show that chemical reactions occur when an aqueous media is exposed to the discharge. Some of these reactions are likely to create hydrogen peroxide which is known to kill a broad range of microorganisms. It is therefore highlighted that collecting electrodes in electrostatic air samplers should rather be dry conductive media.

1. Introduction

Electrostatic precipitation used as a sampling method (sometimes named electrostatic sampling) is one of the most efficient ways to collect particles by charging the aerosols (charger stage) and collecting them due to electrostatics forces (collection stage). Recently it was shown to be a suitable method to collect bioaerosols and several devices were built to investigate their performance [1]. Compared to impaction, advantages of electrostatic sampling are lower impaction stress [2], pressure drop and power consumption [3]. Electrostatic sampling can also achieve cultivable bioaerosol concentration levels as much as nine times higher than the BioStage impactor [4] and five to ten times higher airborne allergen and toxin concentration levels than the BioSampler [5].

However, the DC corona discharges used in most electostatic samplers to charge airborne particles may be germicidal [6-9]. For instance, the atmospheric point-to-plane corona discharge can be used either to collect airborne particles on a culture medium with the purpose to keep bioagents alive [10] or to sterilize a culture medium [9]. In the field of electrostatic sampling, Yao et al. [11] developed a two stages electrostatic sampler and investigated the impact of high electric fields on microorganisms with several media and conditions. Unfortunately the authors restricted their study to the impact of the electric field that is applied in the collection stage of their sampler; the first stage, composed of a charger that is usually used to charge airborne particles prior to their collection, was not studied. This key point should not be omitted in such studies because the reactive species created by the charger are likely to be responsible for the germicidal effect. Direct exposure and remote exposure seem to be both equally effective to kill microorganisms [6]. Since numerous samplers developed to collect bioaerosols include a charger stage, the present study investigates the effect of corona discharges on microorganisms. This study extends the conclusions of Yao et al. [11].

Vegetative bacteria of *E. coli* and *B. subtilis*, spores of *B. subtilis* and fungal spores of *A. fumigatus* were plated on solid media and exposed to the corona discharge in the configuration used by Julák et al. [9] and Sillanpää et al. [10]. To be relevant in the field of air sampling these media included solid culture media [2,4,10,12]. Stainless steel plates were also used to investigate the effect of the discharge over dry conductive media [1,13]. Additional experiments were carried out with a pH indicator gel to investigate the chemical impact of the discharge.

2. Material and Method

The corona discharge was generated from the tip of a needle-shaped electrode made of a hollow stainless steel tube (D_{ext} = 0.5mm; D_{int} = 0.25mm). The needle was placed vertically, 10mm above the surface of the exposed plate. Positive and negative discharges were both investigated. A voltage of 10 kV was applied between the two electrodes with a DC power supply (HT 9, Sefelec) for 10 minutes.

Vegetative cells of *E. coli* (ATCC® 9627™) and *B. subtilis* (ATCC® 23857™) were grown in 250 ml of their culture medium during one night at 37°C and 30°C respectively. The resulting bacterial suspensions had concentrations of about 10^6 CFU/µl. Endospores of *B. subtilis* (ATCC® 23857™) and fungal spores of *A. fumigatus* (ATCC® 1028™) were already in stock suspensions at a concentration of 10^7 CFU/µl. Suspensions were spread out on plates without dilution: 200 µl was plated out on culturing plates; about 10 µl was deposited as little droplets on stainless steel plates and left few minutes at room temperature to dry (Protocols had to be slightly different since stainless steel plates cannot absorb the suspensions). The plates were immediately exposed to the corona discharge. Afterwards cells deposited on stainless steel plates were transferred to appropriate culturing plates. Microorganisms were cultivated overnight. Two Petri dishes were not exposed to the corona discharge and were used as a reference for normal growth.

Bromothymol blue was diluted in hot agar solution (0.5% in weight) to obtain moist solid media such as culture plates. The advantage of these new media is that they can reveal some chemical reactions: Bromothymol blue is yellow when the pH is below 6.0 and blue when the pH is above 7.8. These pH indicator media were exposed to the corona discharge as vegetative cells and spores were.

3. Results

Prior to the experiments on microorganisms, discharges were observed. It appeared as expected that the positive discharge is uniform while the negative discharge is composed of one or two tufts. Next it was verified that the properties of the plates were not altered or deteriorated by exposure to the discharge (surfaces were not altered by arcs and microorganisms can be cultivated onto agar plates previously exposed to a corona discharge).

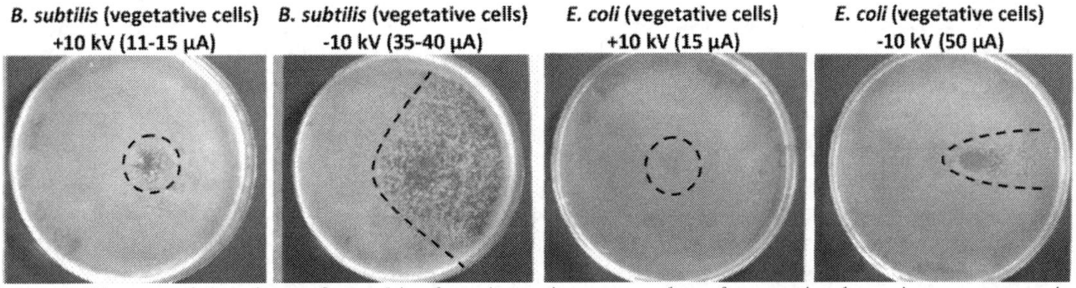

Figure 1. Pictures of the plates after cultivation. A continuous overlay of vegetative bacteria grew except in a zone right under the needle: a circular inhibition zone for the positive corona discharge; a comet-like inhibition zone for the negative corona discharge.

When culturing media were exposed to discharges of 10kV, a continuous overlay for both vegetative bacteria grew and formed a white layer except in a zone right under the needle (see Figure 1). The positive discharge led to a nearly circular inhibition zone and the negative discharge led to a comet-like inhibition zone of a larger surface area. This latter shape is similar to the inhibition zone observed by Scholtz and Julák [14]. The pictures show also that vegetative bacteria of *E. coli* are more resistant to the effects of the discharges than vegetative bacteria of *B. subtilis*. In this configuration, spores of *B. subtilis* and fungal spores of *A. fumigatus* were not affected by the germicidal effect of the discharge. The same vegetative bacteria were not affected by the discharge when they were exposed while being on stainless steel plates. Results obtained with the most sensitive bacteria, namely *B. subtilis*, are shown on Figure 2. The comet-like inhibition zone seen with the negative discharge is not visible here.

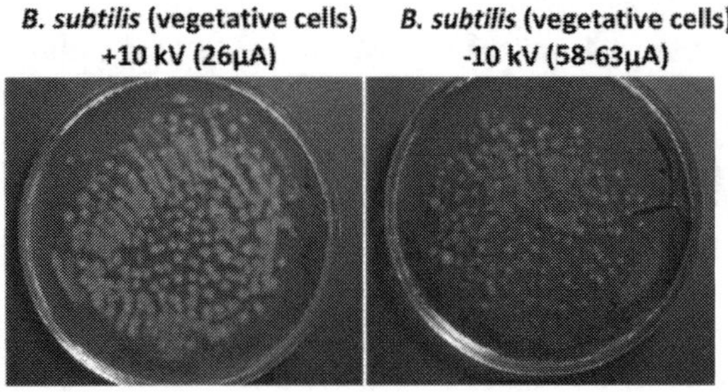

Figure 2: Agar plates where vegetative bacteria exposed on a stainless steel electrode were transferred. The microorganisms grew as spots because they were spotted on the metallic plate as little droplets. White streaks are visible where bacteria were a bit spread out on the cultivation media during the transfer.

Exposure of pH indicator gels to the positive and negative discharges revealed yellow zones (acidic or discolored areas) and blue zones (basic areas) shown on Figure 3. The former correspond to the inhibition zones observed for vegetative bacteria when located in the central region of these gels. Basic areas do not seem to be correlated to an inhibition of the microorganisms' cultivability. Also it should be noticed that pH change is not related to an inhibition of microorganisms' cultivability when located next to the stainless steel counter electrode dipped into the gels.

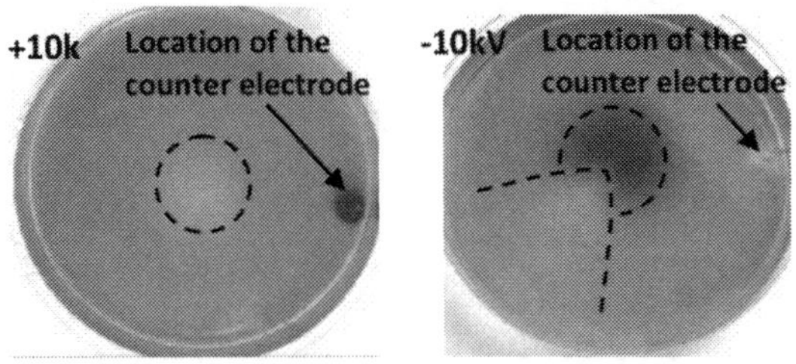

Figure 3. Pictures of the pH indicator gels after exposition: left, exposition to a positive discharge; right, exposition to a negative discharge. Blue areas are basic and yellow areas may be acidic or discolored areas.

4. Discussion

The present experiments show that both positive and negative corona discharges are lethal to vegetative bacteria on culture media. On the contrary, no adverse effect was observed with the same microorganisms on a dry conductive medium. Therefore, the mechanism responsible of the microorganisms' death (or growth inhibition) is likely to be a chemical reaction with the substrate. Electric field strength, ultraviolet radiation and temperature are often considered as candidate to explain the toxicity of electrical discharges. But they seem to play a secondary role [6-7]. Moreover, these processes are independent of the nature of the exposed media so they cannot be responsible for the growth inhibition observed here. Also reactive species generated by the electrical discharge such as ozone do not seem to have a direct role.

The shape of the inhibition zones differs markedly. Exposure to the positive corona discharge resulted in an axisymetric zone, while the negative discharge resulted in elliptical zones occupying a larger area. These differences were found to be related to the plasma region at the tip of the needle. The positive corona discharge is uniform and axisymetric so the spot below is axisymmetric. On the contrary the negative one is composed of one or two tufts. Each tuft results in a comet-like inhibition zone.

Electrical current in culture media implies electrolysis. In that case, one electrode should accept electrons and the pH in the vicinity of the electrode should subsequently decreases. At the same time, the other electrode should donate electrons and the pH should subsequently increases. Experiments with a pH indicator confirmed this expected behavior. However microorganisms grew near the counter electrodes dipped into media regardless of the polarity of the discharge. The pH is therefore not directly related to the culture inhibition. Furthermore, the experiments with a negative discharge revealed that two distinct chemical reactions occur right under the needle (Figure 3): one increases the pH (blue area) while the other seems to decrease it or to discolour the dye (yellow areas). Only the last one is correlated with an inhibition of the microorganisms' growth (or death).

The mechanism of the corona discharge in air is very complex and involves numerous chemical reactions [15]. However a mechanism may be suggested for the negative discharge. It is well-known that the electrons created in the avalanche leave the ionisation region and change into negative ions by attachment to electronegative molecules in the gas (such as O_2 molecules in air) and electron impact dissociation. According to Challenger et al. [16] and Richardson et al. [17] these negative ions combine with water by the following mechanisms:

$$O_2^- + 2H_2O \rightarrow 2H_2O_2 + e^- \qquad 2O_2^- + 2H^+ \rightarrow H_2O_2 + O_2$$

to create hydrogen peroxide that is highly toxic for microorganisms and that can degrade the dye. Meanwhile, electrolysis increases the pH right under the needle in this configuration (blue area). This mechanism is suspected to be responsible for the yellow and blue spots as well as the growth inhibition observed in the middle of the agar plate.

5. Conclusion

Electrostatic precipitators are very efficient samplers that can be used to collect biological airborne agents. But samples analysis requires to maintain the integrity of the microorganisms and this constraint add design rules: chemical reactions should be taken into account. The reported results suggest that dry electrodes prevent some chemical reactions that are harmful for microorganisms and so they preserve the cultivability.

References

[1] Roux J M, Kaspari O, Heinrich R, Hanschmann N, Grunow R 2013 *Aerosol Sci. Tech.* **47**(5) 463-471
[2] Mainelis G, Grinshpun S A, Willeke K, Reponen T 1999, *Aerosol Sci. Tech.* **30** 127–144
[3] Tan M, Shen F, Yao M, Zhu T 2011, *Aerosol Sci. Tech.* **45** 1154-1160
[4] Yao M, Mainelis G 2006, *J. Aerosol Sci.*, **37**, 513-527
[5] Yao M, Zhang H, Dong S, Zhen S, Chen X 2009 *J. Aerosol Sci.*, **40**, 492-502
[6] Laroussi M 2002 *IEEE Trans. Plasma Sci.* **30**(4) 1409-1415
[7] Laroussi M and Leipold F 2004 *Int. J. Mass Spectrom.* **233** 81-86
[8] Stoffels E, Sakiyama Y, Graves D B 2008 *IEEE Trans. Plasma Sci.* **36**(4) 1441-1457
[9] Julák J, Kriha V, Scholtz V 2006, *Czechoslovak J. Phys.*, **56**, 1333-1338
[10] Sillanpää M, Geller M D, Phuleria H C, Sioutas C 2008, *J. Aerosol Sci.*, **39**, 335-347
[11] Yao M, Mainelis G, An H R 2005 *Environ. Sci. Technol.*, **39**, 3338-3344
[12] Mainelis G, Adhikari A, Willeke K, Lee S A, Reponen T, Grinshpun S A 2002, *J. Aerosol Sci.*, **33**, 1417-1432
[13] Sarda-Esteve R, Roux J M, Sciare J, Delapierre G, Nadal M H 2012, *AAAR Conference*
[14] Scholtz V, Julák J 2010 *Journal of Physics: Conference Series* **223** 012005
[15] Kossyi A, Yu Kostinsky A, Matveyev A A, Silakov V P 1992 *Plasma Sources Sci. Technol.* **1** 207-220
[16] Challenger O, Braven J, Harwood D, Rosén K, Richardson G 1996 *Science of the Total Environment* **177** 215-219
[17] Richardson G, Eick S A, Harwood D J, Rosén K G, Dobbs F (2003) *Atmospheric Environment*, 37, 3701–3706

Acoustic effects of single electrostatic discharges

Łukasz Orzech

KOMAG Institute of Mining Technology, Laboratory of Applied Tests,
Pszczynska 37 Street, Gliwice, Poland

e-mail: lorzech@komag.eu

Abstract. Electric discharges, depending on their character, can emit different types of energy, resulting in different effects. Single electrostatic discharges besides generation of electromagnetic pulses are also the source of N acoustic waves. Their specified parameters depending on amount of discharging charge enable determination of value of released charge in a function of acoustic descriptor (e.g. acoustic pressure). Presented approach is the basics of acoustic method for measurement of single electrostatic discharges, enabling direct and contactless measurement of value of charge released during ESD. Method for measurement of acoustic effect of impact of a single electrostatic discharge on the environment in a form of pressure shock wave and examples of acoustic descriptors in a form of equation $Q=f(p_a)$ are described. The properties of measuring system as well as the results of regression static analyses used to determine the described relationships are analysed in details.

1. Introduction

Parameterization of the phenomenon and assessment of hazards caused by electrostatic discharges is directly associated with their proper measurement. Charge transferred during the discharge is one of the parameters of a single electrostatic discharge (ESD). Measuring methods that have been used so far are mainly electrical measuring methods [1], [2], [3], [4]. Analysis of conditions and technical requirements included in the methods for measurement of charge of a single electrostatic discharge, made by the author, has shown limitations and possible technical problems. Measuring method presented in EN 13463-1 Standard does not enable detection of discharges following one after another in a short time (shorter than time constant of electrostatic voltmeter). This leads to overestimation of measurement of discharge from the tested sample. Moreover, this method is sensitive to electrostatic background (saturation of capacitor by approaching the discharging electrode to the tested material without discharge) and coexistence of measuring system and discharge system. Simplicity of measuring system and uncomplicated mathematical relationship $Q=UC$ for determination of charge is the advantage of this method. Von Pidoll [2] suggested another method, based on continuous measurement of current in a circuit during discharge to determine the charge, which has no disadvantages described above.

2. Analysis of the acoustic phenomenon during electrostatic discharge

Acoustic effects, especially in the frequency audible to humans (20 - 20 000) Hz are the result of released energy action, in a form of medium ionization in a strong electric field. The most intensive ionization is in the front of the channel, what manifests with the light effect of discharge, caused by rapid temperature increase up to few thousand degrees (Kelvin). Speed of spark propagation significantly exceeds speed of sound and reaches 10^6 m/s. It causes violent (practically explosive) pressure change in the channel, causing strong acoustic effects [5].

Acoustic shock N-wave (*N-wave, – name comes from similarity of the acoustic waveform, close to the source, to the letter N),* which theoretical and real shape is the result of the phenomenon (e.g. explosion, spark electrostatic discharge) and is presented in below figure.

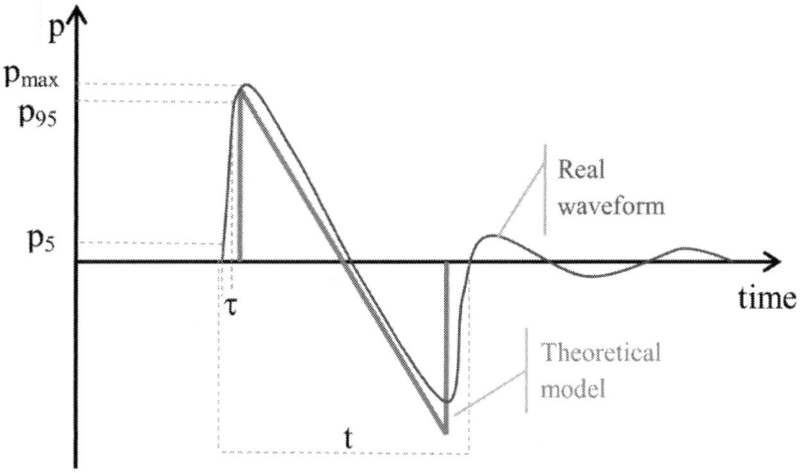

Figure 1. Graphical presentation of N-wave with the main descriptors (own report)

Reference of mechanism of generation of N-wave, presented in a micro scale, to the situation, where the electrodes (discharge electrode and earthed electrode) are placed in a short distance from each other, means that generated wave has a conical shape only at the small distance between these electrodes. In a macro scale, geometrical boundary conditions of such system of electrodes (a distance of up to a few centimetres) cause that the generated shock wave, which propagates, has a spherical shape and audible acoustic effect is its physical effect. Single electrostatic discharges generate short acoustic "thunders" of pulse time process and fuzzy frequency spectrum. Such properties of electrostatic discharge require the use of proper technologies for recording and analysis of measuring signals [6], [7]. The results of laboratory acoustic tests on electric sparks are presented by Klinkownstein R. [8]. He proved the possibility of using the spark discharges as repeatable reference source of acoustic pulses, and confirmed theoretical energy relationships of sound waves for quick-changing processes with fuzzy frequency spectrum in a distant field (free field).

3. Acoustic test of ESD

Part of scheduled tests included the measurements of time process of acoustic pressure during single electrostatic discharge generated by the simulator of single high-energy discharges. The measurements of acoustic pressure were taken with use of two microphones of diameters ½" and ¼" arranged perpendicularly to the axis of discharging electrode. The housings of the microphones were earthed. The distance between the microphones and the electrode axis was equal to 0.5 m. Moreover, time processes of momentary voltage on a capacitor connected in series in the circuit of discharge electrode as well as momentary discharge current in this circuit were recorded.

The measuring system for high-energy single ESD from the discharge generator was created for the testing purposes. The measurements were taken by approaching the electrode of armed discharge generator to the earthed electrode of discharging system with the capacitor of known and constant capacitance, connected in series. When a distance between the electrodes was short enough to overcome the resistance of the medium with charge flashover, EDS occurred. Current of the circuit (of the amount proportional to the released electrostatic charge) charged the capacitor connected in series.

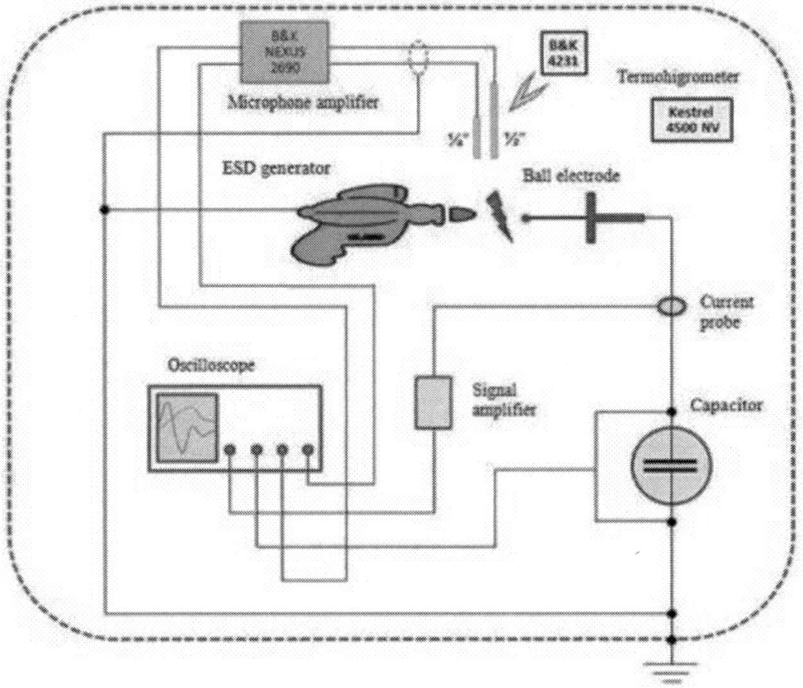

Figure 2. Diagram of the system for realization of first part of measurements (own report)

On the basis of analysis of regression tests of the specified group of acoustic parameters and electrostatic discharges generated by the simulator it was found that uncorrected level of effective acoustic pressure determined for 360 μs (duration of acoustic pulse until decay of wave amplitude) best matches to the acoustic descriptor, which describes high-energy electrostatic discharges generated by ESD simulator. This parameter as well as peak value of amplitude of acoustic process for microphone ½" and its placement at a distance of 0.5 m recreates the amount of charge with accuracy described by R^2 coefficient equal to 0.93.

Identified regression equation, which determines electrostatic charge on the basis of measurement of acoustic parameters, is as follows:

$$Q\ [nC] = 0.4397 \cdot p^2_{a360}\ [dB] - 22.4256 \cdot p^2_{a360}\ [dB] + 33.6560 \cdot p^2_{amax}\ [Pa]$$

where:
Q – electrostatic charge expressed in nC,
p_{a360} – level of effective acoustic pressure for 360μs expressed in dB,
p_{amax} – maximum amplitude of acoustic wave expressed in Pa.

This relationship is presented in Figure 3.

The relationship ($p < 0.001$) enables determining the charge on the basis of measurements of acoustic pressure during single electrostatic discharges (for a given position of microphone) with maximum error determined by a distribution of raw residuals determined from a difference between observed results and the results predicted by the regression model.

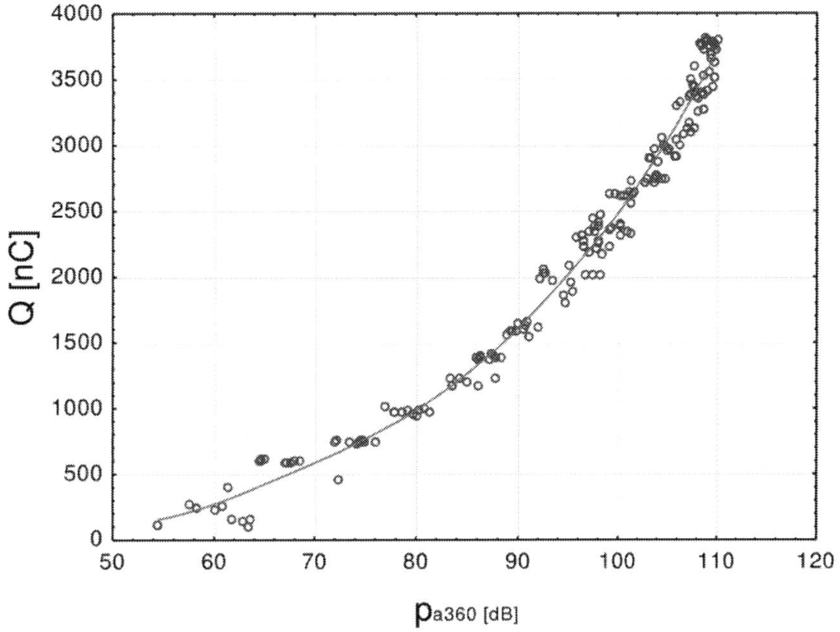

Figure 3. Diagram presenting relationship between electrostatic charge and effective acoustic pressure for high-energy discharges from generator (own report)

4. Summary

In the work, possibilities of using in practice the alternative method of measuring the electrostatic charges from single discharges from the generator of electrostatic discharges, by the measurement of acoustic pressure wave, which is a result of impact of discharges on the medium in which they occur were presented and analyzed. The method is an important element completing the current methods for electrostatic discharge measurements, adding the knowledge, which cannot be acquired by the traditional methods mentioned in the standards, e.g. multiplicity of discharges. Contactless (indirect) measurements of the transferred ESD charge enable eliminating the electrostatic interferences, so they do not affect the measurement results.

References

[1] PN-EN 13463-1. PN-EN 13463-1 Urządzenia nieelektryczne w przestrzeniach zagrożonych wybuchem - Część 1: Podstawowe założenia i wymagania. PKN 2010.

[2] von Pidoll U, Brzostek E and Froechtenigt H 2004 Determining the incendivity of electrostatic discharges without explosive gas mixtures *IEEE Trans. Ind. Appl.* **40** 6 1467-1475

[3] Holdstock P 2008 Does charge transfer correlate with ignition probability? 2008 *J. Phys. Conf. Ser.* **142** 012009

[4] Kacprzyk R 2004 Wybrane zagadnienia badań ładunku i jego zaniku w dielektrykach stałych. (Wrocław: OW PWr, Monografia nr 14)

[5] Weyna S 2005 *Distribution of acoustic energy sources of actual Sources*. (Warsaw: Scientific and Technical Publishing)

[6] Brinkman H H 2000 *Techniques and procedures for the measurement of impulse noise*. Damage risk from impulse noise (Meppen: RTO EN-11)

[7] PN-EN ISO 10843 Methods of describing and measuring individual pulse or series of pulses *Acoustics (Warsaw)* PKN 2002

[8] Klinkowstein R E. 1974 *A study of acoustic radiation from an electrical spark discharge in air*. (Massachusetts: Massachusetts Institute of Technology)

Non-contact surface resistivity measurement for materials greater than 10^9 Ω

Toshiyuki Sugimoto[1] and Koichi Taguchi[2]
[1]Yamagata University, 4-3-16 Jonan, Yonezawa 992-8510, Japan
[2]Napson Corporation, 2-5-10 Ohnodai, Midori-ku, Chiba-city 267-0056, Japan

E-mail: toshi@yz.yamagata-u.ac.jp

Abstract. A non-contact surface resistivity probe for materials with resistivity greater than 10^9 Ω has been developed using surface potential measurement combined with corona charging. The probe is composed of a grid type corona charger and a surface voltmeter located next to the charger. A test material was placed below the probe without contact to the material at a gap of 2 mm. The time variation of the surface potential and the saturation surface potential beneath the charged area are theoretically a function of the surface resistivity; therefore, measurement of the rise time and the saturation potential can be used to predict the surface resistivity in the ranges of lower and higher surface resistivity, respectively. The calibration equation for the probe was determined by solving the circuit equation for a one-dimensional simple circuit model and by extension of the simple model to the actual probe arrangement using experimentally derived constants. The lower surface resistivity, ρ_{sal} (from 10^9 to 8×10^{13} Ω), can be predicted from $\rho_s = 1.0\times10^{11}\times T_m^{1.64}$ using a 63% rise time, T_m. The higher surface resistivity, ρ_{sah} (from 8×10^{13} to 10^{16} Ω), can be predicted using the equation $\rho_s = 3.0\times10^{14}\times V_m^{1.70}$ with the converted surface potential, V_m.

1. Introduction
Surface resistivity is one of the most important properties of materials when contact charging or triboelectric charging of the surface may cause electrostatic problems. In this case, an insulative surface should be coated with a conductive layer that has a surface resistivity ranging from 10^6 to 10^{10} Ω to gently eliminate the surface charge. There are surface resistivity probes with two of four electrodes that contact the test surface to apply voltage and measure the leakage current flowing through the surface. However, this measurement principle requires that an extremely small leakage current be measured for materials in the high surface resistivity range. In addition, the contact condition of the electrodes (difference of real contact area) on the surface may affect the measurement, thereby making it difficult to obtain an accurate measurement. The maximum of the measurement range for a typical surface resistivity probe is 10^{13} Ω. We have proposed a non-contact surface resistivity measurement technique that employs corona charging combined with surface potential measurement [1]. The maximum measurement range can be theoretically extended past 10^{15} Ω. This paper describes the calibration of the probe.

2. Theoretical
Figure 1 depicts the non-contact surface resistivity measurement system, which includes a probe and the test material. The probe is composed of a corona charger and a surface voltmeter. The corona charger has a needle and a grid electrode that provides surface charge to the test material placed below the probe with a gap of δ. The charged potential is controlled by the voltage applied to the grid electrode, V_0.

However, the surface charge on the test material extends over the surface, depending on the surface resistivity; therefore, the change in the surface potential at the outside of the charged area is also dependent on the surface resistivity. The bottom shape of the grid electrode and the body of the potential probe are cylindrical to direct the surface charge from below the grid to below the surface voltmeter. The surface potential measured with the surface voltmeter eventually becomes V_0 if the surface is conductive. The rise time of the surface potential is measured to estimate the surface resistivity as described below.

In a simple one-dimensional model [1], the equivalent circuit equation becomes equation (1) using the surface potential $v(x,t)$ of the test sample:

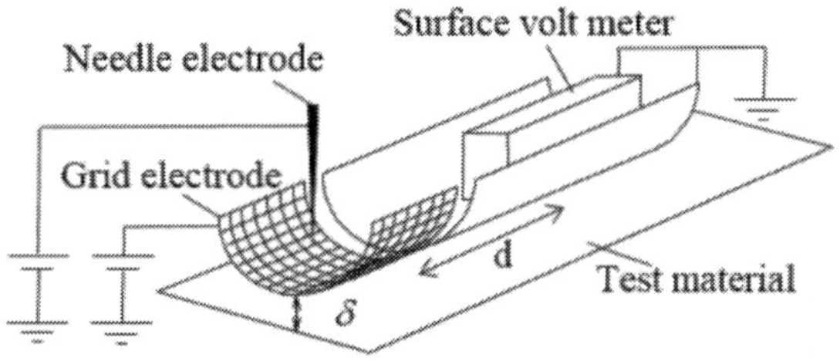

Figure 1. Non-contact surface potential probe.

$$\frac{\partial^2 v(x,t)}{\partial x^2} - \frac{\rho_s}{\rho_v \delta} v(x,t) - \frac{\varepsilon_0 \rho_s}{\delta} \frac{\partial v(x,t)}{\partial t} = 0, \quad (1)$$

where x is the distance from the charge area (below the grid) to the measured point (below the surface voltmeter), t is the elapsed time from charging, ρ_s is the surface resistivity of the test material, ρ_v is the volume resistivity of the air gap, δ is the gap between the test material and the probe, and ε_0 is the permittivity of air. The time variation of the surface potential $v(x,t)$ can be predicted by solving equation (1). The surface potential is a function of ρ_s, x, and t; therefore, the surface resistivity can be predicted by measuring the surface potential at a specific x and t. To obtain the simple solution of equation (1), two different conditions of lower or higher surface resistivity were considered as follows.

For low surface resistivity, $\rho_s << \rho_v\ \delta$, so that the second term of equation (1) can be omitted and $v(x,t)$ becomes equation (2) under the boundary condition $v(0, t) = V_0$:

$$v(x,t) = V_0 erfc\left\{\left(\frac{\varepsilon_0 \rho_s}{4t\delta}\right)^{\frac{1}{2}} \cdot x\right\}, \quad (2)$$

where *erfc* is the complementary error function. At fixed x=d, the normalized surface potential $v(\mathrm{d},t)/V_0$ changes from 0 to 1. For measurement of the surface potential, the 63% rise time, T_m, can be measured when the normalized surface potential $v(\mathrm{d},T_m)/V_0$ becomes 0.63. The theoretical lower surface resistivity ρ_{stl} becomes

$$\rho_{stl} = \frac{4\delta}{\varepsilon_0 d^2}\left\{erfc^{-1}(0.63)\right\}^2 \cdot T_m = A_l \cdot T_m, \quad (3)$$

where A_l is a constant that depends on the dimensions of the probe and the probe set-up. Equation (3) shows that the surface resistivity can be predicted by measuring the rise time, T_m. Equation (3) shows the theoretical surface resistivity derived from the simple one-dimensional model; therefore, the actual surface resistivity of a three-dimensional set-up may be different. We can suppose the constants D and β_l to transform from the one-dimensional theoretical surface resistivity, ρ_{stl}, to the actual surface resistivity, ρ_{sal}, to be as shown in equation (4), considering the large dynamic range.

$$\log_{10} \rho_{sal} = D + \beta_l \cdot \log_{10} \rho_{stl}. \quad (4)$$

Hence, the actual surface resistivity ρ_{sal} can be obtained from equation (5):

$$\rho_{sal} = 10^D \cdot (A_l \cdot T_m)^{\beta_l} = \alpha_l \cdot T_m^{\beta_l}, \quad (5)$$

where both α_l and β_l are constant values determined by calibration experiments using known surface resistivities of test materials.

For the higher surface resistivity range, the surface charge moves extremely slowly and the surface potential v_{sat} becomes smaller than V_0. In this case, the third term of equation (1) can be omitted, so that v_{sat} under the boundary condition $v=V_0$ at $x=0$ becomes

$$v_{sat} = V_0 \exp\left\{ -\left(\frac{\rho_s}{\rho_v \delta} \right)^{\frac{1}{2}} \cdot x \right\}. \quad (6)$$

The theoretical surface resistivity for the higher range ρ_{sth} becomes equation (7) using the surface potential at $x = d$:

$$\rho_{sth} = \frac{\rho_v \delta}{d^2} \left\{ \log_e \left(\frac{V_0}{v_{sat}(d)} \right) \right\}^2 = A_h \cdot V_m^2, \quad (7)$$

where A_h is a constant that depends on the dimensions of the probe and the probe set-up. V_m is the converted surface potential by taking the natural logarithm of V_0/v_{sat}(d). By considering the relationship between the theoretical and actual surface resistivity, as in equation (4), the actual surface resistivity can be expressed in the form

$$\rho_{sah} = \alpha_h \cdot V_m^{\beta_h}, \quad (8)$$

where both α_h and β_h are constants determined by calibration.

3. Experiment and results

Figure 2 shows a photograph of the experimental setup, which includes the non-contact surface resistivity probe and a test material. The size of the grid electrode is 0.9×0.9 mm^2, and the radius of the cylindrical mesh and grounded body is 10 mm. The distance d between the charged and measurement spots is 50 mm. The gap δ, between the probe and the test material is 2 mm. The voltage applied to the grid electrode V_0 was 500 V, while +3.5 kV dc is applied to the needle electrode to start the corona discharge. Table 1 shows test materials prepared on glass substrates. Samples A and B are n-type impurity-doped amorphous silicon. Samples C and D are polysilicon from amorphous silicon by annealing with a laser. Sample E is indium gallium zinc oxide (IGZO). Each of the materials is commercially used as a core material in liquid crystal displays (LCDs). The time variation of the surface potential $v(t)$ was measured and recorded with a personal computer (PC) through an

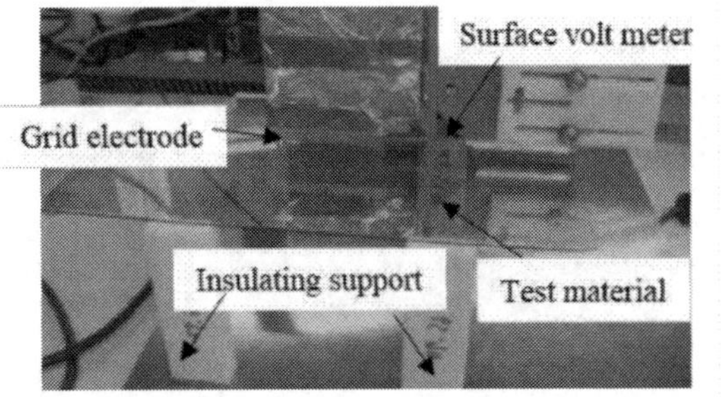

Figure 2. Experimental setup.

Table 1. Prepared test materials

Sample	Surface resistivity (Ω)	Material
A	6.67 x 10^7	n+a Si
B	5.56 x 10^8	n+a Si
C	1.65 x 10^{11}	Poly Si
D	7.37 x 10^{11}	Poly Si
E	6.90 x 10^{14}	IGZO

analog-digital (A-D) converter. The normalized surface potential $v(t)/V_0$ was calculated and the 63% rise time was measured.

Figure 3 shows the time variation of the normalized surface potential $v(t)/V_0$ for each sample. The surface potential increased with time and eventually reached 1. A higher surface resistivity results in a larger rise time. Figure 4 shows the relationship between the surface resistivity and the 63% rise time, T_m. As predicted in equation (5), there is a linear relationship between the rise time and the surface resistivity. From the experimental data, the two constants α_l and β_l in equation (5) were obtained as $\alpha_l = 1.0\times10^{11}$ and $\beta_l = 1.64$.

In the rise time measurement, the measurement time for higher surface resistivity became extremely long. For example, the rise time from measurement of a sample with a surface resistivity over 8×10^{13} Ω became larger than 60 s. Equation (6) indicates the saturation surface potential for a higher surface resistivity sample; therefore, the surface potential calculated from equation (2) should be less than that calculated from equation (6). The calculated potential from equation (6) is theoretically close to the potential of equation (2) at t = 60s. Hence, equation (8) should be effective when the 63% rise time is greater than 60 s. From the experimental data, the two constants α_h and β_h in equation (8) were obtained as $\alpha_h = 3.0\times10^{14}$ and $\beta_h = 1.70$.

The minimum measurement limit of the probe is dependent on the response time of the surface voltmeter. The response time of the probe is 20 ms; therefore, the minimum measurement limit of the surface resistivity is approximately 1×10^{9} Ω. Thus, a much faster response time of the surface voltmeter is required for the measurement of lower surface resistivity. On the other hand, the maximum limit of the measurement is dependent on the voltage resolution of the potential probe. The voltage resolution of the potential probe is a few tens of millivolts; therefore, the maximum detection limit of the surface resistivity is approximately 1×10^{16} Ω. It is important that such a high surface resistivity can be measured without contact to the test material. However, in the high surface resistivity region greater than 10^{14} Ω, precise measurement of the surface resistivity of test samples using a contact type probe, such as two-terminal or four-terminal measurement, is difficult because the leakage current is significantly small. Hence, calibration of the higher surface resistivity range may include uncertainty, so that a standard test sample is necessary for more precise calibration.

4. Conclusion

A non-contact surface resistivity probe was designed using corona charging combined with surface potential measurement and calibrated. The possible measurement range was determined to be from 10^{9} to 10^{16} Ω. Future work will include practical application of the probe for manufacturing qualification in industrial quality control and extension of the measurement range to lower than 10^{9} Ω.

[1] Sugimoto T Abe M and Higashiyama Y 2012 *Trans. IEEE Ind. Appl.*, **48** 1657-61

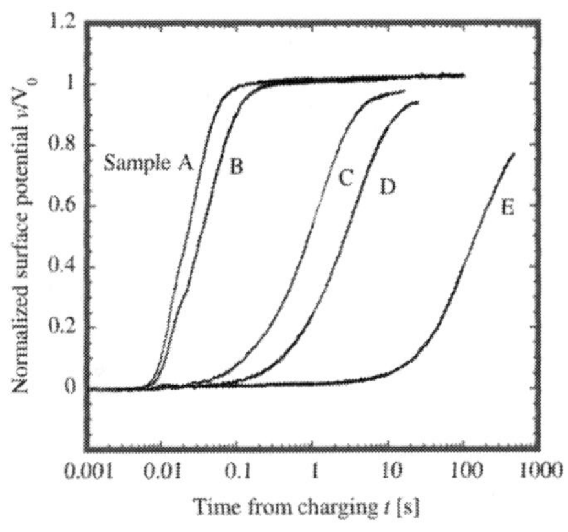

Figure 3. Time variation of v/V_0.

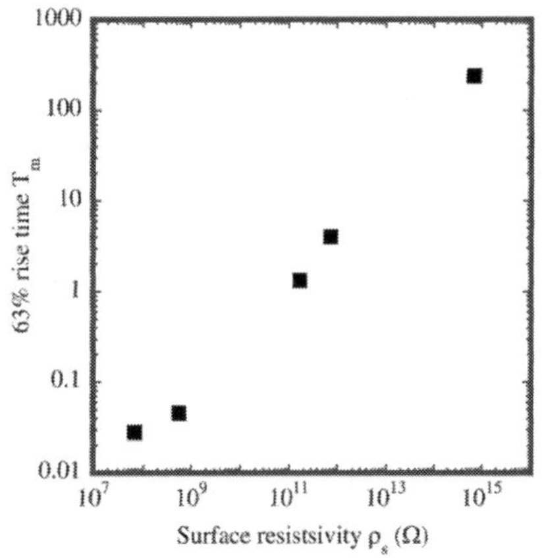

Figure 4. Relationship between ρ_s and T_m.

Current pulses caused by streamers in sphere-sphere electrode system

V A Chirkov, A V Samusenko and Yu K Stishkov
Physics Department, St. Petersburg State University, Ulianovskaia 1, 198504 St. Petersburg, Russia

E-mail: y.stishkov@spbu.ru

Abstract. Streamer is a channel of a low temperature plasma growing due to ionization in the area of the strong electric field at the tip of the channel. Streamer investigation presents a technically highly complicated task due to fleetingness of the process: growing velocity is 10^6–10^7 m/s and characteristic duration is 10^{-8}–10^{-7} s. The electric current pulse registration is a moderate method for investigating so fast process. However, the major part of streamer current investigations refers to low voltage range (about 10^3–10^4 V) and short streamers length range 10^{-2}-10^{-1} cm. Also positive streamers are usually considered and there is a lack of information about current pulses caused by negative streamers. Both positive and negative streamers and their interaction are considered in the present paper. A multibranch streamer corona emerging at voltages above 250 kV and in long gaps (above 40 cm) was investigated.

1. Introduction

Some experimental investigations of streamers contain electric current oscillograms (waveforms). Current oscillograms of positive streamers in "point/wire" electrodes system with 25 mm interelectrode gap and at voltage 25 kV are presented in [1]: maximal current is 8 A and the pulse half-width is about 10 ns. The current oscillograms of positive streamers in "plane with point/plane" system (gap is 14.3 cm; voltage is 98 kV; working gas is a mixture of "15% O_2 / 85% N_2") is presented in [2]: maximal current is 2 A, the duration of the pulse is more than 300 ns. Current oscillograms of positive streamers in "point/plane" electrodes system with 7 mm gap and at voltage 6–9 kV are presented in [3]: maximal current is about 3–30 mA, the pulse duration is 50–200 ns. Current oscillograms of streamers in "plane with point/plane" electrodes system with 13 mm interelectrode gap by voltage 27 kV in N_2 and N_2/O_2 mixture are presented in [4]. In N_2, maximal current is 14 A, the half-width of pulse is about 150 ns. Pulse duration and maximal current decrease with increasing O_2 concentration. Current oscillograms of streamers in "plane with point/plane" system (gap is 35 mm; voltage is 3 kV) are presented in [5]: maximal current is 60 mA, half-width is about 100 ns. Current oscillograms of negative streamers in "plane with point/plane" electrodes system with 30 mm gap and at 12 kV voltage are presented in [6]: maximal current at atmospheric pressure is about 100 mA. Current oscillograms of streamers in "plane with point/plane" electrodes system with 12 mm gap by voltage 5–10 kV and pressure 0.2–0.5 atm. are presented in [7]. Gas mixture of O_2 and 0–5% H_2 is used. Maximal current is 14–40 mA and pulse half-width is 10–30 ns. Current oscillograms of streamers in "point/plane" electrodes system with 10 mm interelectrode gap at 9 kV voltage are presented in [8]: maximal current is 7 mA, half-width of pulse is 1 μs. Current oscillograms of positive streamers burst from spherical electrode (radius 12.5 cm) at about 600 kV voltage are presented in [9]:

maximal current is 10 A and pulses half-width is about 100 ns. Thus, the major part of experimental data on streamer oscillograms refers to voltage values being less than 30 kV, while here a one order higher voltage is considered.

Sphere-sphere electrodes system was investigated. Sphere diameter *D* and interelectrode distance *h* were varied. The first sphere was grounded whereas high voltage pulse was applied to the other one. The voltage pulse form was the standard lightning pulse: fore front duration is 1.2 µs, back front duration is 50 µs. Pulses of the both polarities were considered. Current pulse measurement was implemented by observing voltage oscillogram from a shunting resistor that was series-connected to the grounded sphere. The experiments were implemented in air at the atmospheric pressure.

2. Results

Streamers emerge on the high voltage sphere or on the both spheres depending on the voltage and interelectrode distance. Negative streamer current pulses are short (about 0.2 µs) (figure 1) while those of positive streamers are longer (about 0.5 µs). Positive streamers can initiate negative ones from the opposite (i.e., grounded) sphere due to electric field distortion by space charge emerged in air owing to the discharge. The delayed electric current pulse occurs in this case (figure 2).

The correlation between negative streamer intensity and current pulse magnitude confirms the supposition that it is the first pulse that corresponds to the positive streamer current whereas the second one corresponds to the negative one. Compare results presented in figures 2–3.The data were obtained for the same electrode system but for different voltages: 200 kV and 240–250 kV. Positive streamers have similar length and glowing intensity whereas negative streamers are sufficiently longer and brighter in figure 3 in comparison with figure 2. The corresponding oscillograms show that it is the second current pulse that differs in the figures: the maximal current is 0.2–0.3 A at 200 kV whereas 0.8–1.2 A at 240–250 kV. Also the delay of the second pulse regards to the first pulse decreases: it is 1–1.5 µs at 200 kV and 0.4–1 µs at 240–250 kV. It is worth noting that the "meeting" negative and positive streamers and their electrical contact causes neither current leap nor breakdown (figure 3). Apparently, it is due to the high electrical resistance of negative streamer channels [9].

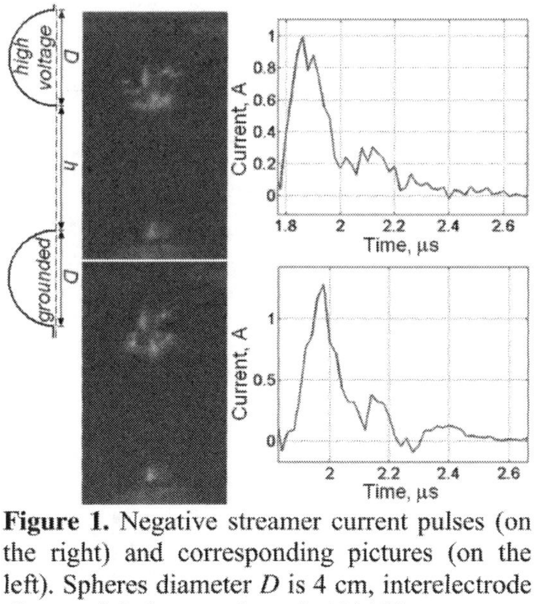

Figure 1. Negative streamer current pulses (on the right) and corresponding pictures (on the left). Spheres diameter *D* is 4 cm, interelectrode distance *h* is 5 cm, voltage is 130 kV.

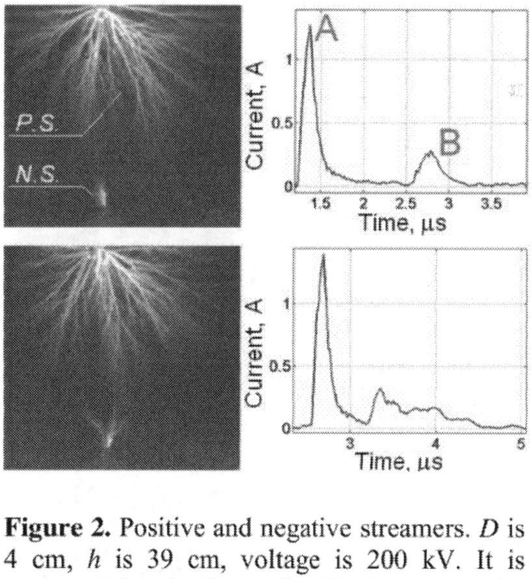

Figure 2. Positive and negative streamers. *D* is 4 cm, *h* is 39 cm, voltage is 200 kV. It is supposed that the first pulse ("A") is caused by positive streamers ("P.S.") from the high voltage sphere, the second pulse ("B") — by negative streamers ("N.S.") from the grounded sphere.

Electric current oscillograms preceding a breakdown are presented in figure 4. The same electrode system and voltage is considered in figure 3 where non-breakdown cases are presented. A breakdown oscillogram starts with a pulse that corresponds to positive streamer pulses in figure 3. The distinction is in the following. Streamers preceding the breakdown provide current 1.3–2 times the current of streamers that is not preceding breakdown. Sometimes the second pulse is distinguished (figure 4, "B"); apparently, it corresponds to negative streamers that start from the grounded sphere: there is similarity with the corresponding pulse "B" in figure 3.

The measured current maximums correspond by the order of magnitude to ones measured at the voltage that is 10 times lower [1, 2, 4]; however, pulses duration is 10 times longer. Apparently, the reason is in the following. The streamer velocity and electric field intensity in the channel remain nearly the same at rising voltage whereas streamer length and growing duration increase.

3. Conclusions

Negative streamer current pulses are shorter than those of positive streamers. In the sphere-sphere electrodes system, positive streamers can initiate the emergence of streamers from the grounded sphere; the delayed electric current pulse occurs in this case. Streamers preceding breakdown provide about 1.5–2 times higher current than those that is not preceding breakdown.

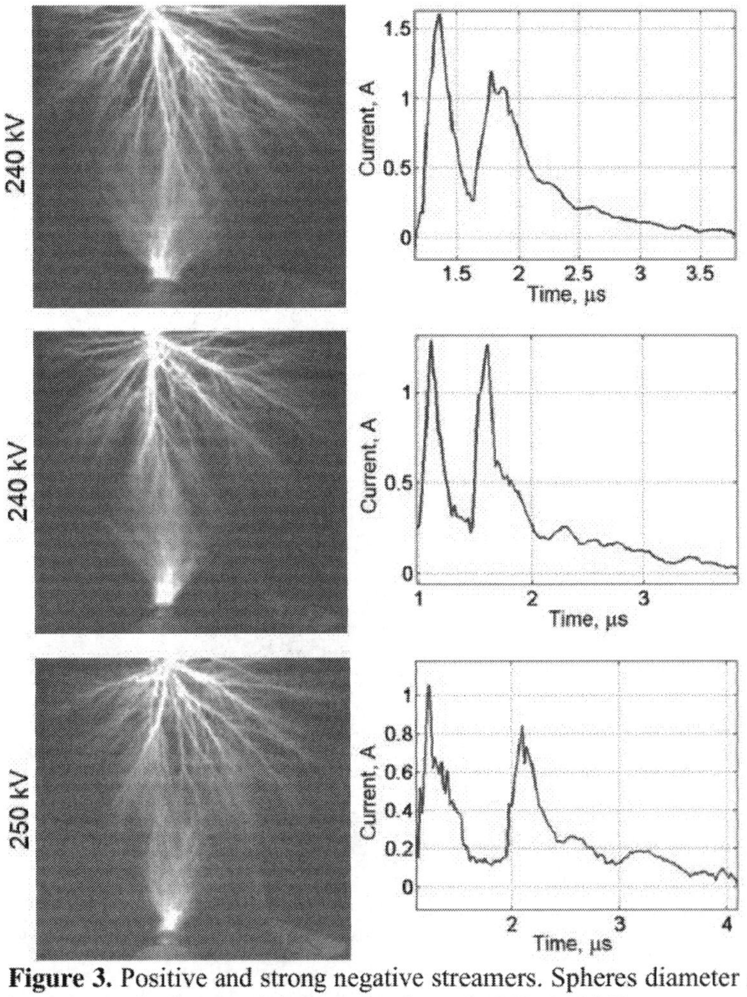

Figure 3. Positive and strong negative streamers. Spheres diameter D is 4 cm, interelectrode distance h is 39 cm, voltage is 240–250 kV.

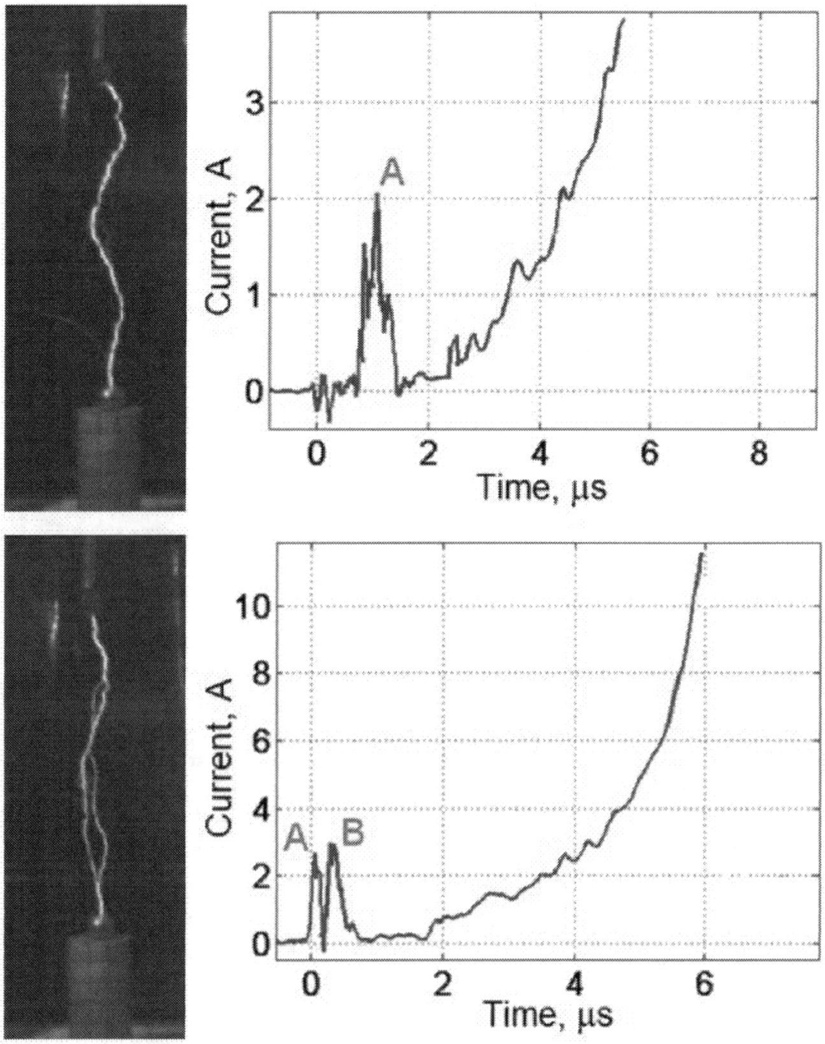

Figure 4. Breakdown. Spheres diameter D is 4 cm, interelectrode distance h is 39 cm, voltage is 250 kV; "A"—positive streamer current pulse, "B"—negative streamer current pulse.

References

[1] Van Veldhuizen E M and Rutgers W R 2002 *J. Phys. D: Appl. Phys.* **35** 2169–2179
[2] Yi W J and Williams P F 2002 *J. Phys. D: Appl. Phys.* **35** 205–218
[3] Eichwald O, Ducasse O, Dubois D, Abahazem A, Merbahi N, Benhenni M and Yousfi M 2008 *J. Phys. D: Appl. Phys.* **41** 234002
[4] Ono R and Oda T 2003 *J. Phys. D: Appl. Phys.* **36** 1952–1958
[5] Gao L, Akyuz M, Larsson A, Cooray V and Scuka V 2000 *J. Phys. D: Appl. Phys.* **33** 1861–1865
[6] Pancheshnyi S, Nudnova M and Starikovskii A 2005 *Phys. Rev. E* **71** 016407
[7] Kúdelcík J, Zahoranova A, Halanda J and Cernak M. 2014 *J. Electrostat.* **72** 417-421
[8] Matsumoto T, Kijima K, Shimoju T, Inada Y, Izawa Y and Nishijima K 2015 *Electrical Engineering in Japan* **190(2)** 10-16
[9] Bazelyan E M and Raizer Yu P 1998 *Spark Discharge* (Boca Raton, FL: CRC-Press)

Characterization of dielectric materials by the extension of voltage response method

Z Á Tamus, D Csábi and G M Csányi

Budapest University of Technology and Economics, Department of Electric Power Engineering, 18. Egry J. str., H-1111, Hungary

E-mail: tamus.adam@vet.bme.hu

Abstract. The voltage response method measures the decay and return voltages on a charged and shorted dielectric and from the slopes of these voltages the specific conductivity and the polarization conductivity of the material can be determined. These two numbers can characterize the main dielectric processes, namely the conductivity and the polarization, however the phenomenon of polarization is the sum of numerous elementary polarization processes with different intensities and time constants. By changing the shorting time of a charged insulation the elementary polarization processes can be examined separately and the whole polarization spectrum can be investigated more precisely. This paper introduces how the voltage response measurement technique is extended by the variation of discharging times and the mathematical model of the measurement is described. Measurements on a model dielectric and on a dielectric material were carried out and the results have been compared by the results of the calculations.

1. Introduction

The voltage response method measures the decay and return voltages on a charged dielectric. The timing diagram and arrangement of the measurement can be seen in figure 1 and 2, respectively. The decay voltage ($V_d(t)$) can be measured after the relatively long ($t_{ch} = 100-1000\ s$) charging period of an insulation, while the return voltage ($V_r(t)$) appears after a few seconds shorting (t_{dch}) of the charged arrangement. The initial slope of the decay voltage (S_d) is directly proportional to the specific conductivity (γ) of the material and the slope of the return voltage (S_r) is directly proportional to the polarization conductivity (β) of slow polarization processes having time constant higher than the discharging time (t_{dch}) [1]. These two figures (S_d, S_r) are successfully used for condition monitoring of the insulation of electric power equipment and cables [2, 3, 4, 5], however the weakness of this method is that the whole polarization spectrum of the material is characterized by only one number (S_r), whereas the distribution of elementary polarization processes by time constant is not uniform. The different ranges of polarization spectrum can be investigated by the changing of charging (t_{ch}) and discharging (t_{dch}) times and in this way the original voltage response method can be extended [6]. In this investigation, the charging time is chosen to constant (1000 s) and the discharging times were increased step by step. The relationship between the variation of charging-discharging times of return voltage measurement and the peaks of return voltages has been published elsewhere [7], however the usage for characterization of dielectric materials by the slopes of return voltages has not been introduced, yet.

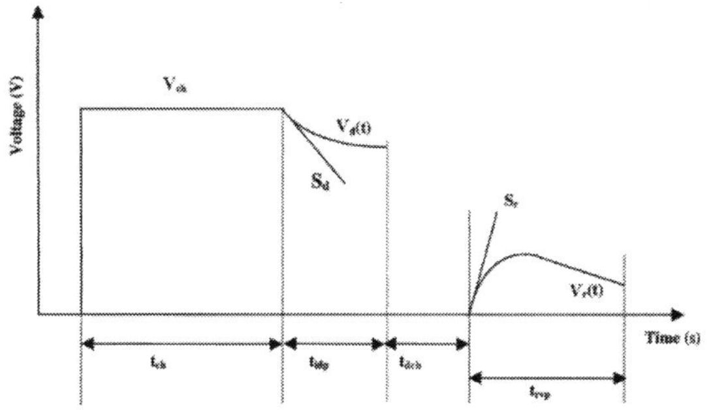

Figure 1. Timing diagram of voltage response measurement

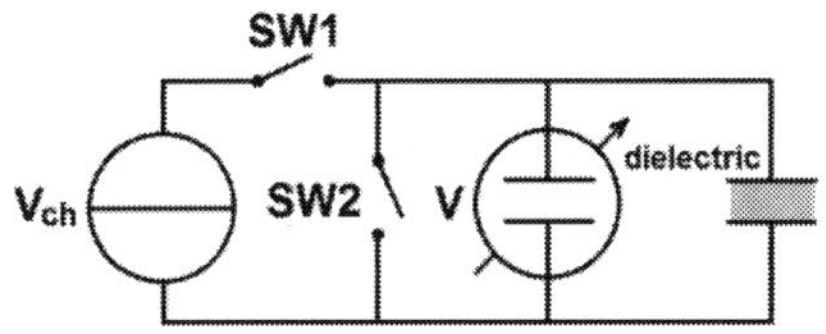

Figure 2. The arrangement of voltage response measurement

2. Model of the measurement

Professor Endre Németh introduced [1] the relationship between the material properties and the result provided by voltage response measurement. The slope of decay voltage is directly proportional to the conductivity of the insulation

$$S_d = \frac{V_{ch}}{\epsilon_0}\gamma, \tag{1}$$

where the V_{ch} is the charging voltage, γ is the specific conductivity of the material and ϵ_0 is the permittivity of free space. Similarly, the slope of the return voltage is directly proportional the polarization conductivity of the material (β):

$$S_r = \frac{V_{ch}}{\epsilon_0}\beta = \frac{V_{ch}}{\epsilon_0}\sum_{i=1}^{n}\beta_{pi}. \tag{2}$$

As equation 2. shows in real dielectrics the polarization conductivity is the sum of the polarization conductivities of the elementary processes (β_{pi}), which remain in activated state after charging and discharging of the insulation. The relation between the polarization conductivity and the polarizability (α_{pi}) of a given elementary process is $\beta_{pi} = \alpha_{pi}/\tau_{pi}$, where τ_{pi} is the time constant of the process. For the calculation of return voltages the extended Debye model of the insulation is used (figure3.). In this model C_0 represents the capacitance of the electrode arrangement without the insulating material (vacuum capacitance). R_0 is the d.c. resistivity of the insulation and the $R_{pi}-C_{pi}$ branches (Debye elements) represent the elementary polarization processes of the insulation. The elementary processes can be characterized by their time constants $\tau_{pi} = R_{pi}C_{pi}$ and intensities, namely the polarizability ($\alpha_{pi} = C_{pi}/C_0$). The relation between the equivalent circuit of an insulation (figure 3) and the parameters provided by the voltage response measurement can be expressed:

$$S_d = \frac{V_{ch}}{\tau_0}, \tag{3}$$

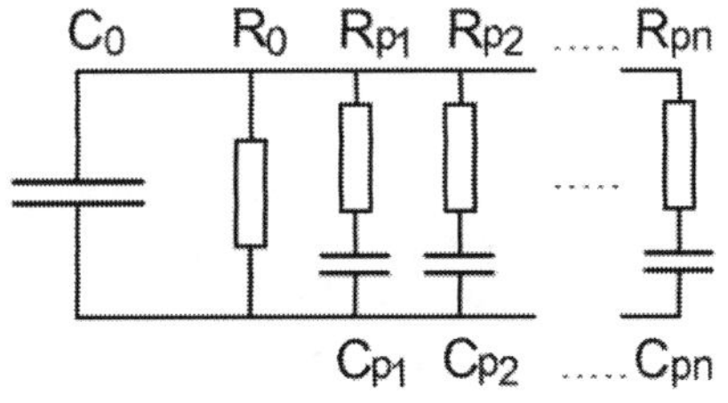

Figure 3. Equivalent circuit of an insulation

where $\tau_0 = R_0C_0$. The initial slope of return voltage (S_r) can be expressed as the sum of the gradients of return voltages generated by Debye (elementary polarization) processes:

$$S_r = \sum_{i=1}^{n} \frac{V_{C_{pi}}}{\tau_{ri}}, \tag{4}$$

where $V_{C_{pi}}$ is the remaining voltage of a C_{pi} capacitor in a Debye element after charging and discharging of the insulation and $\tau_{ri} = R_{ri}C_0$. The relation between the extended Debye circuit and the material properties can be expressed by the susceptibility ($\kappa = \sum_{i=1}^{n} C_{pi}/C_0$), which represents the intensity of polarization process in the material ($\kappa = (\epsilon_r - 1)$) at zero frequency.

3. Computation of return voltages on extended Debye model of dielectrics

The aim of the extension of voltage response measurement is the investigation of the elementary polarization processes and to determine the parameters of the extended Debye model of the measured insulation. For this reason the measurement technique is developed: the charged insulation is not only discharged once, but more shorting is executed, therefore different polarization processes remain activated after the different discharging times and more slopes of return voltages are measured. For the calculations and the measurements, thirteen discharging times between $2 - 200\ s$ were chosen but the charging time was 1000 s in all cases. At the calculations, the dielectric processes were modeled in the $0.1 - 10000\ s$ time constant range and the intensities of the processes were varied till the calculated slopes of return voltages became acceptable close to the measured values.

4. Experimental

4.1. Measurement on dielectric model with two Debye elements

The results of the simulation have been validated on a dielectric model, which contains two elementary polarization processes. Using the symbols of figure 3, the parameters of the model can be seen in table 1. The results of the calculations and the measurement show good agreement (table 2).

Table 1. Parameters of the model dielectric with two Debye elements ($R_0 = 57.0\ G\Omega$, $C_0 = 9.66\ nF$)

i=	1	2
τ_i	5.16 s	41.28 s
R_{pi}	1.72 $G\Omega$	2.10 $G\Omega$
C_{pi}	3.00 nF	19.63 nF

4.2. Measurement on EPR insulation

The measurements were carried out on the insulation of a four core low-voltage ethylene propylene rubber (EPR) insulated low voltage cable. After the measurement the model of the EPR insulation have been determined by the variations of the parameters of an extended Debye model, which contains four Debye elements. The time constants of the elements were chosen to $1, 10, 100$ and $1000\ s$. The calculated parameters of the model are in table 3. The result shows that the slow polarization processes (time constants higher than 1 s) of such a complex dielectric like an EPR insulated cable can be characterized by only four Debye elements (table 4).

Table 2. Slopes of return voltages of the the model dielectric with two Debye elements ($R_0 = 57.0\ G\Omega$, $C_0 = 9.66\ nF$, $T_{ch} = 1000\ s$, $V_{ch} = 1000\ V$)

t_{dch} [s]:	2	4	6	8	10	15	20	30	50	75	100	150	200
Meas.[V/s]	88.0	74.9	64.8	57.3	51.3	39.6	32.6	23.7	14.3	7.6	4.2	1.3	0.4
Calc. [V/s]	87.7	72.4	61.3	53.3	47.3	37.5	31.5	23.9	14.6	8.0	4.3	1.3	0.4

Table 3. Model of the EPR insulation($C_0 = 255\ pF$)

i=	1	2	3	4
τ_i	1 s	10 s	100 s	1000 s
R_{pi}	48.1 $G\Omega$	292 $G\Omega$	1.78 $T\Omega$	10.8 $T\Omega$
C_{pi}	20.78 pF	34.21 pF	56.32 pF	92.69 pF

Table 4. Slopes of return voltages of the EPR insulation ($t_{ch} = 1000\ s$, $V_{ch} = 1000\ V$)

t_{dch} [s]:	2	4	6	8	10	15	20	30	50	75	100	150	200
Meas. [V/s]	24.6	13.4	9.4	7.5	6.3	4.2	3.3	2.5	1.8	1.2	1.0	0.7	0.6
Calc. [V/s]	24.4	12.8	9.9	8.3	7.2	5.1	3.8	2.5	1.6	1.3	1.0	0.7	0.5

5. Conclusion

The voltage response measurement is a tool, which developed for measuring the slow polarization processes of an insulation. The initial technique can characterize the polarization processes by only one figure (S_r) but by the systematic changing of the discharging times, the voltage response method have been developed. By this extension of the method, the parts of the polarization spectrum can be investigated more precisely. From the results provided by the extended method the equivalent circuit of the insulation can be determined. The results have been validated on a model dielectric with two Debye processes. Measurement have been carried out also on the insulation of a low voltage EPR cable and the parameters of a Debye circuit have been calculated. The result shows that the slow polarization processes of such a complex dielectric, like an EPR insulated cable, can be characterized by only four Debye elements.

References

[1] Németh E 1971 *Periodica Polytechnica, Electrical Engineering* **15** 305–322
[2] Németh E 1999 *Science, Measurement and Technology, IEE Proceedings -* **146** 249–252 ISSN 1350-2344
[3] Tamus Z Á and Berta I 2009 *Electrical Insulation Conference, 2009. EIC 2009. IEEE* pp 444–447
[4] Tamus Z Á and Németh E 2010 *Int. Conf. on Condition Monitoring and Diagnosis, CMD 2010* pp 721–724
[5] Tamus Z Á and Szedenik N 2013 *Journal of Electrostatics* **71** 462 – 466 ISSN 0304-3886
[6] Csányi G M and Tamus Z Á 2014 *Electrical Insulation Conference (EIC), 2014* pp 299–302
[7] Ghourab M and Németh E 1992 *Periodica Polytechnica, Electrical Engineering* **36** 121–130

Re-creation of aerosol charge state found near HV power lines using a high voltage corona charger

J C Matthews[1], M D Wright[1], M F Biddiscombe[2], R Underwood[2], O S Usmani[2], D E Shallcross and D L Henshaw[1]

[1] ACRG, School of Chemistry, University of Bristol, UK
[2] NHLI, Imperial College London and Royal Brompton Hospital, UK

E-mail: j.c.matthews@bristol.ac.uk

Abstract. Corona ionisation from AC HV power lines (HVPL) can release ions into the environment, which have the potential to electrically charge pollutant aerosol in the atmosphere. It has been hypothesised that these charged particles have an enhanced probability of being deposited in human airways upon inhalation due to electrostatic attraction by image charge within the lung, with implications for human health. Carbonaceous aerosol particles from a Technegas generator were artificially charge-enhanced using a corona charger. Once generated, particles were passed through the charger, which was either on or off, and stored in a 15 litre conducting bag for ~20 minutes to observe size and charge distribution changes over time. Charge states were estimated using two Sequential Mobility Particle Sizers measuring the size and mobility distributions. Charge-neutral particles were measured 7 times and positive particles 9 times, the average charge-neutral value of x was 1.00 (sd = 0.06) while the average positive value was 4.60 (0.72). The system will be used to generate positive or charge neutral particles for delivery to human volunteers in an inhalation study to assess the impact of charge on ultrafine (size < 100 nm) particle deposition.

1. Introduction

Aerosols in the atmosphere can become electrically charged via collisions with air ions [1] but in most ambient environments remain in a 'charge-neutralised' state in which the number of positive and negatively charged particles is similar [2]. High Voltage Power Lines (HVPL) are a known producer of ions of one or both polarities, which are carried from the line into the atmosphere [3]. Downwind of HVPL, elevated levels of ions of one polarity over the other can cause an increase in aerosol charge state [4].

Particles can deposit in human airways by several mechanisms including diffusion, impaction, interception and sedimentation. Particle deposition is at its least efficient in the range of 200–300 nm [5], and it is at this point that electrical charge may cause enhanced deposition and hence cause an increase in particle deposition within the lung. Human volunteer studies [6] have shown enhanced lung deposition of highly-charged particles, mainly in the alveolar region, but no work has yet been undertaken at charge levels found downwind of HVPL nor for ultrafine (< 100 nm) sizes.

Increased deposition within the lung of charged ultrafine particles has been hypothesised as a possible causal link between HVPL and increased risk of childhood leukaemia [7; 8]. Theoretical modelling and experimental studies are required to test whether charge can increase deposition within the lung. Accurate re-creation of the charge states that could be found downwind of HVPL is essential for such studies.

2. Methods

Figure 1 shows the experimental set up used to verify the charge states of charged particles produced by a Technegas generator (Cyclomedica), a commonly used generator in nuclear medicine imaging studies, which produces carbonaceous particles with a log-normal size distribution of typical sizes 100-150 nm (geometric standard deviation of 1.5-1.9). Technegas particles are usually radio-labelled with Tc-99m, but here only de-ionised water or saline solution was used for safety reasons. It is also possible to remove the sodium from the pertechnetate radiolabel normally used in the Technegas generator [9]. This ensures no salt crystals are present in the particles, reducing mean particle size and minimising hygroscopic growth, both desirable properties in ultrafine particle inhalation studies.

Particles were charged using a custom built corona charger (Figure 2) based on the designs of Qi et al. [10] and Whitby [11]. The corona needle voltage, total flow rate and nitrogen flow rate (on each needle) can be altered to adjust performance. Through experiment with NaCl aerosol and Technegas particles, optimum parameters were found to be 10 lpm total flow, and 2 lpm nitrogen flow and +3.5 kV voltage for each needle. Corona current was stable at ~4 μA per needle, control of corona voltage alone was sufficient for repeatable charging as we observed a constant charge level over an hour's measurement and in repeated characterisation studies.

The vacuum pump and corona charger were connected to two solenoid valves that switched inlet and outlet from HEPA filters to the Technegas generator and a conducting bag. Nitrogen and aerosol flows were started, and where charging was required needles were energised to +3.5 kV and allowed to stabilise, before particles were generated. Following this, solenoid valves were switched for 50 s and particles were drawn from the generator by a vacuum pump through the corona charger (needles either energised to +3.5 kV, or grounded) into a 15 litre conducting bag. All tubing between instrumentation was electrically conducting to minimise particle losses.

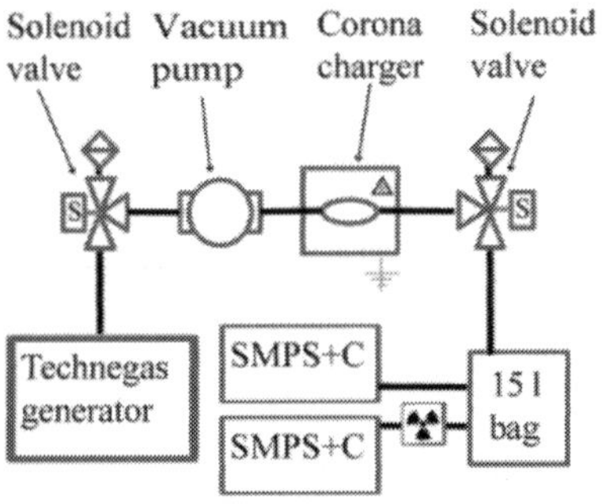

Figure 1. Experimental set up for measuring charge distribution on Technegas particles.

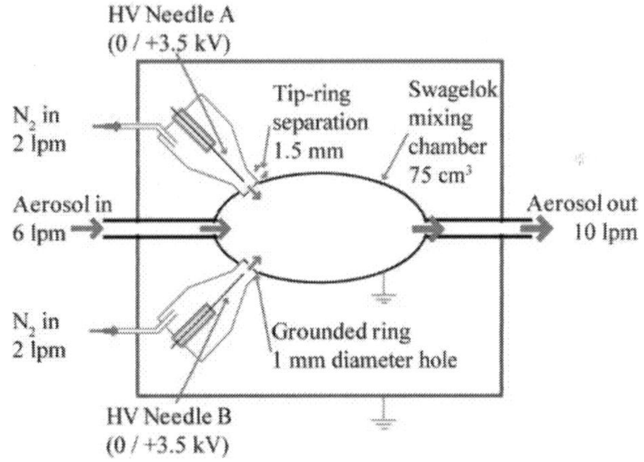

Figure 2. Custom made corona charger based on the designs of Qi et al [10] and Whitby [11].

Particle size and mobility distributions were measured using two Sequential Mobility Particle Sizers (SMPS+C, Grimm Aerosol), one with a neutraliser, over 5 'cycles' of 217 s, for total time ~20 minutes. This allows the charge distribution to be estimated [4]. The charge distribution can be represented by the average number of charges n on particles of diameter d. At a constant charge temperature T this can be described using a single parameter (the charge asymmetry ratio, x) as shown in equation 1, all other values being constant (ε_0 is the permittivity of free space, k the Boltzmann constant and e the charge on one electron) [12]. In total, charge-neutral Technegas particles were measured on 9 occasions, and positive particles on 7 occasions. Each measurement produced 5 size

Journal of Physics: Conference Series **646** (2015) 012044 doi:10.1088/1742-6596/646/1/012044

distributions and 5 mobility distributions as the bag was emptied over 20 minutes, from which charge distributions were estimated.

$$n = \frac{2\pi\varepsilon_0 dkT}{e^2}\ln x \qquad (1)$$

3. Results

Table 1 shows results from the first SMPS+C scan of Technegas particles, passed through the corona charger on (positive) or off (charge-neutral) while Figure 3 shows the change in particle characteristics when holding in the bag for ~20 minutes. Mean (SD) of the charge asymmetry ratio, x, for charge-neutral particles was 1.00 (0.05) implying an equal number of positive and negatively charged particles, and 4.60 (0.72) for positively charged particles. There is some run-to-run variability in the positive charge levels, with higher particle output from the Technegas generator leading to slightly lower charging. However, in all cases average particle charge is much higher than for charge-neutral particles. Particle count is lower for 'positive' compared to 'charge-neutral' particles, suggesting increased loss of particles during charging and bag transfer. Mean diameter (Figure 3b) is similar for positive and charge-neutral particles during the first scan, but charge-neutral particles subsequently grow by coagulation whereas for (unipolar) positive particles coagulation is reduced. Because particle size affects deposition efficiency considerably, for a robust comparison in volunteer studies, particles should be used within this ~200 s time frame.

Table 1: Table showing charge asymmetry ratio (x); the mean number of electrical charges per 100 nm particle; the geometric mean (GMD) and geometric standard deviation (GSD) of particle diameter; and particle count (cm^{-3}) for Technegas particles after being passed through the corona charger either on (positive) or off (charge-neutral).

Charge-neutral					Positive				
Charge Asymmetry Ratio (x)	Mean charges (100 nm)	GMD (nm)	GSD (nm)	Particle Count (cm^{-3})	Charge Asymmetry Ratio (x)	Mean. charges (100 nm)	GMD (nm)	GSD (nm)	Particle Count (cm^{-3})
0.935	-0.06	96.55	1.64	1472000	4.522	1.32	110.19	1.69	1062000
0.986	-0.01	102.59	1.68	1466000	4.640	1.35	110.63	1.71	1023000
0.992	-0.01	132.54	1.71	1676000	5.178	1.44	109.20	1.69	952000
1.094	0.08	96.23	1.61	1814000	5.085	1.43	115.31	1.68	975000
1.017	0.01	111.73	1.64	1722000	4.386	1.30	138.56	1.69	1025000
1.028	0.02	113.09	1.65	1905000	5.918	1.56	97.07	1.65	923000
0.943	-0.05	135.45	1.67	1423000	4.067	1.23	123.70	1.65	1150000
					4.112	1.24	113.91	1.69	1190000
					3.466	1.09	131.84	1.66	1085000

4. Discussion

A charge asymmetry ratio of $x = 1.71$, which implies an average charge per 100 nm particle of 0.47, was observed for particles at a HVPL known to produce a high concentration of ions [4]. Particle charge states in this study are higher than this value, but considerably lower than those measured in the output of metered dose inhalers [13], thus we cannot extrapolate expected charge-related effects to these types of medical devices. Particles produced in this study thus represent an extreme example of what may be found near to HVPL and would therefore provide an upper limit on the possible enhanced ultrafine particulate pollution dose due to HVPL charging effects in real-world conditions.

The corona charger is being used in a human volunteer study in which charged particle deposition in human airways is being investigated. Charged and neutral particle deposition will be measured using particle counters and gamma scintigraphy. As the level of charge exhibited on the particles in this study is slightly higher than that found near HVPL, any observed increase in deposition in the proposed study is likely to be of higher magnitude than that found in populations living near HVPL.

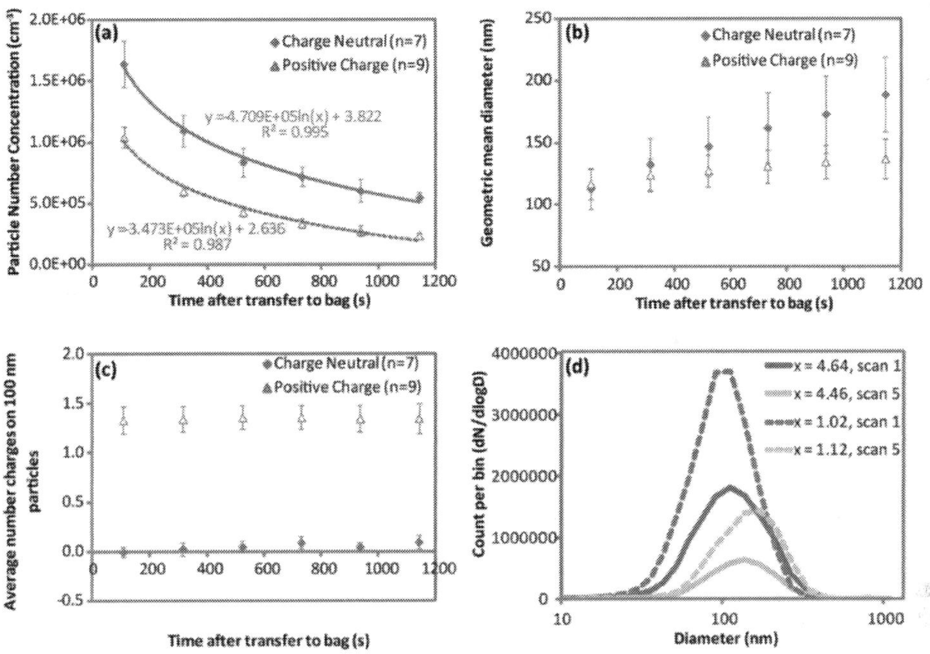

Figure 3. a) Mean and standard deviation of particle number concentration; b) geometric mean diameter; and c) the average number of charges on 100 nm particles on charge-neutral and positively charged particles during holding in the bag for 20 minutes (6 scans). d) Size distributions for charged and charge-neutral particles during the first (~3 minutes after bag filling) and fifth (~16 minutes after bag filling) SMPS+C scans

Acknowledgements
The study was funded by Children with Cancer UK, Registered Charity No. 298405, Grant No. 09/092 and was supported by the NIHR (National Institute of Health Research, UK) Respiratory Disease Biomedical Research Unit at the Royal Brompton and Harefield NHS Foundation Trust and Imperial College London. O.S.U. is a recipient of an NIHR Career Development Fellowship.

References

[1] Fuchs N A 1963 *Geofis. Pura. Appl.* **66** 185-93
[2] Wiedensohler A 1988 *J. Aerosol Sci.* **19** 387-9
[3] Wright M D, Buckley A J, Matthews J C, Shallcross D E and Henshaw D L. 2014. *Atmos. Environ.* **95** 296-304
[4] Buckley A J, Wright M D and Henshaw D L 2008 *Aerosol Sci. Technol.* **42** 1042–51
[5] ICRP. 1994. *Human Respiratory Tract Model for Radiological Protection* ICRP Publication 66 (Oxford: Pergamon)
[6] Melandri C, Tarroni G, Prodi V, De Zaiacomo T, Formignani M and Lombardi C C 1983 *J. Aerosol Sci.* **14** 657–69
[7] Fews A P, Henshaw D L, Wilding R J and Keitch P A 1999 *Int. J. Radiat. Biol.* **75** 1523-31.
[8] Draper G, Vincent T, Kroll M E and Swanson J 2005 *BMJ* **330** 1290-5
[9] Möller W, Felten K, Seitz J, Sommerer K, Takenaka S, Wiebert P, Philipson K, Svartengren M and Kreyling W G 2006 *J. Aerosol Sci.* **37** 631-644
[10] Qi C, Chen D-R and Pui D Y H 2007. *J. Aerosol Sci.* **38** 775-92
[11] Whitby K T 1961 *Rev. Sci. Inst.* **32** 1351-5
[12] Clement C F and Harrison R G 1992 *J. Aerosol Sci.* **22** 481–504
[13] Kwok, Glover and Chan 2005. *J. Pharma. Sci.* **94** 2789

Discrete element method modeling of the triboelectric charging of polyethylene particles: Can particle size distribution and segregation reduce the charging?

Ladislav Konopka, Juraj Kosek

University of Chemistry and Technology Prague, Department of Chemical Engineering, Technicka 5, 166 28 Prague 6, Czech Republic, e-mail: ladislav.konopka@vscht.cz

Abstract. Polyethylene particles of various sizes are present in industrial gas-dispersion reactors and downstream processing units. The contact of the particles with a device wall as well as the mutual particle collisions cause electrons on the particle surface to redistribute in the system. The undesirable triboelectric charging results in several operational problems and safety risks in industrial systems, for example in the fluidized-bed polymerization reactor. We studied the charging of polyethylene particles caused by the particle-particle interactions in gas. Our model employs the Discrete Element Method (DEM) describing the particle dynamics and incorporates the 'Trapped Electron Approach' as the physical basis for the considered charging mechanism. The model predicts the particle charge distribution for systems with various particle size distributions and various level of segregation. Simulation results are in a qualitative agreement with experimental observations of similar particulate systems specifically in two aspects: 1) Big particles tend to gain positive charge and small particles the negative one. 2) The wider the particle size distribution is, the more pronounced is the charging process. Our results suggest that not only the size distribution, but also the effect of the spatial segregation of the polyethylene particles significantly influence the resulting charge distribution 'generated' in the system.The level of particle segregation as well as the particle size distribution of polyethylene particles can be in practice adjusted by the choice of supported catalysts, by the conditions in the fluidized-bed polymerization reactor and by the fluid dynamics. We also attempt to predict how the reactor temperature affects the triboelectric charging of particles.

1. Introduction

Tribolectrification is an undesired phenomenon in a vast majority of industrial processes, mainly in powder processing (polymers, drugs, flour). Although the triboeletric charging of dielectrics in natural and/or industrial systems has been observed for centuries, it is still inadequately understood and it lacks quantitative description. It is difficult to perform well-defined experiments with a good repeatability, since samples used for the measurements usually have random initial charge and the charging dynamics can be altered among other factors by a simple change of the particle surface. Thus mathematical modeling seems to be beneficial to aid in the better understanding of the electrostatic charge related problems.

2. Trapped electron approach

In this work, we follow the approach of Lowell and Truscott [1], who assume the existence of the insulator particle surface electrons that are trapped in the high energy levels of surface orbitals - even if equilibrium surface electron energy states distribution predicts them to be in a lower energy state. The non-equilibrium of surface electron energy states can last even for years, depending on the material of the insulator particle [2]. Due to the high surface resistivity of insulator materials (including polymers), the trapped high-energy (HE) electrons on the particle surface cannot be transferred to the state of lower energy on the same particle. However, the HE electrons trapped on the particle surface can be transported during the particle collision to a vacant low-energy (LE) level on the surface of another particle. This theory was utilized for modeling of triboelectrification in granular systems by Lacks [3]. We shall call this approach the 'Trapped Electron Approach' (TEA) in this paper.

3. Model description

We have developed spatially 2D dynamic model of colliding polyethylene particles involving the contact charging (triboelectrification). The particle dynamics is computed using Discrete Element Method (DEM) incorporating Hertzian [4] description of solid particle contact mechanics and Coulomb's law representing the electrostatic effects. Each particle is represented by a discrete element i in our model. The DEM utilizes Newton's second law to track the motion of each particle i (equation (1)):

$$\frac{d^2\boldsymbol{x}_i}{dt^2} = \frac{d\boldsymbol{v}_i}{dt} = \frac{\sum_j \boldsymbol{F}_{i,j}}{m_i}. \quad (1)$$

Here, $\boldsymbol{x}_i$ and $\boldsymbol{v}_i$ represent particle position and particle velocity, respectively, and m_i stands for the particle mass. The velocity and the position of each particle at each time step are thus computed, since the sum of forces $\boldsymbol{F}_{i,j}$ acting between particle i and another particle j are precalculated for all the particle pairs. Considered forces are Hertzian elastic repulsive force and Coulomb electrostatic force. We utilize the principle of additivity in order to sum up the force effects of all particles acting on particle i.

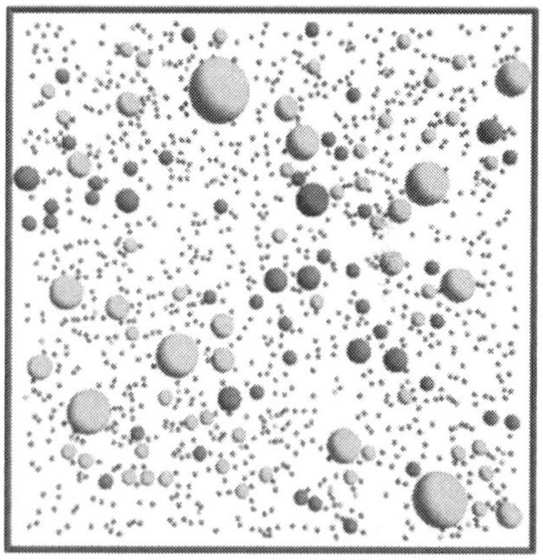

Figure 1. An example of the colliding particles snapshot from the DEM model simulations.

Triboelectric charging of polyethylene particles in our model is implemented as TEA. Inspired by the work of Lacks [3], we simply assume that electrons on a particle surface can be present only in two energy levels; that is a high-energy (HE) state and a low-energy (LE) state. In the beginning of our simulations, all particles are considered to have neutral charge and the same surface density of HE electrons (since it is a material property). We set the initial value of LE surface density to be zero, since LE electrons cannot be transferred. The HE electrons are transferred from a particle to another (or in both ways) during their mutual collisions. We assume that during such collision all the HE electrons present on the contact area of the two colliding particles are transferred according to equations (2) and (3):

$$n_{i,j}^{e} = \rho_i^{HE} A_c \quad (2)$$

$$n_{j,i}^{e} = \rho_j^{HE} A_c\,, \quad (3)$$

where $n_{i,j}^{e}$ and $n_{j,i}^{e}$ stand for the number of electrons transferred from the surface of particle i onto particle j and vice versa, respectively, ρ^{HE} refers to the number of HE electrons present on the surface of given particle. A_c represents the maximum collision contact area, i.e., contact area at the moment, when the distance between the colliding particles centroids reaches its minimum value. The numbers of HE and LE electrons on each particle are recorded in each time step, thus the particle charge at each time step can be calculated using equation (4):

$$Q_{i,t} = \left[(n_{i,0}^{HE} - n_{i,t}^{HE}) - (n_{i,t}^{LE} - n_{i,0}^{LE})\right]q, \quad (4)$$

where $Q_{i,t}$ is the charge of particle i in time t, $n_{i,0}^{HE}$ and $n_{i,t}^{HE}$ denotes the number of HE electrons present on the particle in the beginning of the simulation and in time t, respectively; and $n_{i,0}^{LE}$ and $n_{i,t}^{LE}$ are analogical variables related to LE electrons. In equation (4), q represents the charge of an electron. It is convenient to think of equation (4) as of a simple electron balance:

$$[charge] = [electrons\ donated] - [electrons\ accepted]. \quad (5)$$

Once all HE electrons initially present on the particle surface are depleted (transferred), the charging process and thus also the simulation is finished.

3.1. Model assumptions

In terms of contact mechanics, we assume perfectly elastic, perfectly spherical solid particles with negligible adhesion and no permanent particle deformation. We further assume that the electrons are uniformly distributed on the particle surface.

The charge transfer is considered to be much faster than the collision time of particles. We thus consider the charge accumulation to be in quasi-stationary state, i.e., we treat the charge affecting the Coulomb force in the sum of forces in Equation 1 as a constant. The charge transfer is instant (it takes one time step) and the absolute value of the charge transferred is only a little value, since only a little fraction of the HE electrons is transferred during one collision.

4. Results and discussion

4.1. The effect of particle size distribution on charging

We investigated the charging of spherical polyethylene particles in a spatially 2D box with implemented periodic boundary conditions instead of solid walls. We simulated collisions of particles of various particle size distributions. Although the width of the particle size distribution varied among the simulations, the total mass of the particles and the particle fraction remained constant. The particles were formally divided into several categories according to particle radius and each category contained the same mass of particles. Such a division allows us to deduce on the behavior of real systems like fluidized-bed polymerization reactors, where a constant mass of polyethylene particles is present. For the sake of simplicity, we present results obtained without HE electrons regeneration and without discharge mechanism.

In figure 2 we show the average particle charge at the end of the simulation as the function of particle radius. We can observe a clear trend – big particles tend to gain a positive charge and small particles the negative one. The low absolute value of the average charge for small particles (in comparison to the average charge value for big particles) is expected, because small particles outnumber big particles and the sum of all charges in the system must be inherently zero.

The evolution of the total absolute charge was studied in order to deduce the effect of particle size distribution on the charging dynamics (figure 3). The wider the particle size distribution is the more

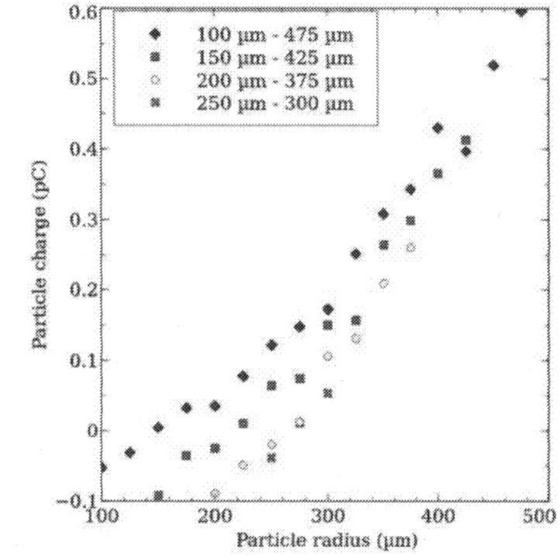

Figure 2. Average resultant particle charge dependency on the particle size for various particle size distributions.

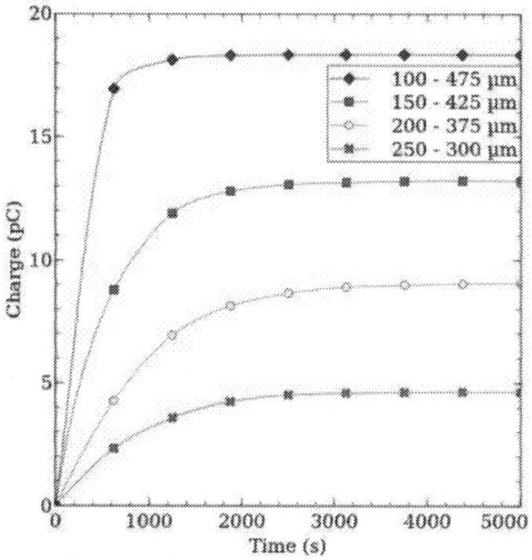

Figure 3. The evolution of the total absolute charge for various particle size distributions.

pronounced and faster is the charging. Also the absolute charge value related to the total surface area of particles in the system (surface charge density) increased if the particle size distribution was wider.

4.2. The effect of particle segregation on charging

In this section we discuss the influence of particle mixing/segregation on the triboelectric charging, since in fluidized-bed polymerization reactor, the particle segregation varies and depends mainly on the flow regime. Our simulations involved polyethylene particles of two particle sizes (1.0 mm and 0.6 mm in diameter) in 2D periodic box. The mass, particle number of each size and particle fraction of the system were constant, however, the level of segregation varied (figure 4). We again follow the time evolution of the sum of absolute particle charges. In figure 4 we show that in the system filled with initially well-mixed particles the charging process was roughly about 30% more pronounced. To explain this behavior, we can think of the system with segregated particles as of the combination of two quasi-isolated systems with a narrow particle size distribution. In case of well-mixed particles, we see the system of wider particle size distribution. Since we know that the charging is intensified in systems with a wide particle-size distribution (figure 3), we expect the system of well mixed particles to charge more than the one of segregated particles – which is observed.

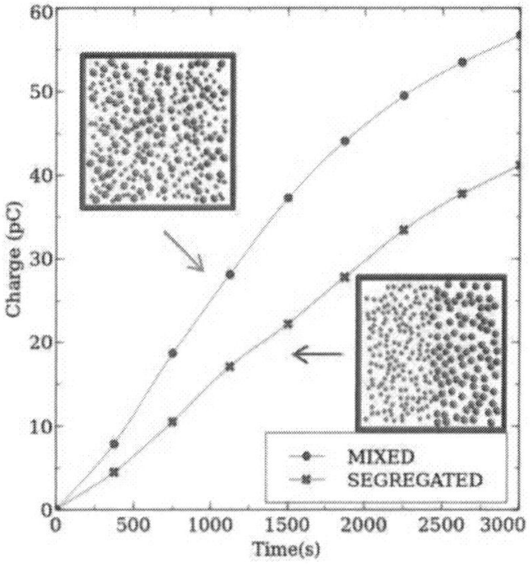

Figure 4. The evolution of the total absolute charge in the segregated and the mixed system, respectively.

5. Conclusions

Although triboelectric charging of polymer powder particles is a significant industrial problem, the phenomenon is regretfully only poorly understood. It is difficult to perform well-defined charging experiments with a good repeatability due to practical complications. As a consequence, mathematical models are unlikely to be quantitatively validated in a short-time perspective, however, such models can provide us with valuable trends and predictions. We present dynamic model of colliding polyethylene particles employing the Discrete Element Method (DEM) approach to describe the particle dynamics. We incorporate the 'Trapped Electron Approach' (TEA) as the physical basis for the charging mechanism. Our results indicate that the wider is the particle size distribution, the more pronounced is the charging process. This observation resonates well with the work of Lacks [3] and with the observation of some natural granular particles charging behavior [5]. We further predict that polymer particles significantly segregated in the system charge less than well mixed particles. We note that both the level of the particle segregation and the particle size distribution of polyethylene particles can be in practice adjusted by the choice of supported catalysts, by the conditions in the fluidized-bed polymerization reactor and by the present flow regime.

References

[1] Lowell J and Truscott W S 1968 *J. Phys. D* **19** 1281
[2] Kron A, Reitberger T, Stenberg B 1997 *Polym. Int.* **42** 131
[3] Lacks D J and Levandovsky A 2007 *J. Electrostat.* **65** 107-112
[4] Hertz H 1882 *Journal of Reine und Angewandte Mathematik* **92** 156-171
[5] Bo T-L, Zhang H, Zheng X-J 2014 *Sci Rep.* 4:5590

Journal of Physics: Conference Series **646** (2015) 012046 doi:10.1088/1742-6596/646/1/012046

Using electrostatic modelling to study cone discharges

W Azizi
Process Safety Specialist, Chilworth Technology Ltd,
Beta House, Southampton Science Park, Southampton SO16 7NS, United Kingdom

E-mail: wahid.azizi@dekra.com

Abstract. Cone discharges, also known as bulking brush discharges, can arise when charged insulating powder accumulates in a heap in silos. They can be an effective ignition source to relatively ignition sensitive powders and therefore represent a possible electrostatic hazard. The current international guidance on control of electrostatic hazards (IEC/TS 60079-32-1 [1]), endorses the usage of electrostatic modelling to estimate the electric field above the powder heap. *"Such model calculations should be based on the charge to mass ratio, bulk density and filling rate of the powder, the relative permittivity and resistivity of the bulked powder as well as the silo geometry."* This study shows a practical demonstration of this modelling technique. It also examines whether the shape of the heap affects the strength of the electric field above the powder heap, and thus the likelihood of cone discharges from occurring.

1. Introduction

Cone discharges are one of five types of electrostatic discharges found in process industries. They occur when highly charged powder lands on a heap of previously settled charged powder. The compression of charges of the same polarity under gravity may lead to a high electric field across the heap. If the breakdown voltage of air is exceeded then a discharge across the surface will occur.

The latest standard (IEC/TS 60079-32-1 [1]) endorses the use of electrostatic modelling to estimate the electric field when assessing the hazards of cone discharges. Electrostatic modelling entails solving the well-known Poisson differential equation [2]. The electric field is then compared with the breakdown voltage of air, taken to be 3 MVm^{-1}, and thus predict the presence of cone discharges. This study gives particular focus on how the shape of the heap affects the strength of the electric field.

2. Process overview, powder properties and shape of heap

The scenario considered is for an earthed metallic silo of 3 m in diameter and 8 m in height, filled up to around the 6 m level at a rate of 5 kgs^{-1}, of powder with which has charge to mass ratio of 1×10^{-6} Ckg^{-1}, relative permittivity of 2, volume resistivity of 5×10^{12} Ωm and bulk density of 500 kgm^{-3}. The charge relaxation time of the powder can be approximated by using Equation (1) [1]:

$$\tau = \varepsilon_0 \varepsilon_r \rho_v \qquad (1)$$

Where τ is the charge relaxation time, ρ_v is the powder resistivity, ε_0 is the Electric Constant and ε_r is the relative permittivity. Substituting the values considered for this scenario into Equation (1) yields a charge relaxation time of 88.5 s.

2.1. Heap Shapes

The heap shapes shown in figure 1 were considered for this study. The angles chosen for the inclined heap, conical heap and inverse conical heap allows a degree of comparability.

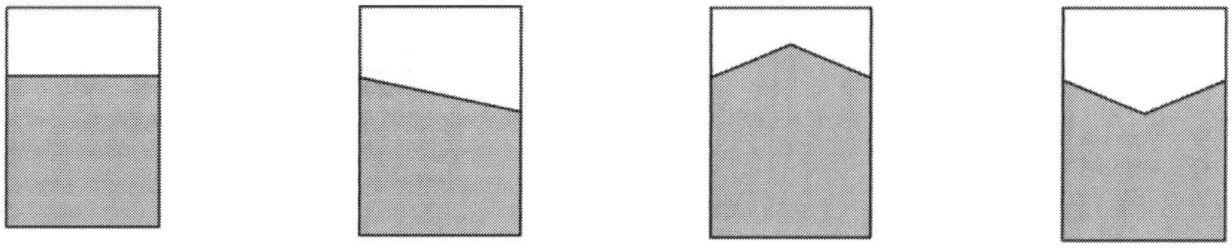

Figure 1a. **Figure 1b.** **Figure 1c.** **Figure 1d.**

Figure 1a. Flat heap, **Figure 1b.** Inclined heap with a repose angle of 30°. **Figure 1c.** Conical Heap with a 60°cone angle. **Figure 1d.** Inverse conical heap with a 300° cone angle.

3. Boundary Conditions

There are two boundary conditions when modelling the electric field in the silo:

1. The silo walls (top, bottom and shell) are at zero potential.
2. The total charge *quantity* in the powder. There are two ways of approaching this. Either, not taking the charge relaxation into account and this is a conservative approach. Or, as carried out in this study, taking into account the charge relaxation of the powder and this is a more realistic approach.

To obtain the total charge in the powder taking into account the charge relaxation, the following method was used. For an ideal material the charge relaxation time is the period when the powder charge will have dropped to 1/e, or about 0.37 of its initial value. For a true exponential decay this would be known as the time constant for the decay. The mass of powder flowing into the silo during this period (i.e. 88.5 seconds) is the product of the *mass flow rate* and the *time constant*, i.e. 442.5 kg. For a charge to mass ratio of 1 μCkg^{-1}, the total charge in this 442.5 kg of powder is the product of the *charge to mass ratio* and the *total mass* of charged powder, i.e. 0.000443 C in this situation.

Assuming a true exponential decay, more than 99% of the starting charge would be lost after 5 time constants, i.e. once a layer of powder has settled in the silo and been there for 5 time constants there is just less than 1% of the charge left which could thereafter be considered at earth potential. Utilising this fact, this study divided the charged powder at the surface into five distinct layers corresponding to the 5 time constants. Here, it is very important to correctly define the depth of each charged layer. The depth of the charged layer itself is dependent of the surface area of the heap. The assessment of how thick the depth will be for various heap shapes is omitted in this paper for the sake of brevity. However the results of the assessment are shown in table 1:

Table 1. Depth of charged powder for various heap shapes and the total charge for each layer.

Layer No. or Time Constant	Exponential decay in percentage (to 1 d.p.)	Flat Heap Powder Depth (m)	Inclined Heap Powder Depth (m)	Conical Top and Inverse Conical Top Powder Depth (m)	Charge ($\times 10^{-3}$ C)
1	36.8 %	0.125	0.0939	0.125	0.443
2	13.5 %	0.250	0.187	0.250	0.163
3	5.0 %	0.375	0.282	0.376	0.060
4	1.8 %	0.500	0.376	0.501	0.008
5	0.1 %	0.625	0.470	0.626	0.004

4. Simulation Results

The electric field was determined numerically by using a finite element software package [3].

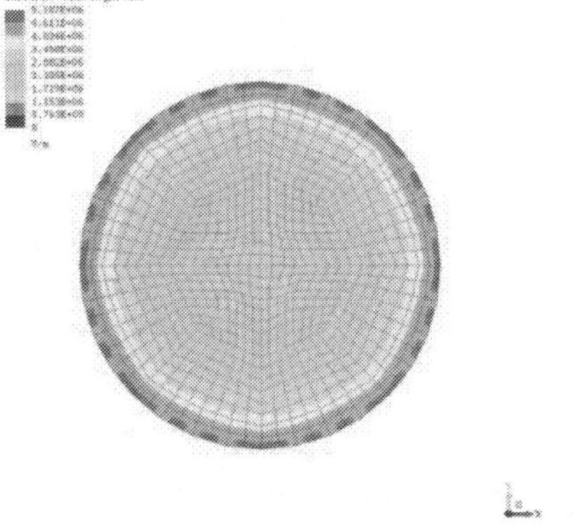

Figure 5a. Electric field magnitude at the surface of the flat heap. The maximum electric field is 5.19 MVm^{-1}.

Figure 5b. Electric field magnitude at the upper part of the silo. Half of the space above the powder surface has been cut away to show regions of interest.

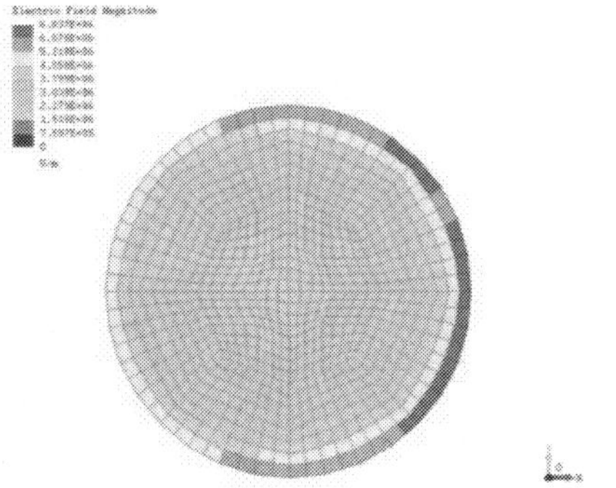

Figure 6a. Electric field magnitude at the surface of an inclined heap. The maximum electric field is 6.84 MVm^{-1}.

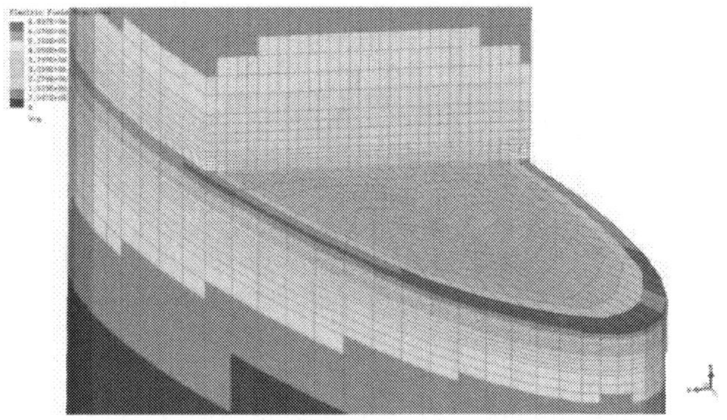

Figure 6b. Electric field magnitude at the upper end of the silo. Half of the space above the powder surface has been cut away to show regions of interest.

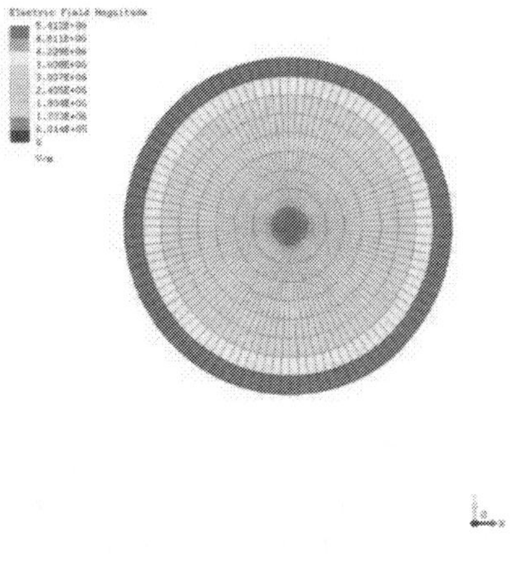

Figure 7a. Electric field magnitude at the surface of a conical heap. The maximum electric field is 5.41 MVm^{-1}.

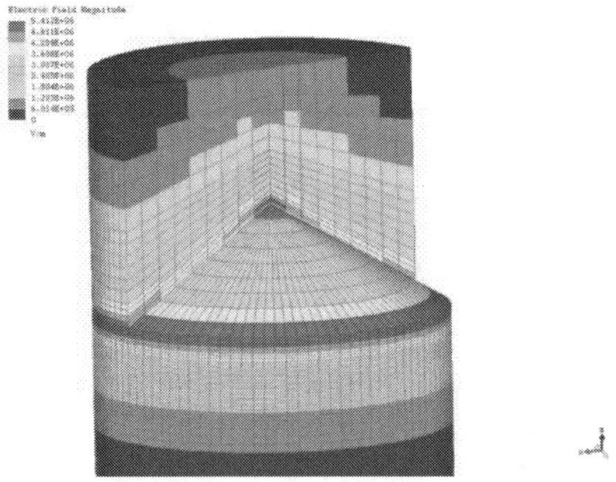

Figure 7b. Electric field magnitude at the upper end of the silo. Half of the space above the powder surface has been cut away to show regions of interest.

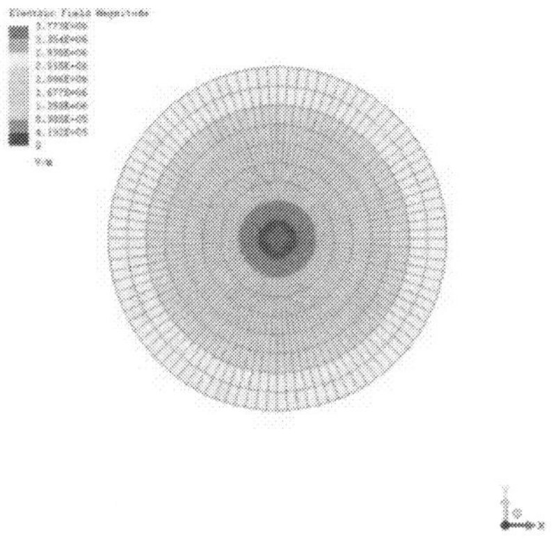

Figure 8a. Electric field magnitude at the surface of an inverse conical heap. The maximum electric field is 2.93 MVm^{-1}.

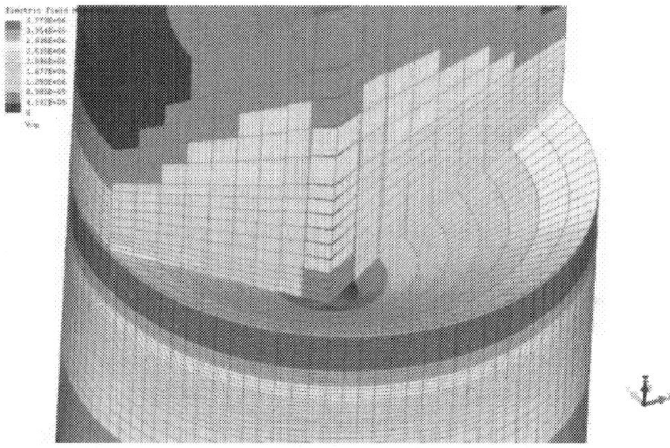

Figure 8b. Electric field magnitude at the upper end of the silo. Half of the space above the powder surface has been cut away to show regions of interest.

5. Discussion and Conclusions

The majority of the silo is at zero electric field. This is because the powder charge relaxation has been taken into account, thus the electric field is concentrated around the top part of the silo where charged powder is concentrated. Out of the four shapes that were considered, the maximum electric field is experienced by an inclined heap, followed by a conical heap, followed by a flat heap and the last is the inverse conical heap. Just to show the significance of this difference; the maximum electric field at the powder surface for the inclined heap is more than twice than the maximum electric field at the powder surface for the inverse conical heap. The simulation has shown that the breakdown strength of air is exceeded in all scenarios except for the inverse conical heap.

The regions of the highest electric field would be where electrostatic discharges should be expected and are therefore important enough to consider in more detail. The precise regions of the highest electric fields vary for the four heap shapes considered. The highest electric field for the flat heap is just below the surface, although the maximum electric field at the surface is not too dissimilar to that just below the surface. The highest electric field for the inclined heap is just below the lower side of the incline. The highest electric fields for the conical heap are at the edge of the heap surface, just below the edge, and just above the cone apex. The highest electric field for the inverse conical heap is below the edge of the powder surface.

Where the highest electric fields occur at the surface or above the surface, then hazardous discharges should be expected. However, if the highest electric fields are just below the surface, then it is very likely that that region will be fuel rich and thus represent a much lower likelihood of causing an explosion.

Given this, it can then be concluded that the safest way to fill a silo is to fill it equally around the circumference such that an inverse conical heap is obtained, and the least safest option would be filling the silo at one side such than an inclined heap is generated. The relationship between the heap shape and the electrical field is worthy of further investigation. The results from this study may be exploited to design silos so as to avoid the problem of cone discharges.

6. References

[1] IEC/TS 60079-32-1:2013 *Explosive Atmospheres – Part 32-1: Electrostatic hazards, Guidance*
[2] Britton L 1999 *Avoiding Static Ignition Hazards in Chemical Operations* (New York: Wiley-AIChE)
[3] Mecway Ltd 2013 *Mecway version 1.1* (New Zealand)

Numerical calculation of dielectrophoretic and electrostatic forces acting on micro-scale particles.

M Praeger, Z Li, J M Smallwood and P L Lewin
Tony Davies High Voltage Laboratory, University of Southampton, UK.
E-mail: mattp@soton.ac.uk

Abstract. Much of the current literature on dielectrophoresis (DEP) relates to micro or nano scale particles; typically in micro-fluidic type experiment geometries. In contrast, this work focusses on the application of DEP forces to larger, micro-scale particles in air. Since DEP scales with particle volume, it can apply a significant force on surprisingly large objects. When using very small particles it is often sufficient to use Pohl's method [1] whereby the particle is considered to be spherical and where it does not interact with the externally applied electric field. For the larger particles used in this work, the spherical approximation does not necessarily hold. DEP forces are therefore calculated using the finite element method (FEM) which permits the use of arbitrary particle shapes. In this model the electric field is solved in the presence of a polarizable particle, the DEP force is then calculated using the Maxwell stress tensor method [2]. The development of this model allows the investigation of the DEP forces acting on non-spherical particles for a specific experimental electrode geometry.

1. Introduction

Since the work of Pohl [1], beginning in the early 1950's, the field of DEP has become increasingly sophisticated. Today it is possible, for example, to use a.c. DEP in a microfluidic chip to sort biological cells by their dielectric spectroscopy properties [3]. Much of the current literature [4] is devoted to the optimum shape of test electrodes [5], to the rotational torques exerted on microparticles [6] and to other subtleties. Despite this, the possibility of making use of DEP simply as an attractive force that may be used for practical purposes also merits consideration. As such, DEP has great potential for exploitation, especially in particlulate based applications. For example, in particle transport, grading by size or dielectric properties or enhanced filtering with the possibility of high yield recovery of trapped particulates. The work presented here has been carried out as part of the SPABRINK project (Self-Printing Advertising Board with Reusable Ink) which aims to apply DEP forces as a means of temporarily fixing micro-scale particles to a surface in order to display an image.

2. Initial verification of the FEM model

Before employing the FEM calculation in more complex geometries it is first necessary to verify that in a simple test model the forces calculated by FEM are in agreement with those obtained using Pohl's method. The hypothetical case of a spherical capacitor with vacuum as the insulating medium is therefore considered. This system consists of a solid perfectly-conducting sphere concentric within a thin, perfectly-conducting spherical shell (see Figure 1); the model parameters are given in table 1. The advantage of this simplified case is that (when no particle is present) it allows the use of an analytical expression for the electric field (directly from Gauss's Law) which is simple to differentiate.

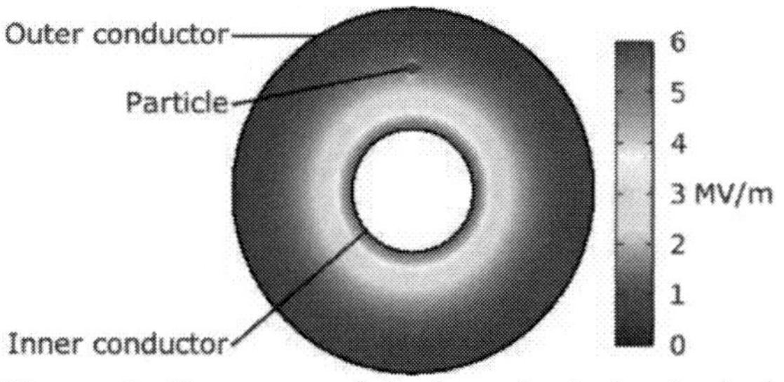

Figure 1. Geometry of the hypothetical spherical capacitor showing E-field strength from FEM.

Table 1. FEM model parameters:

Model Parameter	Value
Inner radius (r_i)	250 μm
Outer radius (r_o)	750 μm
Particle radius (a)	25 μm
Particle radial position (r)	500 μm
Voltage (r_i to r_o)	1000 V
Insulator permittivity	ε_0
Particle permittivity	2.5 x ε_0

Performing this differentiation, Pohl's method gives equation (1) for the DEP force (F) acting on a spherical particle (of radius a, and relative permittivity ε_p) inside the hypothetical spherical capacitor.

$$F = 2\pi a^3 \varepsilon_m \frac{(\varepsilon_p - \varepsilon_m)}{(\varepsilon_p + 2\varepsilon_m)} \frac{-Q^2}{4\pi^2 \varepsilon_0{}^2 r^5}, \qquad Q = \frac{4\pi\varepsilon_0 V}{\left[\frac{1}{r_i} - \frac{1}{r_o}\right]} \qquad (1\ \&\ 2)$$

Here ε_m is the relative permittivity of the insulator, Q is the total charge on the inner sphere and r is the radial distance of the particle from the mutual centre of the two spheres. For comparison with experiment it is often more convenient to consider the applied voltage (V) rather than Q. This can be achieved using equation (2), where r_i and r_o are the radii of the inner and outer conductors respectively. Equations (1 & 2) allow calculation (via Pohl's method) of the DEP force acting on a spherical particle within the hypothetical, spherical capacitor (see figure 2).

A 3d FEM model of the hypothetical, spherical capacitor was constructed in COMSOL (figure 1 shows a 2D slice of the E-field). In COMSOL, the DEP force is calculated by integrating the Maxwell stress tensor $\boldsymbol{T}$ over the surface of the particle [2]; equation (3) defines $\boldsymbol{T}$ in matrix notation.

$$\boldsymbol{n}_1 \boldsymbol{T}_2 = \frac{1}{2} \boldsymbol{n}_1 (\boldsymbol{E} \cdot \boldsymbol{D}) + (\boldsymbol{n}_1 \cdot \boldsymbol{E}) \boldsymbol{D}^T \qquad (3)$$

Where $\boldsymbol{n}_1$ is the outward normal on the surface of the particle, $\boldsymbol{E}$ is the electric field and $\boldsymbol{D}$ is the electric displacement. The subscripts denotes the domain of evaluation, i.e. $\boldsymbol{T}_2$ is evaluated in the media surrounding the particle. The superscript ^{T} denotes the matrix transpose, which makes the final term a tensor product. The markers in figure 2 show the results obtained using the 3d FEM simulation; clearly there is good agreement with Pohl's analytical expression.

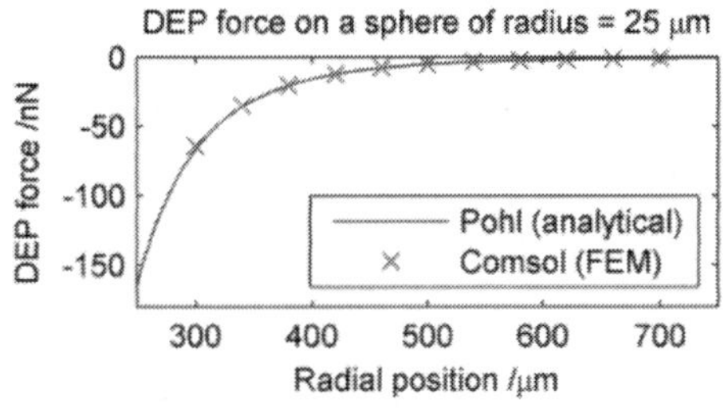

Figure 2. DEP force acting on a 25 μm (radius) particle with relative permittivity of 2.5; plotted as a function of its radial position within a spherical capacitor. The solid line is calculated using Pohl's method via equation (1). The markers show the DEP force as calculated using the FEM model. The DEP force is negative as it acts inward towards the centre of the spherical capacitor.

3. Beyond Pohl's method

The advantage of using the finite element approach is that it allows calculation of the forces acting on arbitrarily shaped particles. To test this, the spherical particle was replaced with an ellipsoid with semi-axes r_x, r_y and r_z. In all calculations presented here $r_x = r_y$, the aspect ratio of the ellipsoid is therefore defined as R = r_z/r_x. Whilst the aspect ratio R is varied, r_x, r_y and r_z will be chosen in order to keep the volume of the ellipsoid equal to that of the spherical particle used previously. Varying R between 1/3 and 3 changes the ellipsoid from oblate to prolate respectively as shown in figure 3a. It is possible to vary the rotation of the ellipsoid as illustrated in figure 3b. Alternatively the position of the ellipsoid can be defined relative to its lowest point (which will be nearest to the charged electrode).

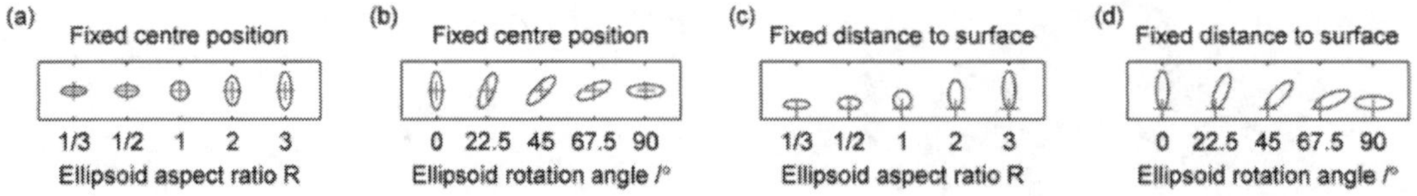

Figure 3. Diagrams showing the effect of ellipsoid parameters. The ellipsoid centre position is fixed whilst in (a) the aspect ratio R is varied and in (b) the ellipsoid is rotated. Figure parts (c) and (d) are similar except that the ellipsoid positions are defined relative to their lowest point.

The FEM model outputs x, y and z components of the DEP force and can also be configured to give rotational torques. In contrast Pohl's method only gives the translational force acting on the centre of mass of the particle. The strongest force (and the one of most interest for the display screen application) is the z-component of the DEP force (towards the centre of the capacitor). This is plotted as a function of ellipsoid aspect ratio and rotation angle in figure 4.

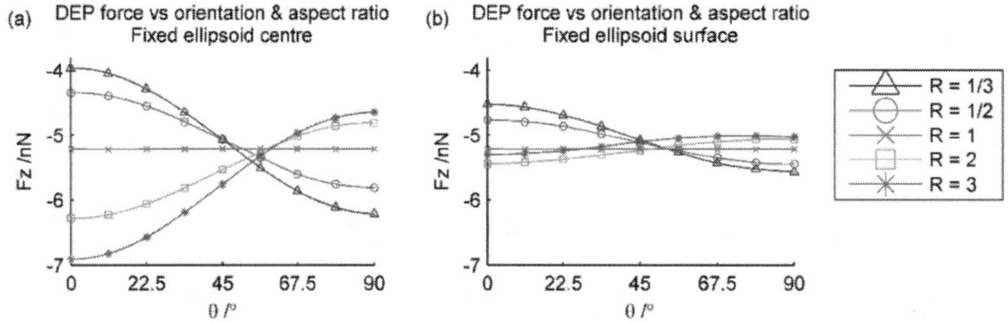

Figure 4. The DEP force acting on an ellipsoid particle within a spherical capacitor, as a function of ellipsoid aspect ratio and rotation angle. (a) When the centre of the ellipse is fixed 500 μm from the centre of the capacitor. (b) When the nearest point of the ellipse is always 475 μm from the centre.

When the centre of the ellipsoid is fixed (figure 4a) it is the prolate spheroid that experiences the strongest DEP force. This occurs when the tip of the ellipsoid comes into closest proximity to the charged inner electrode surface of the spherical capacitor. It is interesting to note that the ellipsoid has to rotate by more than 45° before the oblate shape produces the strongest DEP force.

Intuitively one might expect the prolate ellipsoid to experience the strongest E-field enhancement at its tip and therefore to generate the largest DEP force; however, this is not the case. When the distance between the ellipsoid surface and the charged inner surface of the capacitor is kept constant (figure 4b), it is the oblate spheroid that experiences the strongest DEP force. This is presumably because the rim of the oblate spheroid is narrower than the tip of the prolate spheroid when their volumes are equal.

4. Application to experimental geometry

The FEM approach allows calculation of the DEP forces for non-spherical particles in complex experimental geometries. Figure 5 shows the 3d electric field calculation for an ellipsoid particle in the vicinity of two strip electrodes. This model simulates the experiment geometry used by the display foil application; the interdigitated electrodes produce a divergent electric field near to the surface of the foil.

Figure 6 shows the DEP force acting on the ellipsoid particle as a function of its geometric parameters. The angle Θ is between the unique axis of the ellipsoid and the vertical (z) direction. When $\Theta = 90°$ and $\phi = 0°$ the unique axis of the ellipsoid is aligned across the electrode gap; the strongest DEP force occurs for the prolate spheroid when it is in this orientation.

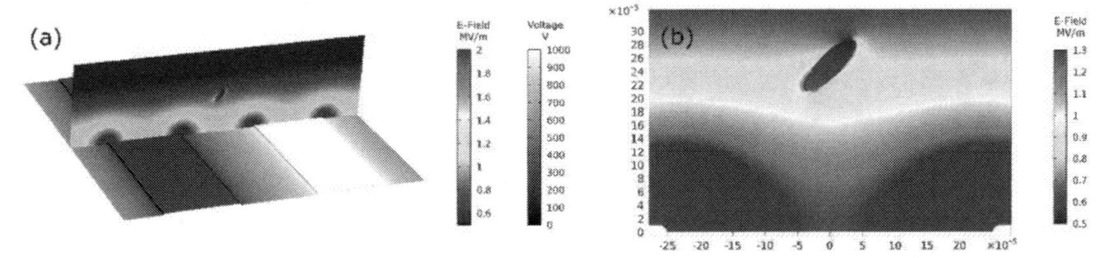

Figure 5. (a) Example of the 3d FEM model used to calculate DEP forces. In the horizontal plane the colour map of the electrodes indicates the voltage. In the vertical plane the colour map indicates the electric field strength. (b) Enlarged view showing the electric field near to the ellipsoidal particle. Particle aspect ratio R=3, and rotation angle is 45°.

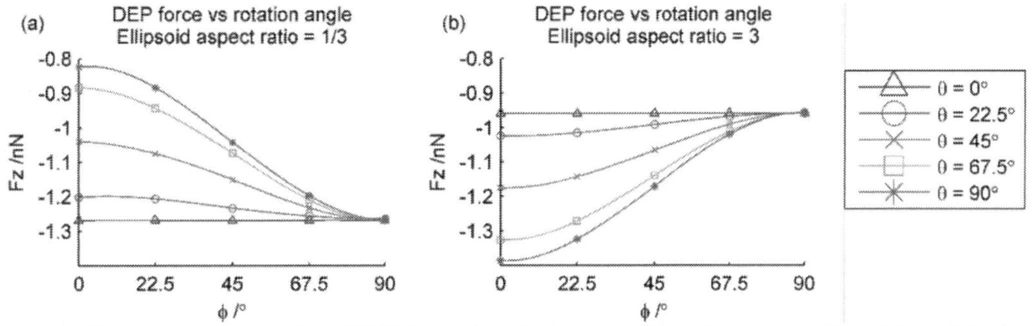

Figure 6. The z component of the DEP force (-ve is downward) acting on an ellipsoid particle in the vicinity of two strip electrodes. (a) Forces on an oblate spheroid. (b) Forces on a prolate spheroid.

5. Conclusions

In the SPABRINK display screen application the particles are made from polymeric materials such as PET ($\varepsilon_r \approx 2.5$). If a density of 1380 $kg.m^{-3}$ is assumed for PET then the weight of the particle produces a force of 0.886 nN. In the 3d model the particle is positioned 250 μm from electrode plane. If the electrodes were inverted so that gravity pulls the particle away from the display surface then the oblate spheroid would fall away from the electrodes when in certain orientations. In all orientations the prolate spheroid would be attracted towards the electrode and will be held in situ by the DEP forces.

This paper has demonstrated the use of FEM to calculate DEP forces in complex physical geometries. In the future this method will be extended to study the forces acting on (and motion of) groups of multiple, irregularly shaped particles; electrostatic charging of these particles could also be incorporated. In this manner FEM could facilitate the engineering of DEP based solutions to a range of practical problems; for example, in filtering and in particulate based industrial processes.

Acknowledgments
This project has received funding from the European Union's Seventh Framework Programme for research, technological development and demonstration under grant agreement no 605299.

[1] Pohl H A 1978 *Dielectrophoresis: The Behavior of Neutral Matter in Nonuniform Electric Fields* (Cambridge: Cambridge University Press)
[2] Wang X J, Wang X B, and Gascoyne P R C 1997 *J. Electrostat* **39** 277-95
[3] Morgan H, Sun T, Holmes D, Gawad S, and Green N G 2007 *J. Phys.* D. **40** 61-70
[4] Pethig R 2010 *Biomicrofluidics* **4** 022811
[5] Ramos A, Morgan H, Green N G, and Castellanos A 1998 *J. Phys.* D. **31**, 2338-2353
[6] Jones T B 2003 *{IEEE} Eng. Med. Biol. Mag.* **22** 33-42

Modelling material dependent parameters of layer type straight coils for fast transient pulses

Z A Tamus, T Orosz, and G M Kiss

Department of Electrical Power Engineering, Budapest University of Technology and Economics, Egry J. u. 18, H-1111, Budapest, Hungary

E-mail: tamus.adam@vet.bme.hu

Abstract. The behavior of coils in case of fast transient pulses is different from that they show at low frequencies. If the dimensions of the coil, i.e. mainly the length of the winding is much shorter than the wavelength of the signal on the coil, a lumped element model can be effectively used taking the capacitances of windings into consideration. In this study a different type of straight, layered coil have been investigated in order to determine parameters of a lumped circuit model of the windings. The frequency dependent parameters are modeled by analytical and finite element calculations and the results are compared to the results of measurements on coils. The finite element method can improve the accuracy of parameter estimation.

1. Introduction

Modelling of coils in case of transient pulses has high importance because transient overvoltages can cause non-uniform voltage distribution between turns, which overloads the turn to turn insulation in electrical equipments e.g. transformers. For measuring high frequency currents and current pulses of partial discharges occurring in insulations simple air-core (Rogowski) coils are used as a current transducer, whose bandwidth is limited by the capacitance of windings [1, 2]. The behavior of coils in case of fast transient pulses is different from their low frequency behavior. While at low frequencies the coil can be modeled as an inductance and a resistance (simple L-R model), at higher frequencies this model have to be improved. If the dimensions of the coil, i.e. mainly the length of the wire is much shorter than the wavelength of the signal on the coil, a lumped element model can also be used to model the coil, however this simple model have to be complemented by a lumped capacitor, which is the result of the stray capacitances. If the wavelength of the signal is shorter than the length of the coil, distributed model has to be used to investigate the behavior of the coil. Treating the lumped parameter model seems much easier than the distributed one, nevertheless the calculation of model parameters is suffer from more difficulty. There are several analytical forms to calculate the inductance of straight coils but the results provided by these formulas are strongly dependent on the shape of the coil and the skin-effect cannot be negligible at the calculation of the inductance. Similarly, the calculation of stray capacitances by analytical method shows high variations in the result depending upon the used formulas. In this study, a single- and multi-layer straight coils are modeled by analytical and finite element methods (FEM) and the results are compared to the results of the measurements.

2. Model parameters

The coil is modeled with a lumped parameter model, which consists of a parallel inductance and capacitance. The resistance of the coil is not calculated because it has no effect on the values of either the capacitance or the inductance. The tested coils are straight single- and multi-layered coil. The diameter (D) of the coil is 22 mm, the height (h) is 17 mm. The insulated diameter (d_{ins}) of the wire is 0.65 mm and the diameter of the conductor (d) is 0.63 mm.

2.1. Turn capacitances

In the analytical approach the turn capacitance of two consecutive tuns is modeled by the capacitance of two parallel wires far from ground potential [1]. The turn capacitance can be expressed:

$$C_t = \epsilon \cdot \pi \frac{D \cdot \pi}{ln \frac{\sqrt{d_{ins}^2 - d^2} + (d_{ins} - d)}{\sqrt{d_{ins}^2 - d^2} - (d_{ins} - d)}}, \tag{1}$$

where d_{ins} is the insulated diameter of the wire, d is the diameter of the conductor, and D is the mean diameter of a given turn. In this equation (1) the fringe effects and related stray capacitances are not taken into consideration. Practically, these effects can be estimated by the increased d_{ins} distance [3] [4]. More general and accurate solution can be achieved by a finite element technique (FEM) from an electrostatic field simulation. The simulation has been made by Agros2D [5]. The turn and layer capacitances can be calculated from the sum of the electrostatic energy:

$$W_m = \frac{1}{2} \cdot \sum C_i \cdot V_i^2 \tag{2}$$

To obtain the turn capacitances C_i, one calculation is enough since the difference between the turn capacitances along a layer can be negligible. The inner i^{th} turn potential is set to $V_i = 1$ V, all the others to 0 V, and the turn capacitance is obtained from (2). The field distribution of the simulation are in Figure 1.

2.2. Layer capacitances

To obtain the layer capacitances, the most common technique is used: the layers are replaced to two fictitious parallel plates [3] [4]:

$$C_l = \epsilon \cdot l \cdot \frac{h + 2 \cdot (d_{ins} - d)}{d_{ins} - d}, \tag{3}$$

where h is the height of the layer, $d_{ins} - d$ is the assumed distance between the two parallel layers and l is the length of the coil. The height of the winding is increased considering the fringe effects [3].

2.3. Equivalent capacitance

The equivalent capacitance is calculated in the following way [1, 4, 6]:

$$C_{eq} = \frac{C_t}{N_l \cdot N_t^2} \cdot (N_t - 1) + \frac{4 \cdot (N_l - 1)}{N_l^2} \cdot C_l, \tag{4}$$

where C_{eq} is the equivalent capacitance of the winding, N_l means the layer number, N_t is the turn number.

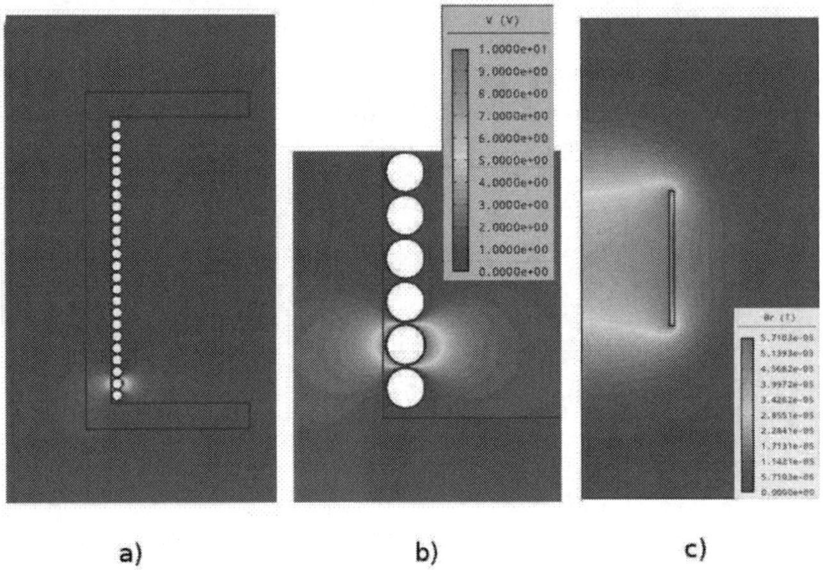

Figure 1. Finite element models of single-layered test coil, the a) and b) images for the turn capacitance calculation, and modeling of the inductance is shown in the c) image.

2.4. Inductance calculation

The inductance of the coil at low frequencies is calculated by the following analytical formula:

$$L = F^2 \cdot N \cdot R, \tag{5}$$

where N is the turn number, R is the mean radius of the coil and F is an empirical correction factor which depends on the ratio of the coil height and mean diameter [7] [8].

3. Measurement of parameters of coils

Measuring of coil parameters have been based on the determination of the resonance frequency of coils. From the resonance frequency the value of lumped capacitance can be calculated if the inductance is the coil is known. The inductance of the coil is measured at a relative low frequency (75 kHz), when the effect of capacitances can be negligible. The measurements have been carried out by a HP4285A precision impedance analyzer. Table 1. shows the results of measurements and the calculated capacitance values.

4. Discussion of the results

The Table 2. contains the result of calculations. Analytical and FEM calculations of the inductance show a good agreement with measured results; however, the difference of analytical calculations from the measurement is higher. The results of turn capacitance calculations are lower than the measured value in both cases, but the FEM calculation provides better results. The explanation is that the turn to turn capacitance in analytical calculations is simplified to the capacitance of two parallel wires [1], therefore this approach neglects the stray capacitances to the farther turns. Comparison of the results of FEM calculations to measurements the values are also lower. However, this accuracy can be enough for designing in most common cases (e.g. Rogowski-coil current transducer), the accuracy can be improved if not only one turn capacitance is calculated in a given layer and the layer to layer capacitances are also calculated.

Table 1. The measured inductance and resonance frequency parameters and the calculated capacitances of the test coils.

Number of layers	Inductance [mH]	Resonance frequency [Mhz]	Capacitance [pF]
1	11.75	30	2
2	99.5	1.75	83
3	169	1.572	61
4	255	1.335	56

Table 2. The calculated parameters of the test coils with analytical and finite element method.

Number of layers	L_{eq} FEM [mH]	L_{eq} Analytical [mH]	C_t FEM [pF]	C_t Analytical [pF]	C_{eq} FEM [pF]	C_{eq} Analytical [pF]
1	11.3	9.96	41.3	15.3	1.65	0.64
2	98.3	81.2	41.3	15.3	64.3	17.3
3	166.9	165.8	41.3	15.3	57.0	15.3
4	258.4	246.6	41.3	15.3	48.1	12.9

5. Conclusion

The high frequency behavior of coils is important in case of fast transient pulses. In this study, straight, single- and multi-layer, air-core coils were modeled by analytical and FEM calculations and the results were compared to measurements on test coils. The results shows the simple analytical formulas do not provide correct results especially at the capacitance calculations. The results of the FEM model are much closer to the measured ones; however, if more precise capacitance values are required for a given application. The improvement of the model is necessary.

References

[1] Orosz T, Tamus Z A and Vajda I 2014 Modeling the high frequency behavior of the rogowski-coil passive l/r integrator current transducer with analytical and finite element method *Power Engineering Conference (UPEC), 2014 49th International Universities* (IEEE) pp 1–4

[2] Polik Z and Kuczmann M 2008 *Przeglad Elektrotechniczny* **12** 230–233

[3] Martinez-Velasco J A 2009 *Power system transients: parameter determination* (CRC press)

[4] Kulkarni S V and Khaparde S 2004 *Transformer engineering: design and practice* vol 25 (CRC Press)

[5] Karban P, Mach F, Kus P, Panek D and Dolezel I 2013 *Computing* **95** 381–408

[6] Massarini A and Kazimierczuk M K 1997 *Power Electronics, IEEE Transactions on* **12** 671–676

[7] Grover F W 2009 *Inductance calculations* (Dover)

[8] Terman F E 1943 *Radio engineer's handbook* (McGrow Hill)

Numerical study of Electrical double layer development: Analysis of the charge genesis.

P Leblanc[1], **J M Cabaleiro**[2], **T Paillat**[1]
[1]Institut P' (CNRS UPR3346 – University of Poitiers – ENSMA) SP2MI – Téléport 2 – 11 Boulevard Marie et Pierre Curie, 86962, France.
[2]CONICET Fluid dynamics Laboratory, Faculty of Engineering, University of Buenos Aires, Av. Paseo Colón 850, Argentina

Abstract. This work presents a numerical simulation of the Electric Double Layer (EDL) development process at a solid/liquid interface for adsorption and corrosion models. First, the study is conducted for static EDL development (without liquid flow) until it reaches a static equilibrium. Afterwards, the EDL is perturbed by a laminar liquid flow leading to flow electrification phenomena (dynamic study). The charge conservation equations of the liquid species have been implemented in an industrial code. A parametric study was performed to consider different chemical reaction scenarios and different models.

1. Introduction

In dielectric liquids, impurities are assumed to be at the origin of the Electric Double Layer (EDL) present at the solid-liquid interface. Due to a physicochemical phenomenon, which is still not well understood, these impurities react with the solid to electrically polarize the interface. One type of charge is accumulated in the solid surface if this one is a dielectric while the opposite charge is distributed in the liquid. Different models, adsorption or corrosion models, had been proposed to help understanding the phenomenon. Different numeric simulations were realized according to each of these two models [1-2]. Unfortunately, all the different simulations have not permitted to confront directly these two models. Using finite volume CFD tool Code-Saturne®, we have developed software that allows simulation, from the same entry data, of both the corrosion and the adsorption models. This paper is focused on a comparative study between the adsorption and the corrosion models on the dynamics of the EDL development versus time without liquid flow (static development) and with flow (dynamic development). The input data will be extracted from literature.

2. Physicochemical Adsorption and Corrosion Models

The two models suppose that in the liquid charge impurities are dissociated in two ionic species, positive and negative. Assuming that only one type of neutral impurity $A_L B_L$ is present in the liquid, the chemical reaction is given according to equation (1):

$$A_L B_L \underset{k_r^L}{\overset{k_d^L}{\rightleftharpoons}} A_L^+ + B_L^- \quad (1)$$

where k_d^L and k_r^L are the kinetic constants of the dissociation reaction. Initially, the liquid is electrically neutral when the contact with the solid medium occurs. Thus, liquid ionic species could react with the solid according to the adsorption or corrosion models (figure 1). The two schemes suppose that the solid is polarized negatively and the liquid positively. This is the major situation observed in experiments.

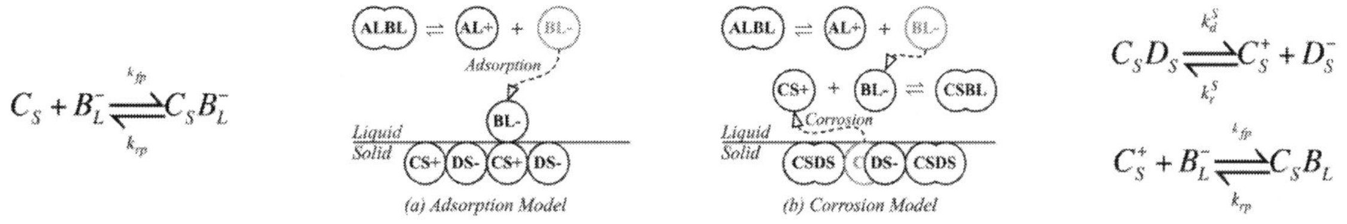

Figure 1. Physicochemical mechanism occurring at the solid-liquid interface for adsorption and corrosion models

All the chemical species (positive, negative and neutral) are transported in the liquid by diffusion, migration and convection (when the liquid flow is induced). The flux transport equations are coupled to the Poisson's equation and the Navier-Stokes equation. To compare the models of adsorption and corrosion, all equations were nondimensionalized according to the Buckingham's theorem. All the dimensionless values were marked with "*" symbol. The simulations carried out with these two models are analysed and compared in this work according to the dimensionless distance z* (z*=z/a with a is half-height) and dimensionless time t* (t*=t/τ_c with τ_c is the Maxwell). Simulations are decomposed into two steps:

- **Step 1 -Static development of the EDL:** without fluid motion, the system reaches an equilibrium depending on the different chemical kinetic constants. When the equilibrium is reached, the electrical current through the interface becomes negligible.
- **Step 2 - Dynamic development of the EDL:** from this static equilibrium, a Poiseuille flow is applied at the entry and the impact of the flow is studied.

The initial concentration of positive and negative ions n_0, (A_L^+, B_L^-) is estimated from the liquid conductivity ($5\times10^{-12} S.m^{-1}$). Thus, the different chemical reactions of ionized impurity in the liquid can be determined considering the space charge (assumed to be initially zero). The main difference between the two models remains in the fact that the corrosion model is based on three successive different chemical reactions while the adsorption model is only based on two reactions. Thus, the initial species concentrations of the solid and liquid, their properties, the chemical kinetics constants are assumed equal for the simulation in both corrosion and adsorption models; the initial data values are summarized in table 1. For both models, the solid reaction is specific. It consists of an adsorption reaction for the adsorption model ($C_S B_L^-$) while it is a corroding reaction for the corrosion model ($C_S D_S$ + $C_S^- B_L^-$). To analyse the results and favour the comparison, we assumed that the chemical reaction between $C_S B_L^-$ and $C_S^- B_L^-$ are equal (k_{fp}, k_{rp}). Solid species are supposed not to limit all the different reactions. Other kinetics constants were chosen according to Touchard's approach [3].

3. Numerical results

3.1. Comparison between adsorption and corrosion models at the Static Equilibrium.

In the goal to compare the dynamics of the different simulation models, we imposed for both simulations the same space charge density at the wall for a fully developed EDL (ρ_{wd}). The major entry data are taken equal for the two models. Then the kinetic constants of the dissociation reaction of the wall corrosion k_d^S, is chosen to reach a value of ρ_{wd} equal to the value simulated with the adsorption model. Therefore dimensionless space charge density ρ_{wd}^* will be equal to 1 in both models. Figure 2 shows the dimensionless space charge distribution in the liquid for a fully developed EDL. Adsorption and corrosion simulations are well superimposed. A small distinction is observed away from the interface. The corrosion model admits the presence of two kinds of impurities and, unlike the adsorption model. It seems that this difference leads to a different distribution of space charge density between both models. Another important difference is observed on figure 3, the dynamics time evolution of the space charge near the interface in the liquid (ρ_{wd}) is plotted. The corrosion model seems about 50% faster. As the corrosion model consists of more successive events, three chemical reactions instead of two in the adsorption model, it could be expected that the equilibrium state would be reached within a longer period of time. However, the opposite effect is observed. The presence of an additional positive charge, C_S^+, in supplement to the positive ions A_L^+, in the liquid, leads to a faster development of the space charge density in the liquid.

Table 1. Physical constants; Initial values of reaction rates.

Constant	Symbol	Value
Initial ions concentration	n_0	$1.582\times10^{-16}molecule.m^{-3}$
Initial neutral ions concentration	n_{0i}	$1.448\times10^{-18}molecule.m^{-3}$
Recombination rate in the liquid	k_d^L	$1.65\times10^{-17}m^3s^{-1}$
Dissociation rate in the liquid	k_r^L	$2.85\times10^{-3}s^{-1}$
Recombination rate in the solid	$k_d^{L'}$	$9.05\times10^{-19}m^3s^{-1}$
Dissociation rate in the solid	$k_r^{L'}$	$1.57\times10^{-4}s^{-1}$

Table 2. Chemical kinetic constants for Adsorption and Corrosion models

Constant	Adsorption model		Corrosion model	
	Liquid	Solid (interface)	Liquid	Solid (interface)
k_{fp}	No reaction	$1.28\times10^{-26}m^3s^{-1}$	$2.324\times10^{-25}m^3s^{-1}$	$1.28\times10^{-26}m^3s^{-1}$
k_{rp}	No reaction	$1.28\times10^{-3}s^{-1}$	$2.324\times10^{-2}s^{-1}$	$1.28\times10^{-3}s^{-1}$
k_d^S	No reaction	No reaction	No reaction	$4.25\times10^{-11}s^{-1}$
k_r^S	No reaction	No reaction	No reaction	$2.81\times10^{-18}m^3s^{-1}$

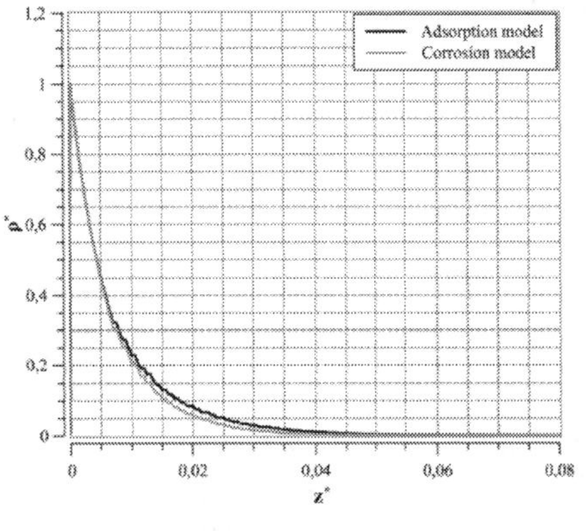

Figure 2. Dimensionless space charge density distribution in the liquid from the interface to the liquid bulk direction for the adsorption and corrosion simulations

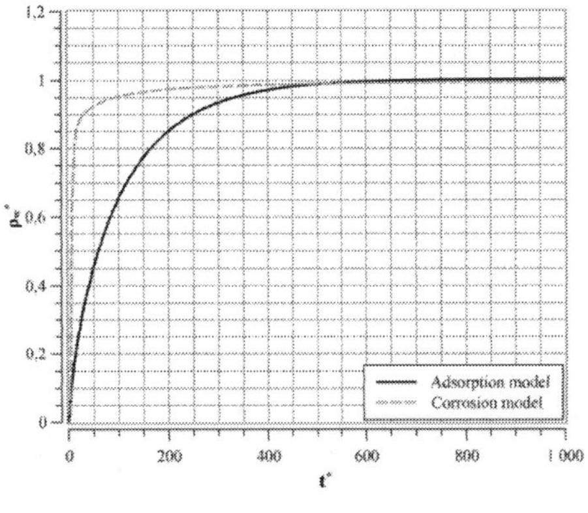

Figure 3. Dimensionless space charge density at the interface versus dimensionless time for the adsorption and corrosion simulations

3.2. Comparison between adsorption and corrosion models at the Dynamic Equilibrium.

Regardless of corrosion or adsorption models, the streaming current shows the same evolution at the different flow velocities (figure 4). However, at the beginning of the flow, the adsorption model leads to a peak current magnitude higher than corrosion model. A reverse behaviour is observed after a few seconds of flow. When the EDL is still fully developed at the exit of the duct, which is the case at the first few seconds of the flow or at low velocity (0.1 $m.s^{-1}$), the streaming current simulated with the adsorption model is usually higher than the one simulated with the corrosion model. When the EDL is partially developed at the exit of the duct, the analysis of the current simulation is quite different. It is important to consider the contact time between the solid and liquid relatively to the time required for the formation of the EDL. Static analyses have clearly shown that the dynamics of the EDL formation for the corrosion model is faster than for the adsorption model. For a given liquid flow velocity and solid length, the EDL space charge distribution developed by corrosion model is higher compared to the one developed by adsorption model and therefore the streaming current.

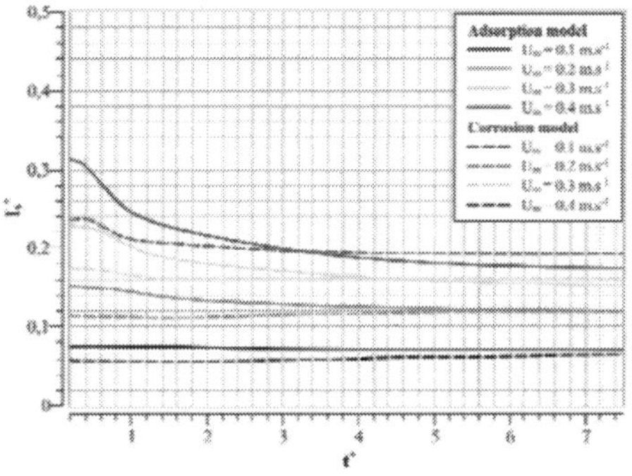

Figure 4. Dimensionless streaming current versus non-dimensional time for the adsorption and corrosion simulations

4. Conclusion

Two scenarios of the interface chemical reactions were simulated. All the numerical simulations are comparable to experimental results for the adsorption and corrosion models. What is notable is the fact that the electrical double layer development, and thus the streaming current, is dependent of the model used. For our entry data, the corrosion model was faster than the adsorption model. Nevertheless, a parametric study is needed to be able to understand the weight of each term and their dynamics. The comparative study is particularly interesting as, based on the experimental results and this numerical simulation, it would be possible to identify the chemical phenomena that control the electrical double layer.

5. Acknowledgments

This work is supported by the French Government program "Investissements d'Avenir" (LABEX INTERACTIFS, reference ANR-11-LABX-0017- 01).

6. References

[1] El_Adawy, Paillat T, Bertrand Y, Moreau O and Touchard G 2010 Physicochemical analysis at the interface between conductive solid and dielectric liquid for flow electrification phenomenon *IEEE Trans. Indus. Appl.* **46**(4) 1593-1600

[2] Cabaleiro J M, Paillat T, Moreau O and Touchard G 2009 Modeling of static development and dynamic behaviour of the electrical double layer at oil-pressboard interface *Electrostatics Joint Conference, USA*

[3] Touchard G, Patzek W and Radke C J 1996 A physicochemical explanation for flow electrification in low-conductivity liquids in contact with a corroding wall *IEEE Trans. Indus. App.* **32**(5) 1051-1057

Electrostatic modeling of cavities inside linear actuators

T A van Beek, J W Jansen, J van Duivenbode and E A Lomonova

Department of Electrical Engineering, Eindhoven University of Technology, Eindhoven, 5600MB, Den Dolech 2, The Netherlands

E-mail: `t.a.v.beek@tue.nl`

Abstract. This paper describes the modeling of two-dimensional electric fields inside linear actuators where cavities in the cooling environment are present. A semi-analytical model based on Fourier series is extended for cavities and incorporates the curvature of corners. The model shows good agreement with finite element analysis. The effect of position and sizes of the cavity are investigated as well as the curvature of the corners of the cavity.

1. Introduction

Increasing demands of the throughput of high-precision positioning systems require more powerful actuators in the positioning stages. To prevent the coils of these actuators from exceeding critical temperatures the coil array is mounted on a, electrically conductive, cooling environment which for safety reasons is connected to protective earth. However, cavities are present in the geometry of the cooling environment because of surface irregularities or mounting holes. The corners of the cavities introduce electric field enhancement, which can cause local dielectric breakdown if the breakdown strength of the insulation is exceeded. Besides the geometrical parameters of the corner, the electric field enhancement introduced by the corner is also determined by its curvature since a sharp corner would lead to an infinite electric field strength at the corner tip [1].

The modeling of the electric field distribution is usually performed by finite element or boundary element methods [2]. However, these methods are slow compared to analytical techniques such as conformal mappings [3] or Schwarz-Christoffel mappings [4]. Curvatures can be taken into account with these modeling techniques, however, due its parametric representation this is only possible for simple domains without loss of accuracy [5].

In this paper, a combination of Fourier modeling and a conformal mapping is proposed as modeling technique for the linear actuator configuration [6]. Globally, the field solution is obtained by Fourier modeling whereas locally the field solution near the curved corner is obtained by a conformal mapping of the curved corner. The derived field solution is validated with a 2-D finite element model.

2. Model formulation

The investigated linear actuator configuration is shown in Figure 1, where a coil is situated between an insulation layer and a cooling environment which inhibits cavities. The dimensions indicated in the figure and the material properties are given in Table 1. The configuration is divided in a number of regions as indicated in Figure 1 such that each region has homogenous material properties. By dividing the investigated configuration a boundary value problem is obtained which is solved based on boundary conditions. Furthermore, the periodicity present in the linear actuator configuration allows a computational reduction to one period and gives a fundamental frequency for the Fourier modeling. Fundamental frequencies for region I and II are given by

$$\omega^{I} = \frac{2\pi}{w_{per}} \quad \text{and } \omega^{II} = \frac{2\pi}{w_{cav}}. \tag{1}$$

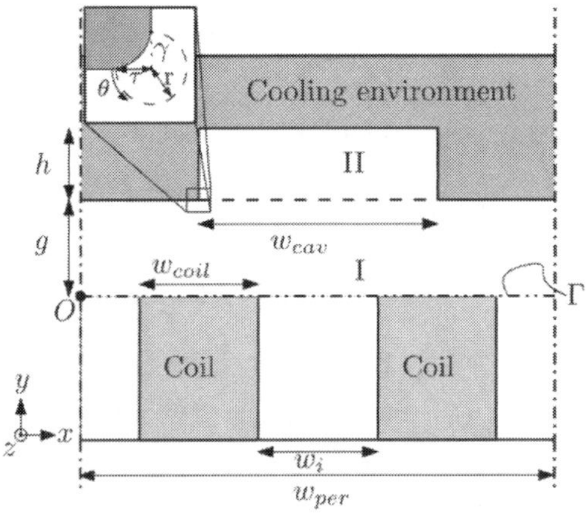

Figure 1: The linear actuator configuration in a Cartesian coordinate system with origin O

Table 1: Dimensions and material properties used in the linear actuator configuration

w_{coil}	10 [mm]	Width of the coil
w_{cav}	30 [mm]	Width of the cavity
w_{per}	40 [mm]	Width of the period
w_i	10 [mm]	Width of intercoil gap
g	5 [mm]	Height of the insulation gap
h	5 [mm]	Height of the cavity
ε_I	1	Relative permittivity of region I
ε_{II}	1	Relative permittivity of region II
V_p	1 [V]	Coil voltage wrt PE
V_e	0 [V]	Cooling environment voltage wrt PE
τ	100 [μm]	Curvature radius
N_I	91	Number of harmonics region I
N_{II}	21	Number of harmonics region II

2.1. Semi-analytical solution
Based on the electrostatic Maxwell equations and the assumption that the electric and the magnetic fields are decoupled, then $\nabla \times \vec{E} = 0$ and the electric field strength can be expressed by

$$\vec{E} = -\nabla\phi, \tag{2}$$

where ϕ is the electric scalar potential. Since it is assumed that no free charges are present in the materials, Gauss's law reduces to

$$\nabla \cdot \varepsilon\vec{E} = 0, \tag{3}$$

where ε is the relative permittivity of a material. By substituting Eq. 2 into 3 the Laplace equation is obtained for the electric scalar potential. Using the separation of variables the Laplace equation is solved. As a solution, a product of two function is obtained, one dependent on the x-direction and one on the y-direction.

$$\phi_k = a_0^k + \sum_{i=1}^{N}(a_i^k cos(\omega_i^k x) + b_i^k sin(\omega_i^k x))(c_i^k e^{\omega_i^k y} + d_i^k e^{-\omega_i^k y}), \tag{4}$$

where ω_i^k is the i-th spatial harmonic of the k-th region and N the number of spatial harmonics.

2.2. Boundary conditions
The field quantities in the different regions of the semi-analytical model are dependent on the boundary conditions. Regions I and II have a boundary at $y = 0$ and $y = g + h$, respectively. The electric scalar potential at these boundaries is defined by the environment. At $y = 0$, along path Γ, the electric scalar potental is subjected to a Dirichlet boundary condition which mimics the potential distribution of the coil and is given by

$$\phi_I\Big|_{y=0} = \begin{cases} V_p & \frac{1}{2}w_i \geq x \geq w_{coil} + \frac{1}{2}w_i \\ -V_p & \frac{1}{2}(w_{per} + w_i) \geq x \geq w_{per} - \frac{1}{2}w_i \\ 0 & \text{else.} \end{cases} \tag{5}$$

At $y = g + h$, the electric scalar potential is defined by the cooling environment and given by

$$\phi_{II}\Big|_{y=g+h} = 0. \tag{6}$$

The boundary condition between region I and II is a combination of a Dirichlet and a continuous boundary condition. This combined boundary occurs between regions with unequal width and implies that both regions have a different fundamental frequency. Observing the interface between region I and II, it is clear that for region II a continuous boundary is present at the interface with region I. Therefore, for region I a continuity is required at the interval with region II. For the remainder of the interface of region I, a Dirichlet boundary is present which is imposed by the cooling environment. The boundary conditions between region I and II are given by

$$\phi_I = \begin{cases} \phi_{II}\big|_{y=g} & 0.5\,w_{per} - dx \le x < 0.5\,w_{per} + dx, \\ 0 & \text{else,} \end{cases} \tag{7}$$

$$\varepsilon_{II}\frac{\partial \phi_{II}}{\partial y} = \varepsilon_I \frac{\partial \phi_I}{\partial y}\bigg|_{y=g}. \tag{8}$$

The combined boundary conditions in Equation 7 and 8 can be solved using mode-matching [7].

2.3. Curvature of corners

The semi-analytical model as described in Section 2.1 is unable to model the curvature of rounded elements since the system is solved in a Cartesian coordinate system. This results in solutions that are globally accurate but locally inaccurate in the neighbourhood of a rounded element. To incorporate the curvature of rounded element a separate model calculates the electric potential due to an unbounded profile. The calculation of the electric potentials due to the unbounded profile are performed by a conformal mapping. The conformal map of the unbounded profile is given by

$$z = k[(w+a)^{\frac{1}{\alpha}} + (w-a)^{\frac{1}{\alpha}}], \tag{9}$$

where

$$z = x + iy \qquad \text{corresponding to the real geometry,} \tag{10}$$

$$w = u + iv \qquad \text{where v is the electric potential,} \tag{11}$$

with $k = \frac{1}{2}$, $a = 2^{\alpha-1}$ and $\alpha = \frac{2}{3}$ for a straight corner. The total potential distribution, ϕ_t, is obtained by combining the solutions of the semi-analytical model, ϕ_{sa}, and the unbounded profile, ϕ_{unb}, and is given by

$$\phi_t = \lambda\tau^{\alpha}\phi_{unb} \qquad \text{for } r \le \tau, \tag{12}$$

$$\phi_t = \phi_{sa} + \lambda\tau^{\alpha}\phi_{unb} \qquad \text{for } r > \tau, \tag{13}$$

where λ represents the singularity factor, as in [6], and τ the curvature radius. The singularity factor is obtained by performing a weighted line integral in the semi-analytical model from border to border of the sharp corner. The singularity factor is given by

$$\lambda = \frac{4}{3\pi}\tau^{-\alpha}\int_0^{\frac{3\pi}{2}} \phi_{sa}(\tau,\theta) sin(\alpha\theta) d\theta, \tag{14}$$

where θ is the angle from border to border of the corner. In this model the electric field strength is obtained numerically.

3. Verification of the combined model

To validate the results of the combined model, the linear actuator configuration in Figure 1 is modeled with 2-D finite element method using Cedrat FLUX2D [8] for the parameters given in Table 1. Near the corner, the electric field strength is compared along path γ, parallel to the y-axis and starting from the original position of the sharp corner and ending at the cooling environment. The electric field strength in the y-direction along path γ is shown in Figure 2a for both the combined model and the Finite Element Model (FEM). In this figure Q is the normalized position along γ with respect to the curvature radius. The average discrepancies between the model and FEM for curvature radii, τ, of 0.1, 1 and 3 [mm] are 19, 17 and 10 [%], respectively. Furthermore, it is observed that for a curvature radius of 3 [mm] the discrepancy increases to above 20 [%] near the starting point of path γ. This is due to the conformal mapping presented in Equation 9, which assumes a constant potential on its mapped electrodes. For small curvatures this is a valid assumption but for larger curvatures this is not the case and the discrepancy at points far from the corner increases for increasing curvature radii.

The selected harmonic numbers, N_I and N_{II}, are sufficient for these simulations since tripling the number of harmonics resulted in less than 1 [%] change in the first 9 harmonic coefficients.

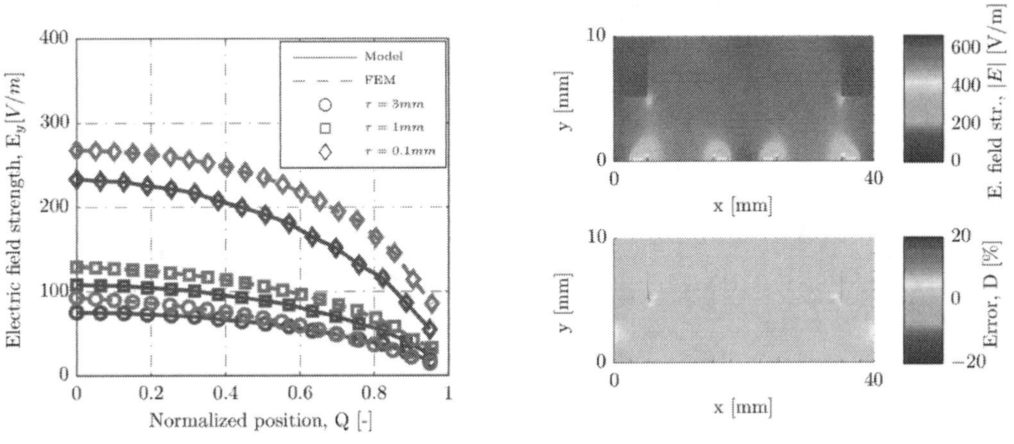

(a) (b)

Figure 2: (a) Electric field of the combined model and FEM along path γ for radii. (b) Electric field and the discrepancy between FEM for the combined model for $w_{cav} = 30[mm]$ and $\tau = 0.1[mm]$.

In Figure 2b the potential distribution and the discrepancy with the finite element model of a periodical section of the linear actuator configuration is shown. The discrepancy, D, is calculated as

$$D = 100\frac{|E_t| - |E_{FEM}|}{max(|E_t|)}, \tag{15}$$

where E_t and E_{FEM} are the electric field strength of the combined model and FEM, respectively. The discrepancies near the corners are below 15 [%], this is lower than the discrepancies as obtained from Figure 2a. This is due to meshsize of the results, which is larger than the curvature of the corner. Furthermore, the discrepancies, up to 20 [%], on the sides of the cavity are caused by a slight difference in cavity width between the combined model and FEM (30.02 vs 30 [mm]) to allow numerical convergence of the semi-analytical model.

4. Conclusion

In this paper, a conformal mapping and semi-analytical model combined presents the electric potential distribution of a periodical section inside a linear actuator where cavities are present. The local potential distribution near rounded elements is taken into account by a conformal mapping of an unbounded profile. Globally, the electric potential distribution inside the linear actuator is modeled using the Fourier modeling technique extended with mode-matching. The field solutions of the combined model are verified with 2-D finite element analyses for different curvature radii of the corner of the cavity. For large curvature radii the average discrepancy is low, 10 [%], compared to smaller curvature radii, 19[%]. However, for large curvature radii the discrepancy far from the corner increases to above 20 [%]. Therefore, this model is suited to investigate linear actuator configurations with relative small curvatures.

References

[1] Langton N and Smith M 1977 *Electrical Engineers, Proceedings of the Institution of* **124** 277–284 ISSN 0020-3270
[2] Lerch R, Kaltenbacher M, Landes H, Peipp R and Simkovics R 2002 *Sensors, 2002. Proceedings of IEEE* vol 2 pp 1233–1238 vol.2
[3] Brenna M, Martinelli F and Zich R 2004 *Mathematical Methods in Electromagnetic Theory, 2004. 10th International Conference on* pp 277–279
[4] O'Connell T and Krein P 2010 *Energy Conversion, IEEE Transactions on* **25** 606–618 ISSN 0885-8969
[5] Delillo T K *SIAM Journal on Numerical Analysis* **31** pp. 788–812
[6] Krahenbuhl L, Buret F, Perrussel R, Voyer D, Dular P, Peron V and Poignard C 2011 *Journal of Microwaves, Optoelectronics and Electromagnetic Applications* **10** 66 – 81 ISSN 2179-1074
[7] Gysen B, Meessen K, Paulides J and Lomonova E 2010 *Magnetics, IEEE Transactions on* **46** 39–52 ISSN 0018-9464
[8] Cedrat 2014 *FLUX 11.2.2 User's Guide* (Meylan: France)

Development of super-clean diesel engine and combustor using nonthermal plasma hybrid aftertreatment

Masaaki Okubo

Department of Mechanical Engineering, Osaka Prefecture University, 1-1 Gakuen-cho, Naka-ku, Sakai 599-8531, Japan

E-mail: mokubo@me.osakafu-u.ac.jp

Abstract. One of important and successful environmental applications of atmospheric-pressure corona discharge or plasma is electrostatic precipitator (ESP), which have been widely used for coal- or oil-fired boilers in electric power plants and particulate matter control emitted from industries such as glass melting furnace system, etc. In the ESPs, steady high voltage is usually applied to a pair of electrodes (at least, one of these has sharp edge). Unsteady pulsed high voltage is often applied for the collection of high-resistivity particulate matter (PM) to avoid reverse corona phenomena which reduce the collection efficiency of the ESPs. It was found that unsteady high voltage can treat hazardous gaseous components (NO_x, SO_x, hydrocarbon, and CO, etc.) in the exhaust gas, and researches were shifted from PM removal to hazardous gases aftertreatment with unsteady corona discharge induced plasmas. In the paper, recent results on diesel engine and industrial boiler emission controls are mainly reviewed among these our research topics.

1. Introduction

One of important and successful environmental applications of atmospheric-pressure corona discharge or plasma is electrostatic precipitator (ESP), which have been widely used for coal- or oil-fired boilers in electric power plants and particulate matter control emitted from industries such as glass melting furnace system, etc. In the ESPs, steady high voltage is usually applied to a pair of electrodes (at least, one of these has sharp edge). Unsteady pulsed high voltage is often applied for the collection of high-resistivity particulate matter (PM) to avoid reverse corona phenomena which reduce the collection efficiency of the ESPs. It was found that unsteady high voltage can treat hazardous gaseous components (NO_x, SO_x, hydrocarbon, and CO) in the exhaust gas, and researches were shifted from PM removal to hazardous gases aftertreatment with unsteady corona discharge induced plasmas [1-3].

In our laboratory, we have tried to develop nonthermal-plasma environmental applications technology since 1998. Main research topics are exhaust gas cleaning as follows: diesel engine emission control, boilers emission control, NO_x removal for glass melting furnace, VOC (volatile organic compounds) control from factories, odor and dioxin controls from garbage incinerators, CO_2 and PFC (perfluoro compounds) emission and moisture controls in various environments, and indoor electric air cleaner. All our papers are listed in [4]. In the paper, recent results on diesel engine and industrial boiler emission controls are mainly reviewed among these our research topics.

2. Marine Diesel Engine Emission Control [5-8]

A photograph of the target diesel engine (6DK-20e, Daihatsu Diesel MFG Co. Ltd., Japan) is shown in **figure 1**. In marine vessels, diesel engines are used not only as the main propulsion power sources but also for electrical sub power generation. In this study, the target is the latter of these. The specifications of the engine are as follows: a maximum or 100% output power of 1071 kW, four cycles, six cylinders with a cylinder bore of 200 mm and a stroke of 300 mm, a constant rotation rate of 900 rpm, and the total mass of the engine and dynamo of 16 ton. Marine diesel oil (A-heavy oil, sulfur = 0.075 mass%, nitrogen = 0.01 mass%, heat quantity = 45.4 MJ/kg) is used as the fuel. The exhaust flow rate is 3920 Nm^3/h for 50% load or output power, 5526 Nm^3/h for 75%, and 6815 Nm^3/h for 100% (N denotes the standard state of 0°C, 0.1 MPa).

Figure 1. Photograph of the targeted diesel engine (6DK-20e, max. power = 1071 kW) [5]

Figure 2 shows the experimental setup for the aftertreatment system in the marine diesel engine (6DK-20e, output power = 1071 kW, Daihatsu Diesel MFG Co., Ltd, fuel: A-heavy oil or marine diesel oil). In this system, the combination of adsorption and the NTP (nonthermal plasma) is used for reduction of NO_x. The exhaust gas emitted from the engine is first separated and approximately 13% of the bypassed gas the flow rate of which is controlled by the flow regulation valve passes through a set of ceramics diesel particulate filters (DPFs). More than 95% of the PM in the exhaust gas is

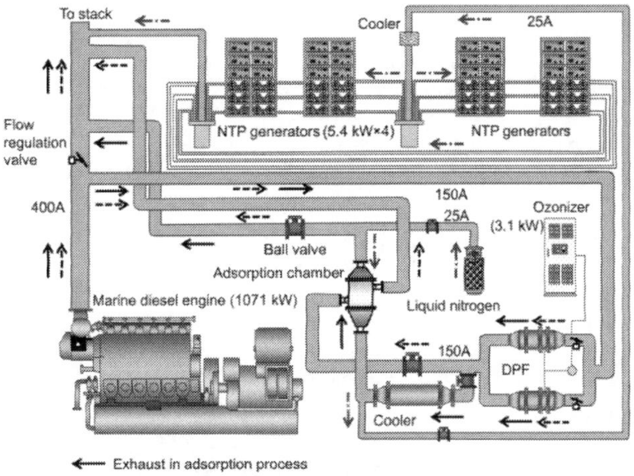

Figure 2. Schematic diagram of experimental setup for the aftertreatment system in the marine diesel engine [5]

captured in the DPFs. The accumulated PM in the DPF is treated by nonthermal plasma-induced ozone (O_3) injection technology [6, 8]. After this process of the PM removal, NO_x in the exhaust gas is treated by NTP combined with repeated adsorption and desorption processes. In the adsorption process, for an effective NO_x adsorption to adsorbents, the exhaust gas is cooled from 270°C down to approximately 45°C with the water-cooling type cooler. An adsorption chamber that contains adsorbents (1.2–2.4 mm-sized granular pellets of MnO_x–CuO compound) adsorbs NO_x of 92% in maximum. Finally, the exhaust gas is flowed out from a stack. In the desorption process, after the valves are controlled, N_2 gas of lower flow rate of 11.1 ~ 22.2 Nm^3/h supplied from liquid N_2 cylinder passes through the chamber. The direction of the N_2 gas is opposite to that of the exhaust gas in the adsorption. Simultaneously, the chamber is heated up by the waste heat of the exhaust gas. As a result, high-concentration (typically 1% = 10000 ppm) NO_x flows out from the chamber, and introduced to the NTP reactors of total 21.6 kW in maximum.

Figure 3 shows a time-dependent NO_x emission before and after the gas passes through the aftertreatment system for operation cycles 2 ~ 4, 9, and 10. Exhaust gas is continuously treated during these repeated cycles. In this graph, the mass flow rate of NO_x is taken in the vertical axis evaluated by NO_x analyzers based on the molecular mass of NO_2 with the unit of $g(NO_2)/h$. Furthermore, the untreated NO_x is indicated by white circles with lines, and treated NO_x is indicated by black circles

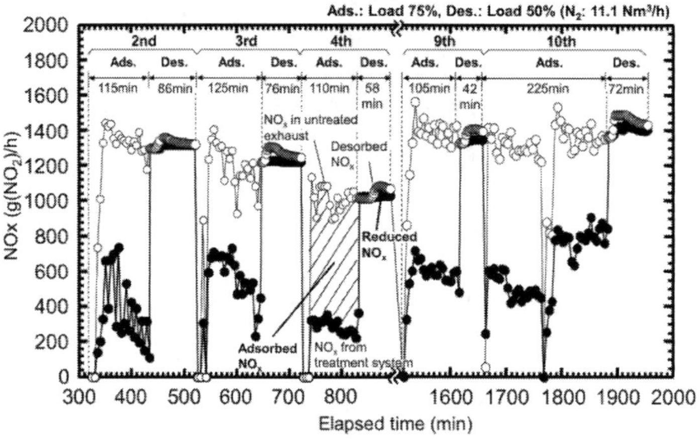

Figure 3. Time-dependent NO_x emission from the aftertreatment system for the marine diesel engine

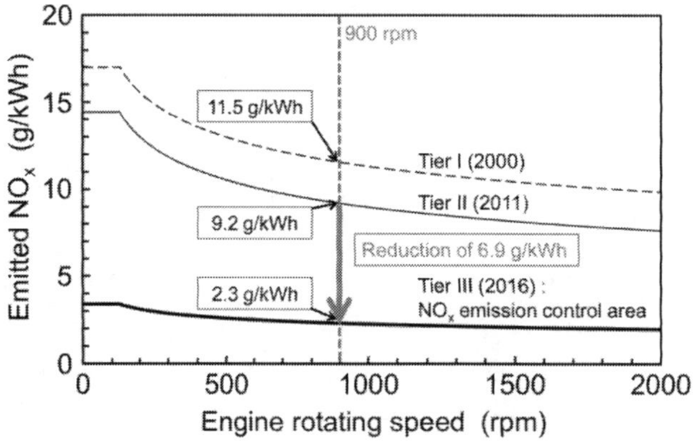

Figure 4. IMO emission standards and the required NO_x reductions for Tiers I, II, and III [5]

with lines. NTP is applied only in the desorption processes, and the input power to the NTP generator is 21.6 kW in the cycles 2 ~ 4 and 12.0 kW in the cycles 9 and 10. It is known this graph that large amount of NO_x is treated. At 10th cycle, the highest system energy efficiency among these cycles of η_{system} = 161 g(NO_2)/kWh is achieved. Application of the highest recorded energy efficiency to the requirement of the IMO (International Maritime Organization) emission standards from Tier II to III corresponds to only 4.3% (= 6.9 / 161 × 100) of the engine output power as explained in **figure 4**.

3. Industrial Boiler Emission Control [9-13]

The number of small boilers using city natural gas, heavy oil, and waste oils has been increasing year by year in Japan, and more stringent regulations for NO_x emission are being anticipated in order to reduce environmental NO_x concentration. A suitable flue gas treatment system for small boilers will be required soon. Several studies have been conducted on laboratory-scale nonthermal plasma–chemical hybrid processes for the removal of NO_x from gases emitted from various stationary sources. The authors propose a plasma–chemical hybrid process consisting of an indirect nonthermal plasma process and wet-chemical treatment. The principle of NO_x removal is as follows:

$$NO + O_3 \rightarrow NO_2 + O_2 \quad (1)$$
$$2NO_2 + 4Na_2SO_3 \rightarrow N_2 + 4Na_2SO_4 \quad (2)$$

Figure 5 shows an overview of the plant. The flue tube boiler (Takao Iron Works Co., Ltd.) has an original rotary burner for gas and/or oil and is operated by using city natural gas (13A) at 157 Nm^3/h and A-type heavy oil at 171 L/h and waste oils. The boiler has a steam generation rate of 2.5 t/h. Two sets of silent discharge-type ozonizers (EW-90Z, Ebara Jitsugyo Co., Ltd.) equipped with a pressure swing adsorption (PSA) oxygen generator are employed to generate O_3.

Figure 5. Plasma hybrid clean boiler system

Figure 6 shows the structure of the discharge section of the plasma reactor in the ozonizer. A coaxial NTP reactor or silent discharge-type reactor is used. O_3 gas is generated after O_2 gas by PSA passes through the 30 plasma reactors inside the ozonizer and then injected into a flue gas duct for NO oxidation. The O_3 gas flow rate is constant, 0.9 Nm^3/h. When the discharge power is 1.5 kW and the PSA power is 1.6 kW, 90 g/h of O_3 is generated in maximum with a concentration of 4.7% and the energy efficiency of 29 g/kWh. High-concentration O_3 gas is injected into the duct and diluted in order to oxidize almost all the NO to NO_2 according to reaction (1). The flue gas is then introduced into the scrubber in order to reduce NO_2 to N_2 according to reaction (2).

Experimental results and discussion are shown next. In order to confirm NO_x removal performance, the mixed oil of WVO and heavy oil and city natural gas are tested as the boiler fuels.

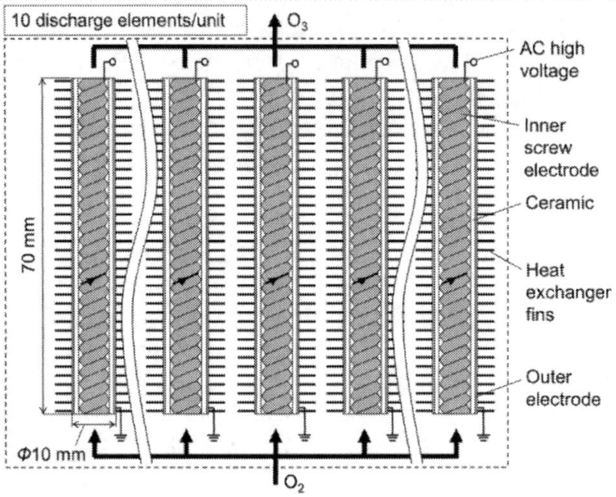

Figure 6. Structure of the discharge section in the ozonizer

Figure 7 shows time-dependent NO and NO_x concentrations at MP1 (measurement point 1 at the exit of the boiler) and MP3 (measurement point 1 at the exit of the exhaust gas cleaning) [13]. The NO_x levels are measured by NO_x analyzers. The experiment is carried out with daily start and stop operation. The average flow rate of city natural gas is 52 Nm^3/h and that of flue gas is 718 Nm^3/h at an O_2 concentration of 3.3–4.0%. The average rates of O_3 injection and Na_2SO_3 supply are 86 g/h and 63 mol/h. NO_x concentrations at MP1 are within 45–50 ppm. This small difference of NO_x each day is due to the difference in the O_2 concentrations at MP1 between 3.3% and 4.0%. The NO_x concentrations at MP3 are within 5–8 ppm, which are equivalent to a NO_x removal efficiency of more than 85% over an operating time of 23 h. The NO_x removal performance achieved is satisfactory and stable. Concerning the specific gravity of the scrubbing solution, it increases gradually and is stable at approximately 1.08 which is an allowable value as a control index.

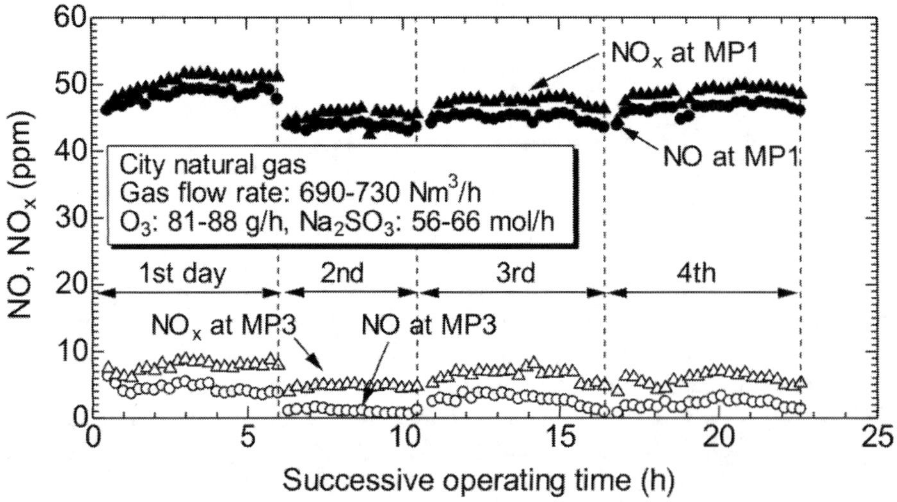

Figure 7. Time-dependent NO and NO_x concentrations at MP1 (boiler exit) and MP3 (exit of aftertreatment) during a four-successive-day operation in firing city natural gas

4. Conclusion

Recent results on the development of super-clean diesel engine and combustor using nonthermal plasma hybrid aftertreatment are reported. We achieved highest removal efficiency of NO_x reduction greater than 100 g(NO_2)/kWh when the nonthermal plasma is combined with other chemical processes (adsorption or wet chemical scrubbing). The efficiency is approximately ten-times higher than that in the case with plasma application only.

An adsorption-plasma combined NO_x aftertreatment system for marine diesel exhaust gas is developed. An experiment using a pilot-scale aftertreatment system for a marine diesel engine with an output power of 1 MW has been carried out using an NTP generator with a power of 21.6 kW. The characteristics of NO_x adsorption/desorption and NO_x reduction by NTP are analyzed using the experimental data. Finally, the energy efficiency of NO_x reduction by NTP is evaluated. This system demonstrates an excellent system energy efficiency for NO_x removal of 161 g(NO_2)/kWh. Because high-concentration NO_x is treated by NTP with the help of NO_x adsorption and desorption, the present aftertreatment system can achieve higher energy efficiency.

Further, the pilot-scale investigation of a plasma–chemical hybrid NO_x removal process is carried out using a low emission boiler system comprising multi-fuel boiler and a chemical scrubber. The amount of NO removed is almost the same as the amount of the corresponding ozone concentration to oxidize NO to NO_2 (1:1 stoichiometric ratio) regardless of the flue gas flow rate. NO_x removal efficiency of more than 85% is achieved over an operating time of 23 h when firing city natural gas.

It is confirmed that the present nonthermal plasma hybrid aftertreatment is very effective to these targets and applicable for industrial use.

Acknowledgments

The author thanks Dr. H. Fujishima, Dr. T. Kuroki (Osaka Prefecture University), Dr. T. Kuwahara (Nippon Institute of Technology), and Dr. K. Yoshida (Osaka Institute of Technology) for their support in the researches. The author also thanks kind reproduction permission for **Figures 1**, **2** and **4** from Springer Science and Business Media. This work was supported by the Regional R&D Resources Utilization Program in JST and JSPS KAKENHI Grant Number 24246145 and 249848.

References

[1] Yao S, Fushimi C., Madokoro K, Yamada K 2006 *Plasma Chem Plasma Process.* **26** 481
[2] Hołub M, Kalisiak S, Borkowski T, Myśków J, Brandenburg R 2010 *Pol. J. Env. Stud.* **19** 1199
[3] McAdams R, Beech P, Shawcross JT 2008 *Plasma Chem Plasma Process.* **28** 159
[4] Okubo M 2014 A paper list by Google Scholar, http://www.mokubo.com/personal/scholar
[5] Kuwahara T, Yoshida K, Kuroki T, Hanamoto K, Sato K, and Okubo M 2013 *Plasma Chem. Plasma Process.* **34** 65
[6] Kuwahara T, Yoshida K, Hanamoto K, Sato K, Kuroki T, Yamamoto T, and Okubo M 2012 *IEEE Transactions on Industry Applications* **48(5)** 1649
[7] Kuwahara T, Yoshida K, Hanamoto K, Sato K, Kuroki T, Yamamoto T, and Okubo M 2012 *J. Japan Institute of Marine Engineering* **47** 379
[8] Okubo M, Arita N, Kuroki T, Yamamoto T 2008 *Plasma Chem. Plasma Process.* **28(2)** 173
[9] Fujishima H, Takekoshi K, Kuroki T, Tanaka A, Otsuka K, and Okubo M 2013 *Applied Energy* **111** 394
[10] Fujishima H, Yoshioka Y, Kuroki T, Tanaka A, Otsuka K, and Okubo M 2011 *IEEE Transactions on Industry Applications* **47(5)** 2210
[11] Fujishima H, Kuroki T, Ito T, Otsuka K, Yamamoto T, Yoshida K, and Okubo M 2010 *IEEE Transactions on Industry Applications* **46(5)** 1707
[12] Fujishima H, Tatsumi A, Kuroki T, Tanaka A, Otsuka K, Yamamoto T, and Okubo M 2010 *IEEE Transactions on Industry Applications* **46(5)** 1722
[13] Kuroki T, Fujishima H, Tanaka A, Otsuka K, Okubo M 2014 *Proc. 14th Int. Symp.High Pressure Low Temperature Plasma Chemistry (Hakone XIV)*, Topic 06, Oral-06-28.

Atmospheric-pressure ionization: New approaches and applications for plasmas in contact with liquids

DB Go[1]
Aerospace and Mechanical Engineering
Chemical and Biomolecular Engineering
University of Notre Dame
Notre Dame, IN, 46556, United States

E-mail: dgo@nd.edu

Abstract. Historically, gas discharges have been difficult to stabilize at atmospheric pressure, and this has confined them to operation at low pressure under vacuum conditions. However, recent advances in plasma technology have enabled stable high pressure gas discharges up to and even exceeding atmospheric pressure. One significant advantage of operating at atmospheric pressure is that the plasma can be brought into contact with non-conventional substrates, especially soft materials such as plastics, biological tissue, and aqueous solutions. This last example is of prime interest as plasma/liquid interactions have a number of important implications in applications ranging from water purification to plasma medicine. In this paper, recent work studying the impact of electrons in the plasma inducing reactions in aqueous solutions is discussed. These studies include measurements of the bulk solution as the electrons induce long-lived species as well as interfacial measurements directly at the plasma/liquid interface to probe the behaviour of electrons traversing from the plasma into the liquid.

1. Introduction

Gas discharges or plasmas are ionized gases, and for the most part they are operated under vacuum conditions. Recently, however, new technologies using small dimensions have led to stable atmospheric pressure plasmas termed microplasmas [1]. In addition to shepherding in advances in lighting [2], gas sensing [3], and nanomaterials synthesis [4], one key advantage of operating at atmospheric pressure is that it enables the plasma to interact with non-traditional substrates. That is, plasma interactions with soft materials, ranging from plastics to biological tissue, have been uniquely enabled by atmospheric pressure operation.

One substrate that has been of significant interest over the past decade is water or aqueous solutions, where the plasma comes in direct contact with the liquid [5],[6]. One of the key features of plasmas is that they produce an abundance of reactive species including electrons, ions, radicals, metastables, and photons. The plasma therefore becomes an excellent delivery system, dissolving these reactive species into the water leading to a wide range of reactions. Recent review articles (*e.g.*, [7]) have focused on reactive oxygen species (ROS) and reactive nitrogen species (RNS) highlighting the importance of these species in water purification and biological applications. However, one key species in both the plasma and liquid phase is the electron. This article will overview electron-driven

[1] To whom any correspondence should be addressed.

chemistry in plasma/liquid systems, summarizing recent studies aimed at understanding this important mechanism.

2. Historical Perspective and Recent Applications

In the mid-twentieth century, the concept of using plasmas as electrochemical reactors, often called glow discharge electrolysis, was pioneered by Hickling and his collaborators [8]. Borrowing from conventional electrochemical cells, in glow discharge electrolysis, also called contact glow discharge electrolysis, one of the electrodes was replaced by a plasma such that the electrolyte itself acted as the counter electrode under direct current (DC) conditions. Figure 1 shows two basic plasma configurations along with a conventional electrochemical cell. In a conventional electrochemical cell, two electrodes are submerged in an electrolyte solution and a DC bias is applied between them, typically on the order of ~1-10 V. In glow discharge electrolysis, either one (or sometimes both) of the electrodes can be replaced with a suspended electrode. Applying a very high voltage (~1-10 kV) causes the interstitial gas between the suspended electrode and solution to breakdown, forming a plasma. Unlike many plasma/liquid configurations where the plasma is 'remote' and formed only by electrodes external to the solution, in glow discharge electrolysis the electrolyte solution itself is an important conductive element in the entire circuit. Thus, while in remote configurations the charged electrons and ions in the plasma only pass between two external electrodes, in glow discharge electrolysis the charged particles must traverse into the solution in some manner in order to produce current. When the plasma is the cathode (Figure 1b), these charged particles are electrons entering the anodic solution. Conversely, when the plasma is the anode (Figure 1c), the charged particles are positive ions entering a cathodic solution.

Like conventional electrochemistry, the current in plasma electrolysis is maintained by reactions occurring at the electrode interfaces – reduction at the cathode and oxidation at the anode. However, in the absence of a solid electrode, the reactions occur at the plasma/liquid interface, reduction for a cathodic plasma and oxidation for an anodic plasma. This is fundamentally different from conventional electrochemistry. Hickling and his collaborators, for the most part, focused on anodic plasmas. Today, much of the interest is in using cathodic plasmas in order to understand the behaviour of electrons interacting at the solution interface. One of the most promising areas that has recently emerged is using these types of contact plasmas for the electrochemical synthesis of colloidal nanoparticles [9]. In a cathodic configuration, free plasma electrons interact at the solution interface, inducing reduction reactions. If the solution cations are metallic, the metal cation is reduced. For silver ions (Ag^+) from solutions of silver nitrate ($AgNO_3$), the silver cation (Ag^+) is reduced to Ag^0. In conventional electrochemistry, these reduced species adsorb onto the metal cathode, coating it with silver metal – a process called electroplating. However, if the metal cathode is replaced with a plasma cathode, there is no solid surface on to which the Ag^0 can absorb, and they agglomerate to form nanoparticles. Similarly, if the initial solution contains gold chloride ($HAuCl_4$), gold (Au) nanoparticles can be efficiently produced.

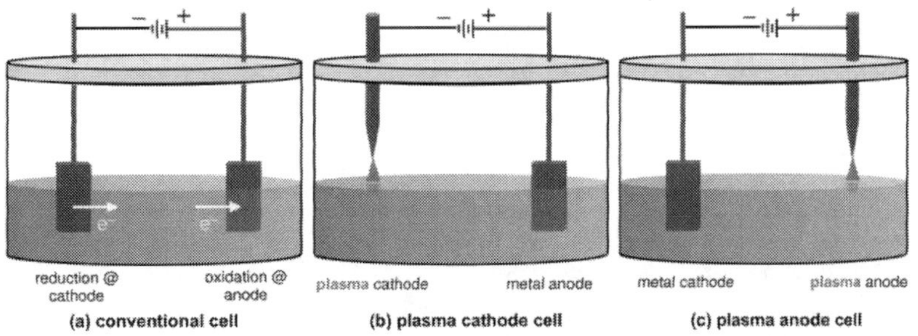

Figure 1. Schematic of electrochemical cells: (a) conventional cell, (b) plasma cathode cell, and (c) plasma anode cell.

3. The Question of Electron-Driven Reactions

Ostensibly, the simplest picture of electron-driven reactions for nanoparticle synthesis is the direct reduction of the cation by the plasma electron. For example, in the case of silver, it would be a single electron reaction ($Ag^+ + e^- \rightarrow Ag^0$), whereas for gold reduction it is a three-electron reaction ($Au^{3+} + 3e^- \rightarrow Au^0$). However, a number of recent studies have raised the question of whether this is indeed the case. For example, Chen et al. recently showed that it is more likely the H_2O_2 formed in solution by the various reactive species that reduces the gold cation [10]. Thus, it is not clear if the reduction is always by plasma electron transfer, sometimes by plasma electron transfer, or never by plasma electron transfer with the reduction always through some other pathway.

A second question that is often raised is the nature of the electron transfer. Specifically, the question arises as to whether the electron initiates the reduction right at the plasma/liquid interface or if the electron penetrates into the solution, effectively becoming solvated before reacting with solution species. The solvated electron is one of the most powerful reduction agents known in chemistry, with a reduction potential of -2.89 V [11], and modern techniques typically use radiation sources (electron or gamma) to ionize the solution and causing a subsequent electron avalanche. These liberated electrons eventually thermalize in the solution as solvated electrons before reacting away [12]. The prevailing question is whether the relatively 'low' energy electrons escaping the plasma are injected into the solution. Recently, a number of studies have begun to narrow in at resolving these questions and unraveling the complex nature of electron chemistry at plasma/liquid interfaces.

3.1. Bulk Solution Studies

Initially studies on electron-driven chemistry focused on effects that would persist for long times and ultimately affect the bulk properties of the solution. One of the first of these studies was conducted by Richmonds et al., who studied the classic ferri/ferrocyanide redox couple [13] using a split electrochemical cell (H-cell) with separated cathodic and anodic chambers. Ferricyanide, $Fe(CN)_6^{3-}$, is well known to undergo a single electron reduction to form ferrocyanide, $Fe(CN)_6^{4-}$. Importantly, ferricyanide absorbs strongly in the blue at ~420 nm such that the color of the solution gradually turns from yellow to clear as the ferricyanide is reduced. By quantifying the absorbance using ultraviolet-visible (UV-vis) spectrosopy, Richmonds et al. were able to show that there was some inherent inefficiency. In theory, the reduction reaction should proceed Faradaically such that the moles reduced is proportional to the plasma current and the plasma processing time. While they found that the amount reduced did proceed linearly at constant current, it was less than that predicted by Faraday's law.

One obvious culprit is that some percentage of the electrons injected by the plasma go directly to water electrolysis, rather than reduction of the reagent. In acidic solutions, electrons will directly recombine with protons to produce hydrogen gas (H_2) via

$$2(H^+)_{aq} + 2e^- \rightarrow (H_2)_g, \tag{1}$$

which is the well known hydrogen evolution reaction. This cathodic reaction at the plasma interface is complemented by an oxidation reaction at the anode that produces oxygen (O_2) gas via

$$H_2O \rightarrow (O_2)_g + 4(H^+)_{aq} + 4e^-. \tag{2}$$

These electrolytic effects can be measured in multiple ways, either via changes in the local pH of the solution or by detecting the subsequent gas products (*e.g.*, H_2). The author's group took advantage of this to confirm that in fact the plasma does drive hydrogen evolution via electrolysis [14]. Using a split H-cell, with separated cathodic and anodic chambers, and a 0.001 M solution of hydrochloric acid (HCl) with 1.0 M potassium chloride (KCl) for conductivity, they both measured the pH change in the solution and the $(H_2)_g$ using an isotope mass spectrometer. As expected, the reduction proceeded linearly, but like Richmonds et al. [13] it was less than that predicted by Faraday's law.

Follow up studies focused on understanding these electrolytic processes further. In particular, it is well known that other reactive species formed in the plasma also dissolve into the solution, changing the bulk chemistry. Therefore, there is potentially an inherent competition at the plasma/liquid interface – chemistry driven by electrons (electrolytic) and chemistry driven by dissolved species. In order to resolve this inherent competition, the author's group studied electrochemical processes in solutions of sodium chloride (NaCl) as opposed to the acidic solutions studied earlier [15]. Because electrolytic reactions in NaCl lead to excess OH^-, the pH increases. However, in nitrogen/oxygen-rich environments, such as air, the plasma forms nitric oxides (NO_x), and these plasma products dissolve in to the solution to form nitrous (HNO_2) and nitric (HNO_3) acid, which lowers the pH. By measuring pH variation in different plasma environments (pure Ar, O_2, N_2, and air), Rumbach et al. [15] revealed conditions where electrolytic reactions dominate the chemistry and conditions where dissolved plasma species dominate the chemistry. Importantly, they also showed that the electrolytic and that dissolved species pathways can couple with each other to fundamentally alter the chemistry in the solution.

3.2. Interfacial Studies

While these bulk studies confirmed the electrolytic activity driven by cathodic plasmas, they did not directly address one persistent question – do the plasma electrons solvate before reacting or do they react right at the plasma/liquid interface? The author's group has developed a novel approach to directly measuring solvated electrons at the interface of the plasma and liquid, taking advantage of the absorption behavior of solvated electrons [16]. It is well known in the radiolysis community that solvated electrons absorb strongly in the red [17], with a peak near 700 nm that shifts further into the red as the solution temperature increases [18]. Therefore the strategy was to probe the interface with a red laser and measure the absorption signal as shown in Figure 2. If the electrons *do* solvate at the interface, then there would be a measured absorption. However, a number of technical challenges make this measurement particularly difficult. In the absence of anything that reacts directly with electrons, such as a cation like Ag^+, solvated electrons eventually diffuse to each other and react via

$$2(e^-)_{aq} + 2H_2O \rightarrow 2(OH^-)_{aq} + (H_2)_g, \qquad (3)$$

which is referred to as 2nd-order recombination. At the concentrations expected from a typical cathodic plasma operating at ~1-10 mA, the timescale for recombination is on the order of μs, such that the change in signal due to absorption is only $I/I_0 \sim 0.999999$, which is exceedingly low. Thus any sources of random noise, such as through electromagnetic pick-up, or systematic effects that cause spurious reflections, could swamp the measured absorption signal. To overcome these limitations, an alternative configuration was used where the laser passed through the solution at an angle less than the critical reflection angle, such that all the laser light that is not absorbed is captured by the photodetector, mitigating spurious reflections. This total internal reflection configuration, shown in Figure 2, also included modulating the plasma between low current and high current states at a known frequency of 20 kHz. Locking in the photodetection at this same modulation frequency filtered out any incoherent noise generated by the experiment, allowing for the very high signal-to-noise ratio required to detect solvated electrons.

Figure 3 shows a series of absorption measurements taken in aqueous solutions of NaOH using diode lasers of various wavelengths. Consistent with the absorption spectrum measured from radiolysis [18] shown as a solid line, there is peak in the red near 700 nm confirming that plasma electrons do in fact solvate in the solution. Interestingly, the spectrum appears to be blue-shifted from those measured via radiolysis. There are various effects that could cause this shift, including a Stark effect due to the high electric field at the interfacial double layer, space charge in the double layer, or even stratification due to localized heating by the plasma. What the blue shift does suggest is that the plasma generates solvated electrons under conditions distinct from other methods such as radiolysis, and these conditions impact their behaviour.

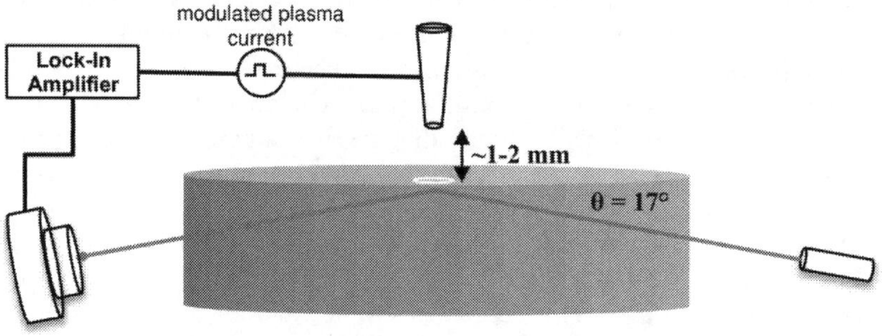

Figure 2. Total internal reflection absorption spectroscopy configuration where the plasma is generated using a stainless steel capillary with a diameter ~100 μm suspended ~1-2 mm above the solution surface biased negatively relative to a submerged platinum electrode. The plasma current is modulated at 20 kHz and lock-in amplification is used to mitigate sources of noise, including spurious reflections from the liquid interface mechanically deforming beneath the plasma.

Rumbach et al. [16] also confirmed that these electrons are solvated by introducing additional species that are known to react with electrons, scavenging them and reducing the absorption signal. That is, for the reaction

$$(e^-)_{aq} + S^n \rightarrow S^{n-1}, \quad (3)$$

as the concentration of scavenger S increases, reaction (4) dominates reaction (3) and extinguishes the absorption signal. By varying the concentration of cationic (H^+), anionic (NO_2^-, NO_3^-), and neutral scavengers (H_2O_2), they showed that the signal decreases as $I_0/I_0 \propto [(S)_{aq}]^{-1}$, confirming that the electrons are solvated *before* reacting away. Adding even greater confidence to their results, Rumbach et al. [16] extracted reaction rate coefficients from these scavenger experiments, and they were consistent with known values from the radiolysis literature.

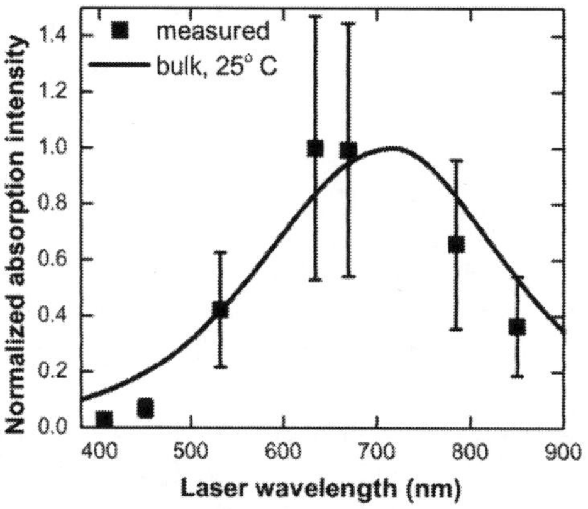

Figure 3. Measured plasma-generated solvated electron absorption spectrum using a solution 0.163 M $NaClO_4$. Adapted from [16].

Ultimately, these recent interfacial measurements confirm that cathodic plasmas inject electrons into the liquid phase. This has important implications, as the solvated electron is one of the most powerful reducing species with a reduction potential of -2.89 V. Thus while much is known about the impressive chemical effects of dissolved plasma species, it is clear that plasma electrons can have a significant impact on liquid chemistry, with implications on applications ranging from plasma medicine to water purification.

4. Conclusions

It is well understood that the way that plasmas and gas discharges interact with liquids is very complex. Recent developments in using plasmas for nanomaterials synthesis have raised questions about how plasma electrons, in particular, interact with liquids. It is now becoming more evident that the plasma can act as a source of solvated electrons, and that these solvated electrons will subsequently drive reduction chemistry in the liquid. However, it is also clear that the many other plasma species also induce liquid chemistry that can often times be in direct competition with the electron-driven chemistry. Looking toward the future, it is necessary to continue to parse and understand all these reaction paths in order to design plasma/liquid systems that are truly optimized toward a specific application, whether that be materials synthesis, water purification, or biomedical in nature.

Acknowledgments

The author wishes to acknowledge his collaborators and all those who contributed to this work: P Rumbach, D M Bartels, R M Sankaran, M Witzke, S Martinez, N Griggs, and A Talbot. The author acknowledges that this material is based upon work supported by the Air Force Office of Scientific Research under AFOSR Award No. FA9550-11-1-0020 and the Army Research Office under ARO Award No. W911NF-14-1-0241.

References

[1] Becker K H, Schoenbach K H and Eden J G 2006 *J. Phys D: Appl. Phys.* **39** R55
[2] Eden J G *et al.* 2013 *IEEE Trans. Plasma Sci.* **41** 661
[3] Gianchandani Y B, Wright S A, Eun C K, Wilson C G and Mitra B 2009 *Anal. Bioanal. Chem.* **395** 559
[4] Mariotti D and Sankaran R M 2010 *J. Phys D: Appl. Phys.* **43** 323001
[5] Bruggeman P and Leys C 2009 *J. Phys D: Appl. Phys.* **42** 053001
[6] Samukawa S *et al.* 2012 *J. Phys D: Appl. Phys.* **45** 253001
[7] Graves D B 2012 *J. Phys D: Appl. Phys.* **45** 263001
[8] Hickling A 1971 Electrochemical processes in glow discharge at the gas-solution interface *Modern Aspects of Electrochemistry No. 6* (New York: Springer) pp 329-373
[9] Richmonds C and Sankaran R M 2008 *Appl. Phys. Lett.* **93** 131501
[10] Chen Q, Kaneko T and Hatakeyama R 2012 *Appl. Phys. Exp.* **5** 086201
[11] Armstrong D A, Huie R E, Lymar S, Koppenol W H, Merényi G, Neta P, Stanbury D M, Steenken S and Wardman P 2013 *BioInorg. React. Mech.* 9 59-61
[12] Tabata Y 1990 *Pulse Radiolysis* (Boca Raton, FL: CRC Press)
[13] Richmonds C, Witzke M, Bartling B, Lee S W, Wainright J, Liu C-C and Sankaran R M 2011 *J. Amer. Chem. Soc.* **133** 17582
[14] Witzke M, Rumbach P, Go D B and Sankaran R M 2012 *J. Phys. D: Appl. Phys* **45** 442001
[15] Rumbach P, Witzke M, Sankaran R M and Go D B 2013 *J. Amer. Chem. Soc.* **135** 16264
[16] Rumbach P, Bartels D M, Sankaran R M and Go D B 2015 *Nat. Commun.* **6** 7248
[17] Hart E J and Boag J W 1963 *J. Amer. Chem. Soc.* **84** 4090
[18] Bartels D M, Takahashi K, Cline J A, Marin T W and Jonah C D 2005 *J. Phys. Chem. A* **109** 1299

Reduction of NOx and PM in marine diesel engine exhaust gas using microwave plasma

W Balachandran, FInstP, N Manivannan, R Beleca, and M Abbod
Centre for Electronic Systems Research
Brunel University, Kingston Lane, Uxbridge, UB8 3PH, UK

Email: Wamadeva.Balachandran@brunel.ac.uk

Abstract. Abatement of NOx and particulate matters (PM) of marine diesel exhaust gas using microwave (MW) non-thermal plasma is presented in this paper. NOx mainly consist of NO and less concentration of NO_2 in a typical two stoke marine diesel engine and microwave plasma generation can completely remove NO. MW was generated using two 2kW microwave sources and a saw tooth passive electrode. Passive electrode was used to generate high electric field region within microwave environment where high energetic electrons (1-3eV) are produced for the generation of non-thermal plasma (NTP). 2kW gen-set diesel exhaust gas was used to test our pilot-scale MW plasma reactor. The experimental results show that almost 100% removal of NO is possible for the exhaust gas flow rate of 60l/s. It was also shown that MW can significantly remove soot particles (PM, 10nm to 365nm) entrained in the exhaust gas of 200kW marine diesel engine with 40% engine load and gas flow rate of 130l/s. MW without generating plasma showed reduction up to 50% reduction of PM and with the plasma up to 90% reduction. The major challenge in these experiments was that igniting the desired plasma and sustaining it with passive electrodes for longer period (10s of minutes) as it required fine tuning of electrode position, which was influenced by many factors such as gas flow rate, geometry of reactor and MW power.

1. Introduction

Marine diesel engine exhaust gas contains toxic gases such as NOx($NO+NO_2$) and SO_2 and it very important to convert these gases into harmless composition to avoid the detrimental effect these gases can cause to environment (e.g.: acid rain) as well as human health (e.g.: Aggravation of asthma and Inflammation)[1]. New regulations are being also imposed by UN IMO to control the emission levels of NOx and SOx from marine exhaust [2]. While SOx emission can be limited by using fuel without sulphur, an abatement method should be adapted to remove NOx from the exhaust gas. Particulate matters (PM) are also another harmful composition found in marine diesel engine exhaust [1]. Some of the health effects caused by PM are premature death in people with heart or lung disease, nonfatal heart attacks, irregular heartbeat, aggravated asthma, decreased lung function and increased respiratory symptoms [1]. Soot particles (unburnt carbon) are the major contributor to PM and needs to be removed from the exhaust before getting into the atmosphere.

In this paper, the results obtained in the experiments performed to abate NOx and PM using MW plasma is presented. Non-Thermal Plasma (NTP) is an emerging technology for abatement of NOx of gaseous emissions from diesel engine exhaust. The fundamental nature of non-thermal plasma is that the electron temperature is much higher than that of the gas temperature, including vibration and rotational temperature of the molecules. The electron energy is consumed in the ionisation, excitation and dissociation of the molecules and finally in the formation of active free radicals OH•, O•, N• and H•. These radicals oxidise NO to NO_2, which reacts with water vapour present in the exhaust gas forming aerosol HNO_3 which can be safely disposed into after adjusting its PH level using seawater. It was reported in our previous study that kinetics of NTP allows removal of NOx very effectively to almost 100% [3]. There are three methods for generating non-thermal plasma in atmospheric gas pressure; electron beam (EB) radiation, Microwave (MW) radiation and Electrical Discharge (DC, AC and pulsed field). Out of these there MW is very cost effective and had been exploited in many applications [4][5] and hence experimented in this work.

Section 2 presents the MW reactor experimental set-up and the results are presented in Section 3 along with brief discussion. Finally conclusion is drawn in Section 4.

2. Experimental setup with microwave plasma

Figure1: Pilot scale Microwave plasma at Brunel University: 1- 2kW gen-set generator, 2 – Testo gas analyser at inlet/outlet of NTPR, 3 – Spelman 30 kV DC power supply, 4 – Multi-mode MW cavity, 5 – Sensors data logging system, 6 – Flow meter; 7 – Gas extraction to atmosphere

Figure 1 show the MW plasma experimental setup used in Brunel University laboratory. This set-up includes multi-mode microwave cavity, gen-set diesel generator, Testo-350 gas measurement systems (x2), 30kV high voltage supply, gas flow rate meter and data logging system. Two 2kW microwave generators operating at 2.45GHz were used to supply required microwave energy into the microwave cavity through two slotted waveguides. Full details of the cavity design can be found in [6]. The gen-set generator exhaust contains higher percentage of NO than NO_2, which is very similar to the exhaust from a typical 2-stoke diesel marine engine. Passive electrodes (needle electrode, saw tooth electrode and cross saw tooth electrodes) were used to ignite plasma in the MW reactor and it produced excellent results in the NOx abatement for moderate flow rate (60 – 70 l/s) from the gen-set generator.

3. Results and discussion

Two sets of experiments were carried out with pilot scale MW plasma. The first one was performed in the Brunel Laboratory for the exhaust from a 2kW gen-set diesel electricity generator and the second test was for the exhaust from a 200kW diesel engine in the Southampton pilot scale test site. The main purpose of the first set of experiments was to understand the performance of NOx removal whereas the second set of experiments was intended to perform for the removal of PM (soot).

NOx Removal

A number of test were performed at different operating condition of exhaust of the gen-set diesel generator (temperature, flow rate and load) and each of the test required fine tuning of MW power and position of cross saw tooth in the reactor to ignite and maintain the plasma. The typical results are shown in figure 2. As shown in figure 2(a), when MW plasma is ignited, the concentration NO went down to almost zero and hence the concentration of NOx reduced significantly while there was no reduction happened to NO. This results were obtained when exhausts gas temperature was between 40-700C and exhaust flow rate was 60l/s. Removal efficiency for this test was 100% of NO and 65% of NOx for the microwave power of 1kW (i.e.: 2x 500W). When the flow rate was increased for same supplied MW power, the removal efficiency dropped in proportion. The removal efficiencies for three flow rates are shown in the figure 2(b). As can be seen in this figure NO (and NOx) removal efficiency has decreased for the increased gas flow rate, as it would require more plasma energy to treat higher gas flow rate. It was also noted from another set of tests that for relatively higher gas temperature (> 1000C), the concentration of NO was increased while gas passed through MW plasma. This resulted in negative removal efficiency of NO and NOX (<-100%) and it was due to oxidation of N radicals into NO. Therefore the temperature of the exhaust gas is very important in the NO abatement process.

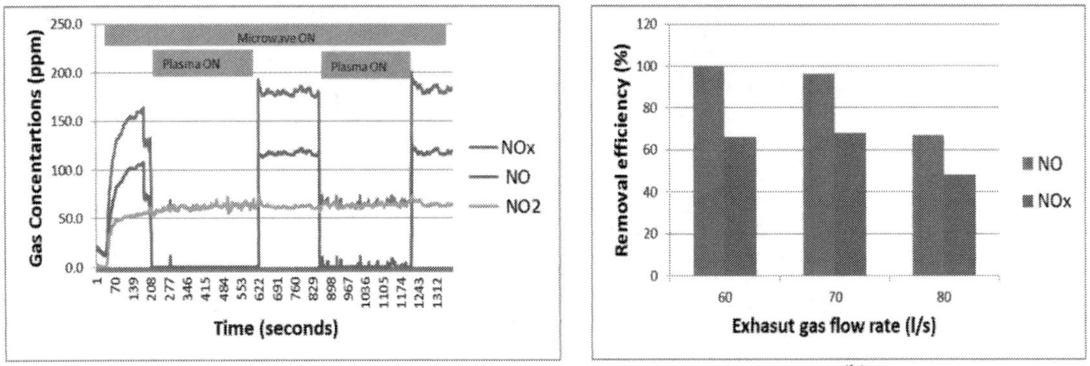

(a) (b)

Figure 2: Results for NOx removal (a) NOx concentrations when plasma is ignited (b) NOx removal efficiency vs exhaust flow rate for the gas temperature between 40-70 ^{0}c

PM Removal

A number of experiments were carried out to investigate the PM removal capacity of MW plasma at the Southampton pilot scale test site. Exhaust from 200kW diesel engine at 40% engine load was passed through MW plasma and measurements at the outlet were taken using TSI PM Analyser. Two scenarios of PM removal with MW were experimented, with and without plasma. These results are shown in figure 3. Figure 3(a) shows particle of various size form 11.5nm to 365.2nm. As can be seen in this figure, when microwave is ON the concentrations of particles has dropped significantly, mainly the small sized particles. When plasma stuck (at around 11:38am) for a very short period, the removal of particles was very high. Figure 3(b) shows the removal of total particles for Microwave ON and plasma ON. It is estimated that form number of repeated PM removal tests that removal efficiency with microwave without plasma was 50% and as high as 90% when MW plasma was generated. The precipitation of the particles is due to the distribution of electric field within the reactor generated by the MW. When the plasma is formed the ionisation of the gas molecules enhanced the electric field as a consequence the particles removal efficiency increased significantly.

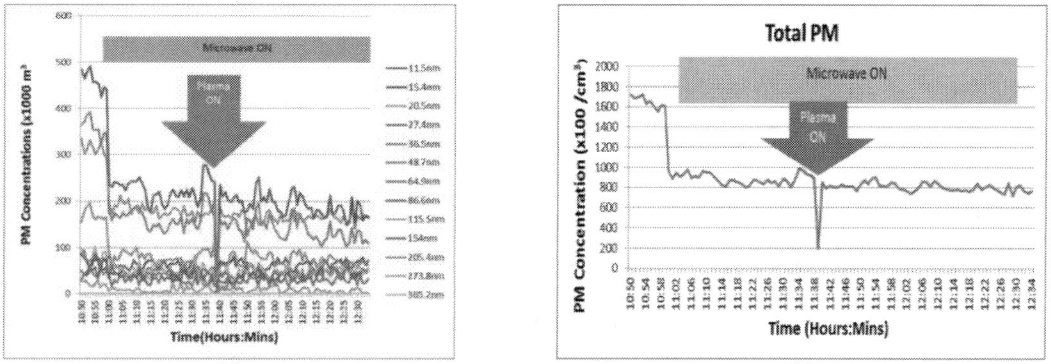

(a) (b)

Figure 3: PM removal using MW plasma (a) various sizes of PM (b) Total PM

3. Conclusions

Removal NOx from the exhaust was shown experimentally using microwave plasma for the exhaust flow rate up to 60l/s and for the exhaust temperature between 40-70oC. It was also shown that removal of PM (soot particles) can also be performed using MW plasma very effectively. The challenges still remaining are ignition and maintaining of non-thermal MW plasma for larger volume of multi-mode MW cavity reactor.

Acknowledgements

Authors wish to acknowledge EU for their financial support under FP7 program (EU contract number: 284745) to carry out the research presented in this paper.

References

[1] U.S Environmental Protection Agency, "What Are the Six Common Air Pollutants?," U.S Environmental Protection Agency, 22 December 2014. [Online]. Available: http://www.epa.gov/airquality/urbanair/. [Accessed 1 Feb 2014].

[2] International Maritime Organization, "IB664E - MARPOL Annex VI & NTC 2008, 2013 Edition," *2013* International Maritime Organization, London, UK.

[3] Manivannan N , Balachanran W, Beleca R and Abbod M 2014 *Journal of Clean Energy Technologies* **2** 233-6

[4] Chandrasekaran S, Ramanathan S and Basak T 2013 *Food Research International* **52** 243–61

[5] Remya N and Jih-Gaw Lin J 2011 *Chemical Engineering Journal*, **166** 797-813

[6] Manivannan N, Balachandran W, Beleca R and Abbod M. 2015 *International Journal of Environmental Science and Development* **6** 151-4

Electrostatics 2015 IOP Publishing
Journal of Physics: Conference Series **646** (2015) 012054 doi:10.1088/1742-6596/646/1/012054

Electrical and mechanical characteristics of nanosecond pulsed sliding dielectric barrier discharges with different electrode gaps

K D Bayoda[1], N Benard[1] and E Moreau[1]
[1]Institut PPRIME (CNRS UPR 3346, Université de Poitiers, ISAE-ENSMA)
SP2MI - Téléport2 Bd Marie & Pierre Curie BP 30179, 86962 Futuroscope, France.

E-mail: kossi.bayoda@univ-poitiers.fr

Abstract. This study proposes the characterization of a surface sliding discharge that extends over a length of 80 mm. The gas ionization is caused by series of high voltage pulses with nanosecond rising and decaying times while ion drift is forced by a negative DC component. Different plasma diagnostics such as electrical measurements, iCCD visualizations and strioscopy have been performed. They highlight that a threshold mean electric field between both air-exposed electrodes is required to fully establish a sliding discharge. Compared to a single nanosecond pulsed dielectric barrier discharge, the sliding discharge results in an energy consumption increase. Moreover, the pressure wave induced by the discharge is strongly impacted.

1. Introduction

Non-thermal surface plasma discharges can be used to control actively airflow over profiles. However, realistic applications at real aircraft or MAV scale require usually a large surface plasma in order to enhance the interaction region. Dielectric Barrier Discharges (AC-DBD) supplied by ac sine high voltage, are used for such applications at low Reynolds number and reduced scale [1, 2]. In this case, the flow control is based on the electrohydrodynamic force produced inside the discharge, resulting in an electrical wind-based wall jet. Recently, a new kind of plasma actuator has proven its effectiveness for airflow at high Reynolds number [3, 4]. In this case, the high voltage applied between both electrodes has a nanosecond pulsed waveform and the plasma extension is limited to a few millimeters. The flow control mechanism is no more caused by an induced body force but it seems result from fast heat deposition whose gas signature is a hemispherical region of high pressure propagating at sound speed. In view of overcoming the limited plasma extension, a new design based on three electrodes has been perfected to ignite a sliding discharge [5, 6]. Here the second air-exposed electrode that is supplied by a dc voltage allows us drifting electrical charges originating from the active electrode. In the present paper, an experimental study compares the electrical and mechanical characteristics of a single nanosecond surface dielectric barrier discharge (DBD) and a sliding discharge with electrode gap varying from 40 up to 80 mm. First, iCCD visualizations of both discharges are proposed. Secondly, the corresponding electrical measurements of discharge currents and energy consumptions are compared. Finally, shadowgraph characterizations of the induced pressure wave are performed

2. Experimental setup

The design used to produce a sliding discharge consists of three aluminum electrodes flush mounted on both sides of a 2-mm thick PMMA plate, as described in Figure 1. The single DBD design, composed of only two electrodes mounted on both sides of the dielectric plate (only electrodes (1) and (3) in Figure 1) is considered as the reference case (Gr). With this last design, a typical nanosecond discharge is generated by connecting electrode (1) to a pulsed high voltage power supply (FID). The high voltage waveform is shown in Figure 2 (1 kHz, 16 kV_P peak voltage). The rising and decaying time are about 10 ns. To extend the plasma layer, a third electrode (3) is placed at 3 different positions from electrode (1) (gaps of 40, 60 and 80 mm). These designs are called G40, G60 and G80. The third electrode is supplied by a negative DC high voltage. Current measurements are performed on electrode (1) and (3) by means of Bergoz current transformers, allowing to compute the total input energy. Details of the electrical system can be found in [8]. Plasma imagery diagnostics are also carried out with the help of iCCD camera (PI-MAX-1024i) coupled with an UV lens (100F/2.8), showing the plasma morphology. Finally, experiments of shadowgraph visualizations of the induced pressure wave using the optical bench described in [7] are performed.

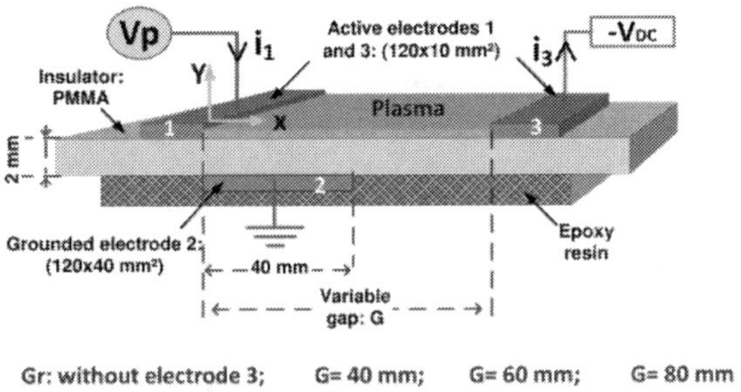

Figure 1: Schematic of the plasma actuator.

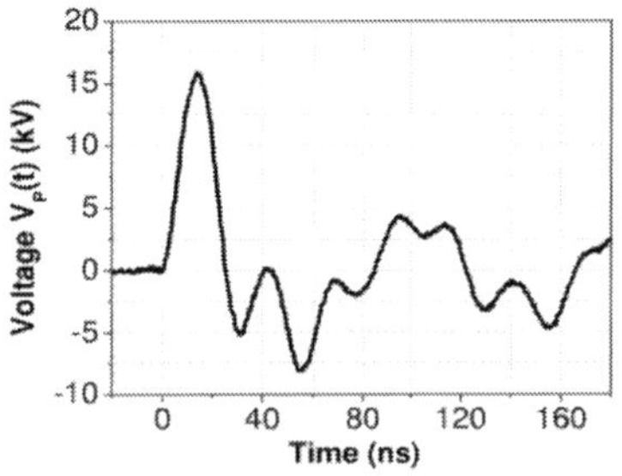

Figure 2 : Pulsed voltage versus time.

3. Results

3.1. iCCD visualisations of the discharges

Figure 3 shows top view iCCD images of the discharge for the four configurations. For all the cases, these images reveal that streamers propagate from electrode (1) over the actuator surface, as observed in [7, 9]. For the reference case (Gr), the pulsed voltage is fixed at 16 kV_P (peak value). The plasma layer extends up to 16 mm from electrode (1) (Figure 3). When the third electrode is added and supplied by a negative DC voltage, the charged species slide over the dielectric surface. As demonstrated in [8], the presence of the third electrode, when it is not supplied, does not influence the pulsed discharge behaviour. For a gap between electrodes (1) and (3) equal to 40 mm (G40), a sliding discharge is obtained for V_{DC} = -18 kV (V_P = 16 kV_P). By increasing $|V_{DC}|$ up to 34 kV_{DC} and V_P maintained at 16 kV, it is possible to extend more the plasma layer, over a length of 60 mm (G60 in Figure 3). For G80, a sliding discharge is initiated by increasing not only the DC voltage (-40 kV, corresponding to the DC power supply limitation) but also the pulsed voltage (V_P = 20 kV_P).

3.2. Electrical diagnostics

Figure 4-A and B illustrate the current measured at electrode (1) and (3) respectively, for all the cases. The currents $i_1(t)$ is composed of two peaks as observed in previous studies [7, 10]; that correspond to the ionization caused by the pulsed high voltage. When the sliding discharges are ignited, the positive peak value of $i_1(t)$ increases from 1.6 A/cm for the reference case (Gr) up to 2.4 A/cm for G40 and G60 (in both cases, V_P = 16 kV_P). For G80 (V_P= 20 kV_P and V_{DC} = -40 kV), the current peak reaches 2.7 A/cm.

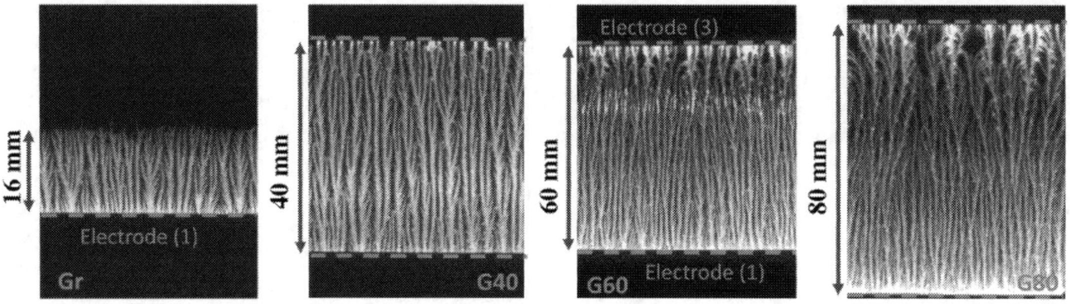

Figure 3 : ICCD images of the plasma layer for a single DBD (Gr) and the sliding DBD with gap equal to 40 mm (G40), 60 mm (G60) and 80 mm (G80).

Moreover, the establishment of the sliding discharge can be also highlighted by current $i_3(t)$ (Figure 4-B), that reflects the amount of charges collected by electrode (3). For G40, G60 and G80 configurations, the i_3 amplitudes are equal to 0.64, 0.38 and 0.15 A/cm, respectively. These variations in i_3 amplitudes show the influence of the mean surface electric field between electrode (1) and (3), defined as $V_{PD}/cm = (V_P - (-V_{DC}))/G$ where G is the gap between electrodes (1) and (3). Indeed, this value affects the propagation speed of the ionized species, as discussed in [8]. From current and voltage curves, the total energy consumed by the discharge is evaluated. As a reminder, it has been demonstrated previously that the input energy is strongly linked to the pressure wave intensity [3, 7]. The energy consumed by the discharge is computed as the addition of the pulsed and DC contributions. Figure 4-C presents the energy variation as a function of the mean electric field for $V_P =$ 16 and 20 kV, for G40, G60 and G80 configurations. On one hand, the total energy consumed by the three-electrode actuator is constant (0.24 mJ/cm for V_P = 16 kV) and equal to the pulsed contribution when $V_{PD}/cm < 6.5$ kV/cm because there is no sliding discharge as indicated by iCCD images (data not shown). On other hand, when V_{PD}/cm exceeds 6.5 kV/cm, the sliding discharge is established and total energy increase gradually (1.25 mJ/cm for G60). This result is consistent with AC sliding discharges results which are reported in [5] where the same threshold of 6.5 kV/cm has been observed.

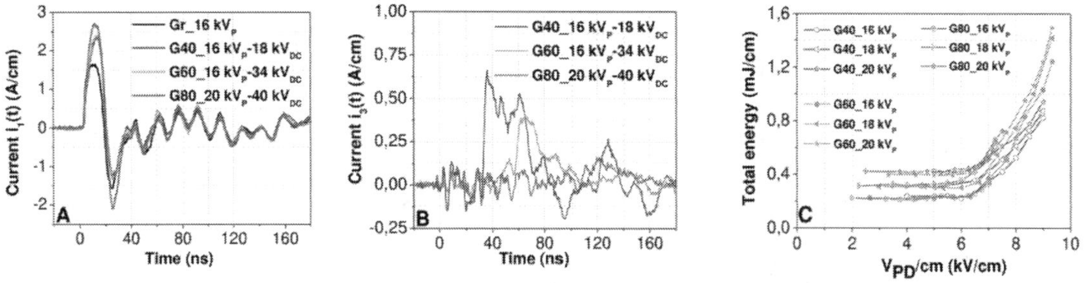

Figure 4 : Current waveforms measured at electrode (1) (A), at electrode (3) (B) and total energy consumption as a function of the mean surface electric field (C).

3.3. Optical characterization of the induced pressure wave

The last experiment is addressed to the pressure wave characterization through shadowgraph images. For all the configurations, Figure 5 shows side views of the air density gradient (pressure wave topology) at t = 30 µs after the discharge ignition. By investigating the position of the wave front versus time, a propagation of the wave at 351 m/s has been observed, as in [3, 7, 10]. For the reference case, the pressure gradient consists of a hemispherical wave and a planar one. The first one, originating from the edge of electrode (1), is attributed to the release of electrical energy transformed into a sudden localized heat source. The planar wave is suggested to be caused by the plasma extension over the dielectric surface. The present investigation confirms this hypothesis. Indeed, when the sliding discharge is ignited, the planar wave extends from electrode (1) to electrode (3), as the surface discharge. Finally, one can remark the presence of a second hemispherical wave located at electrode (3) with the three sliding discharges, certainly due to a heat release induced by the current i_3. Concerning the pressure wave intensity, the shadowgraph images allow us to estimate it qualitatively through the grey level of the images. For instance, on the reference image (Gr in Figure 5) it can be observed that the rear half part of the spherical wave is less bright than the forward and front parts, indicating that the pressure gradient of the wave is maximum downstream electrode (1). For the sliding discharges, the intensity of the second spherical wave at electrode (3) is weaker than the one occurring at electrode (1). This can be explained by the lower amplitude of current $i_3(t)$ compared to $i_1(t)$.

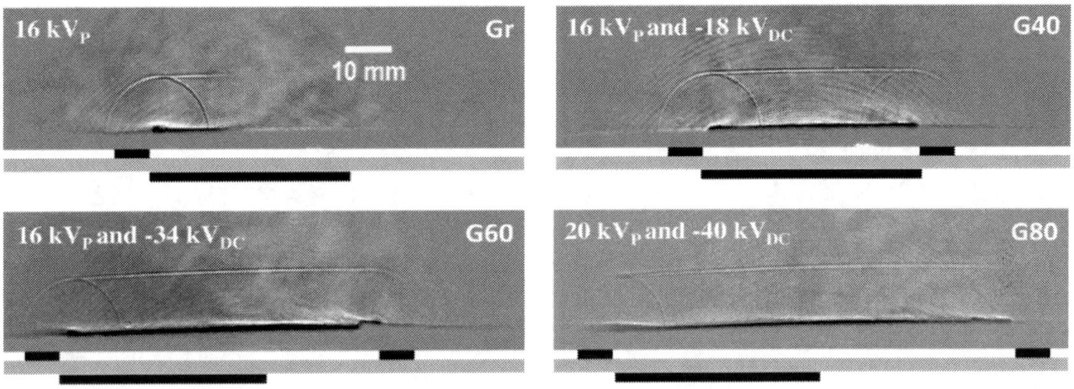

Figure 5 : shadowgraph images of the induced pressure wave for the studied configurations

4. Conclusion

A single nanosecond dielectric barrier discharge based on two-electrode design and a sliding one that uses three electrodes are compared by electrical and optical diagnostics. The sliding discharge can be established over the dielectric surface, in the region located between both air-exposed electrodes, if the mean electric field is higher than a threshold value equal to 6.5 kV/cm. Moreover, the energy consumed by the sliding discharge is three or four times the one needed for a single DBD. The sliding discharge induces two spherical pressure waves which are linked by a planar one. Consequently, that should enhance the interaction region between the incoming flow and plasma.

5. References

[1] Moreau E 2007, *J. Phys. D*, **40** 605-636.
[2] Benard N and Moreau E 2014, *Exp. Fluids*, **55**,1846.
[3] Roupassov D, Nikipelov A, Nudnova M and Starikovskii A 2009, *AIAA J*, **47** 168-185.
[4] Wu Y, Yinghong L, Hua L, Jun L, 2014 *Proc. AIAA (Atlanta)*, **45**. 2118.
[5] Sosa R, Kelly H, Grondona D, Marquez A, Lago V and Artana G 2008, *J. Phys. D: Appl. Phys.*, **41** 035202.
[6] Song H; Li Y; Zhang Q, Jia M; Wu Y, 2011, *J. Plasma. Sci. Technol.*, Vol. **13** 5.
[7] Paniel E, Rabat H and Hong D, 2014, *IEEE Trans. Plasma Sci.*, **42**.
[8] Benard N, Zouzou N, Claverie A, Sotton J, and Moreau E, 2012, *J. Appl. Phys*, **111** 033303.
[9] Bayoda K D, Benard N and Moreau E. 2015, *J. Electrostaics*, **74**, 79-84.
[10] Dawson R and Little J 2013 *J. Appl. Phys*, **113**, 103302.

Evaluation of the time-resolved EHD force produced by a plasma actuator by particle image velocimetry – a parametric study

N Benard[1], M Caron[1] and E Moreau[1]
[1] Institut PPRIME - CNRS - University of Poitiers – ENSMA, 86962, Futuroscope, France

E-mail : nicolas.benard@univ-poitiers.fr

Abstract. Surface plasma discharges, and more specifically dielectric barrier discharge, find several applications in aerodynamic due to their capability to produce a local flow at the dielectric wall. The present study proposes a numerical method to estimate the EHD force by using experimental velocity information. Here, this method is used in a parametric study and the obtained results are compared with force balance measurements. It is shown that the EHD volume force increases in space and in amplitude for increasing voltage and increasing AC frequency. Furthermore the force distribution expands in an homothetic manner.

1. Introduction

Non-thermal surface plasma discharges have gained an increasing interest over the last decade due to their capability for manipulating flows at low speed flow conditions. A variety of actuators have been designed over the years [1] and all of them have demonstrated their own merits in terms of aerodynamic performance increase [2]. These surface actuators produce a tangential wall jet limited to maximal velocity of about 7 m/s, but the control mechanism is most of the time related to excitation or damping of the natural flow instabilities suggesting that the induced mean flow velocity is not necessarily the main actor for an effective control. The studies performed over the years have confirmed that the optimization of these discharges in terms of efficiency and electromechanical conversion is definitively a harsh challenge. Despite significant effort from the community, several technical obstacles remain and have to be tackled. Among them, the experimental characterization of the electrohydrodynamic (EHD) body force (in space and in time) produced by such a discharge is a key point because no optimization can be conducted without a fine understanding of the body force production mechanisms and also because experimental data could be of great help for an accurate numerical modeling of the electro-mechanical transfer occurring in a non-thermal plasma.

Here, by inversion of the Navier-Stokes equations [3,4,5], the spatial distribution of the mean EHD force is evaluated based on the mean velocity field. Velocity measurements are conducted for different voltages and frequencies. Then, the spatial distribution of the EHD force and its integrated value can be compared for an increasing applied voltage and an increasing frequency.

2. Experimental set-up

The DBD actuator consists of two aluminum electrodes asymmetrically mounted on a 3-mm thick dielectric plate made of PMMA. All its dimensions are indicated in figure 1. A signal amplifier (Trek, model 30/40 mA) is used to apply a gain of 3,000 V/V to the input voltage provided by a function generator. The input signal is a sine waveform applied to the top electrode while the bottom electrode is grounded and encapsulated.

The mean velocity produced at the center of the actuator (z=0) is measured by a PIV acquisition system composed of dual-head Nd–YAG laser (Evergreen Big Sky, Quantel), a CCD camera with 4920×3280 px² resolution (LX 16M, Lavision), a trigger unit and a multi-core PC running Davis 8 PIV software. The camera is equipped with a 105 mm to record a field of view of 40×14 mm (spatial resolution of 8.1 μm). The flow is seeded with droplets of dielectric oil (Ondina 919). The velocity fields are obtained by a cross correlation algorithm with adaptive multi-pass (resolution of one vector every 65μm). The time-averaged velocity field is obtained by recording 6000 images for each configuration.

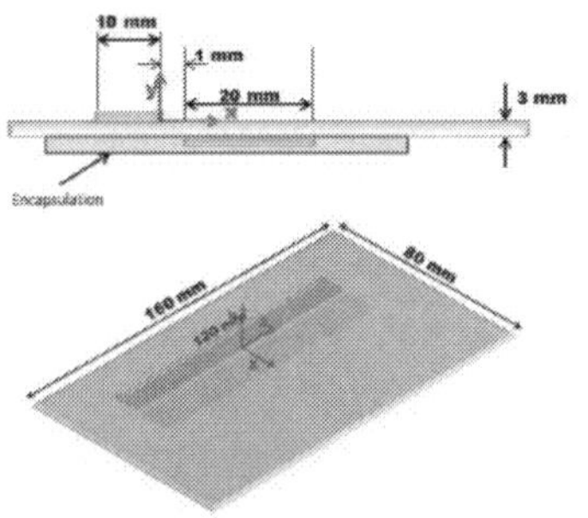

Figure 1. Schematic illustration of the surface DBD plasma actuator.

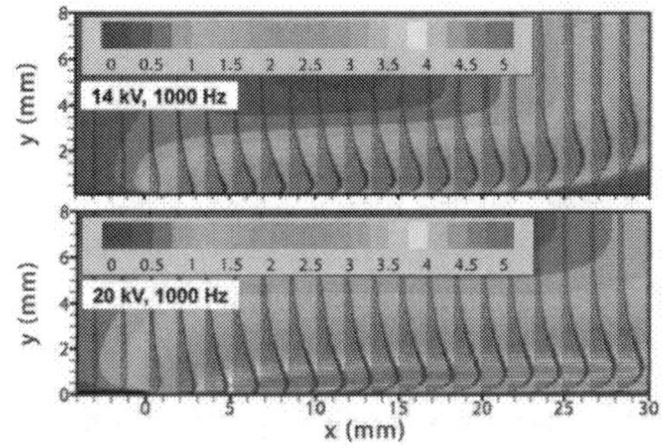

Figure 2. Illustration of the local flow produced by a surface plasma actuator operated at different voltages

3. Results and Discussion

3.1. Estimation of the EHD body force

The mean flow produced by the discharge obeys the incompressible 2D flow momentum equations. At each data point of the measured velocity field, the mean EHD volume force f can be expressed as:

$$\mathbf{F} - \nabla p = \rho \mathbf{U} \cdot \nabla \mathbf{U} - \mu \nabla^2 \mathbf{U} \qquad (1)$$

where p is the pressure field, **U** is the mean velocity field while ρ and μ are the gas density and gas dynamic viscosity, respectively. The right hand side of equation (1) can be evaluated at each measurement location from the velocity measurements as the two cases shown in Figure 2. It is considered here that the pressure contribution is minor regarding the volume force produced by the discharge. Subsequently, the left hand side of equation (1) approximates the EHD volume force. A fortran code (F90) is developed to discretize and then solve equation (1) by finite difference method with second order spatial schemes.

The distribution of the volume force is illustrated in figure 3 for an applied voltage of 20 kV and AC frequency of 1000 Hz. The force is high on the edge of the top electrode, but it also spreads above the dielectric surface. The total volume force is mainly due to f_x, the influence of f_y being restricted to the region above the top electrode. Here, the total force propagates in the first 7-10 mm where the density of the charged species under the influence of the electric field is high.

The EHD body force can be computed as the integral surface of the volume force. The definition of the integration surface needs a careful attention in order to limit the contribution of background noise. Here, the body force is evaluated by integrating the volume force over a control surface that starts from a fixed point located 3.5 mm upstream of the top electrode edge and that propagates gradually in the x and y directions following a rectangular surface (n×2n where n increases by steps of 65 μm). A convergence plot such as the one presented in figure 4 is finally obtained. The force magnitude for which a further increase in the spatial domain does not change the EHD force by more than 1% is retained as final converged value of the EHD body force.

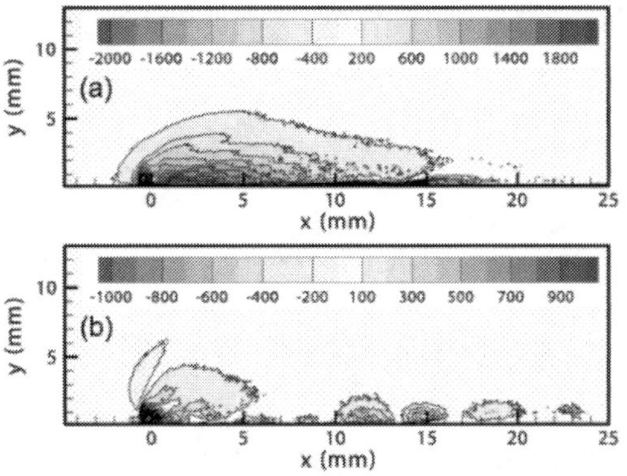

Figure 3. Spatial distribution of the horizontal (a) and vertical (b) components of the evaluated EHD volume force. Force in N/m^3.

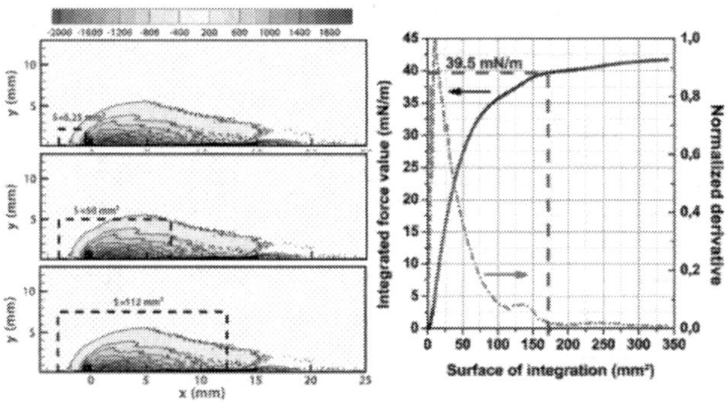

Figure 4. Illustration of the control surface evolution and typical convergence plot for the evaluation of the mean EHD body force.

3.2. Influence of the voltage amplitude and AC frequency

Some illustrations of the volume mean force distribution are proposed in figure 5. These plots indicate an increase in space and in amplitude of the volume force when higher voltage amplitude and frequency are applied. By increasing the voltage and the AC frequency, the density of the ionized species is promoted. This is accompanied by an increase of the produced force in space and amplitude as shown here. However, the topology of the volume force is not dependent on the amplitude and the frequency of the applied high voltage but the body force distribution seems to enlarge by an homothetic manner.

In order to fully validate the evaluation of the EHD body force by experimental measurements coupled to numerical approaches, the results of the present investigation are compared to those reported in [6] where the mean EHD body force was experimentally measured by a precision force balance. In most of the cases, the EHD body force is slightly overestimated when it is evaluated by velocity derivation probably because force balance measurements suffers for parasitic drag penalizing the thrust produced by the discharge. However, the evolution of the EHD body force for an increasing voltage, frequency or consumed power are in agreement with the ones obtained by a precision balance (figure 6). The EHD body force increase as a power of 2 with the applied voltage amplitude while the increase is linear for an increasing AC frequency [1] [2]. Here the saturation regime is not reached, and then the produced EHD body force varies linearly with the electrical consumed power.

4. Conclusion

The spatial distribution of the volume force as well as the EHD body force have been evaluated from the velocity fields produced by a surface DBD plasma actuator. Measurements for different voltage amplitudes and frequencies have been conducted. The evaluation method has been validated by comparison with direct experimental measurements by a force balance. The present study gives useful information regarding the spatial distribution of the mean EHD force and the obtained spatial distribution of the EHD volume force can be used in computational fluid dynamic codes as a realistic source that models plasma actuator.

Acknowledgment

Results obtained in SINAPS project supported by the French ANR program and DGA under reference ANR-13-ASTR-0005. A part of the equipment used in this work was funded by the French Government program "Investissements d'Avenir" (LABEX INTERACTIFS, reference ANR-11-LABX-0017-01)

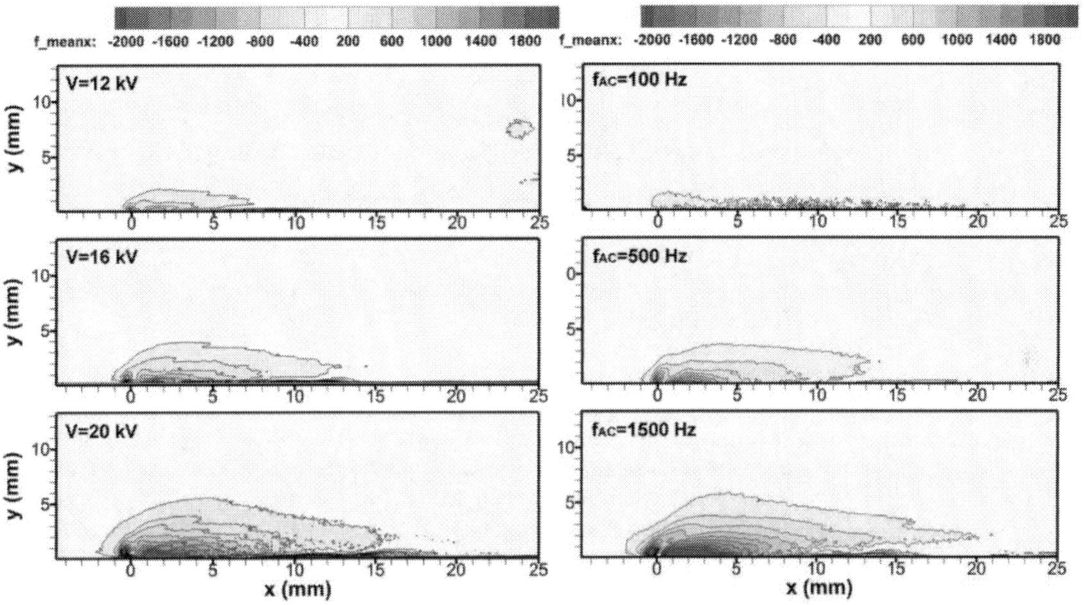

Figure 5. Influence on the EHD volume force of the voltage amplitude while AC frequency is 1000 Hz (a) and applied AC frequency while the voltage amplitude is 20 kV (b). Force expressed in N/m^3.

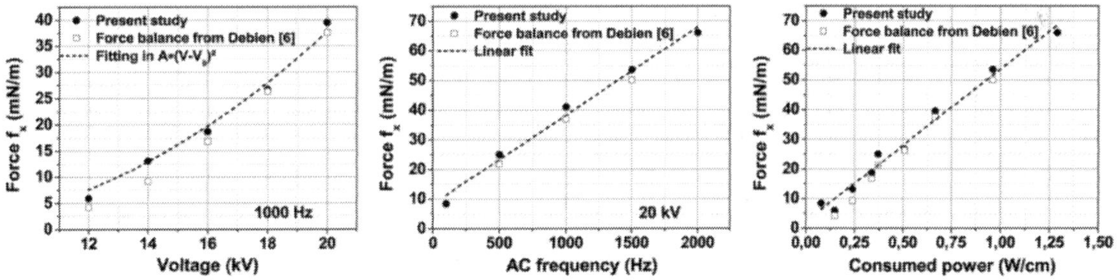

Figure 6. Influence on the EHD body force of the voltage amplitude, AC frequency and consumed electrical power.

References

[1] Benard N and Moreau E 2014 *Experiments in Fluids* **55** 1846
[2] Corke T C, Enloe C L, and Wilkinson S P 2010 *Ann. Rev. Fluid Mech.* **42** 505-529
[3] Kotsonis M, Ghaemi S, Veldhuis L and Scarano F 2011 *J. Phys. D: Appl. Phys.* **44** 045204
[4] Kriegseis J, Schwarz C, Tropea C and Grundmann S 2013 *J. Phys D: Appl. Phys.* **46** 055202
[5] Benard N and Moreau E 2013, *J. Phys D: Appl. Phys.* **46** 245201
[6] Debien A 2012 PhD Dissertation, University of Poitiers

CO_2 reduction using adsorption followed by nonthermal plasma treatment

Kenji Nakajima, Kazuya Takahashi, Masanari Tanaka, Tomoyuki Kuroki and Masaaki Okubo

Department of Mechanical Engineering, Osaka Prefecture University, 1-1 Gakuen-cho, Naka-ku, Sakai 599-8531, Japan

E-mail: sx101072@edu.osakafu-u.ac.jp

Abstract. Carbon dioxide (CO_2) is one of the main substances linked to global warming, and its emission should be reduced. In this study, a CO_2 reduction treatment using an adsorbent and a nonthermal plasma flow is investigated. This treatment comprises a physical adsorption process and nitrogen (N_2) plasma reduction process. In the physical adsorption process, CO_2 is adsorbed by the adsorbent. In the N_2 plasma reduction process, the adsorbed CO_2 is reduced to CO by a nonthermal plasma flow that is generated by a plasma reactor with a circulating N_2 plasma flow. The generated CO can be reused as a fuel. We estimate this experimental results by calculating conversion efficiency of CO_2 to CO. In the N_2 plasma reduction process, the CO concentration reaches approximately 1%, regardless of the number of experiments, and conversion efficiency reaches at most 5.3%.

1. Introduction

In recent years, global warming has become a worldwide problem. Carbon dioxide (CO_2) is one of the main substances linked to global warming, and its emission should be reduced. However, the proposed CO_2 removal technologies such as chemical adsorption technology using aqueous alkanolamine solutions require a large cost [1]. Therefore, CO_2 treatment technology having a low cost and high efficiency is required. One promising technology is the conversion of CO_2 to CO by a plasma treatment [2] because CO can be used as fuel for combustion engines.

In this study, a CO_2 reduction treatment system using an adsorbent and a nonthermal plasma flow is investigated. In this experiment, CO_2 is adsorbed by the adsorbent, then the adsorbed CO_2 is reduced to CO by a nonthermal plasma flow. We develop a highly efficient CO_2 reduction treatment system using an adsorbent and a nonthermal plasma flow at the laboratory scale.

2. Experimental setup and method

A schematic diagram of the experimental apparatus is shown in figure 1. In the physical adsorption process, valves F, H, I, and J are opened. An adsorbent with a volume of 1.22 L is placed in the adsorbent chamber that is downstream from the plasma reactor. As an adsorbent, the molecular sieve zeolite 13X (APG-III) [3] is used. CO_2 gas at a concentration of 10.0% with a flow rate of 10 L/min is prepared by a 99.5% CO_2 cylinder and an air cylinder, then introduced into the experimental apparatus, and exhausted through valve H. In this process, CO_2 is adsorbed by the adsorbent. The upstream and downstream CO_2 concentrations from the adsorbent chamber are measured with a gas-detecting tube every 10 min. When adsorption is completed, measurement is finished. It is known that adsorption is

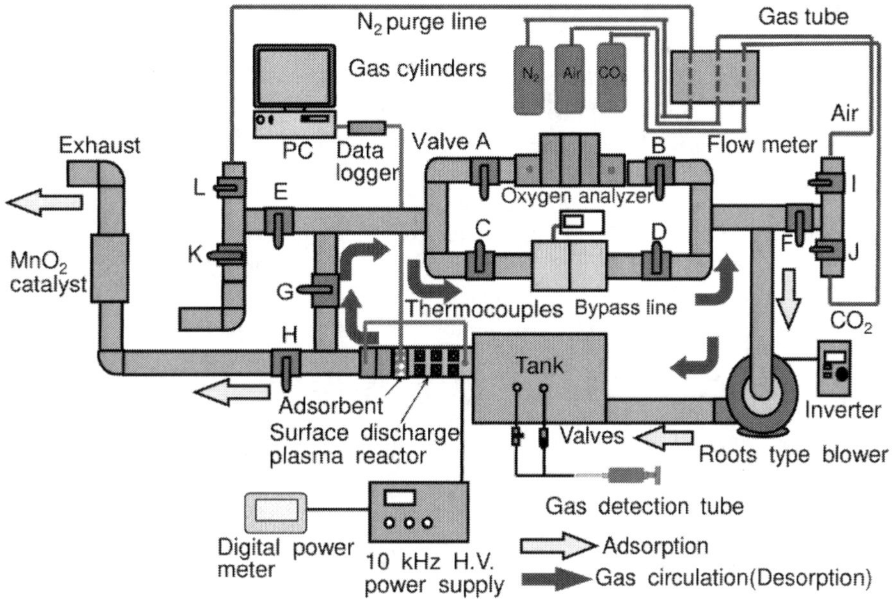

Figure 1. Schematic diagram of the experiment apparatus.

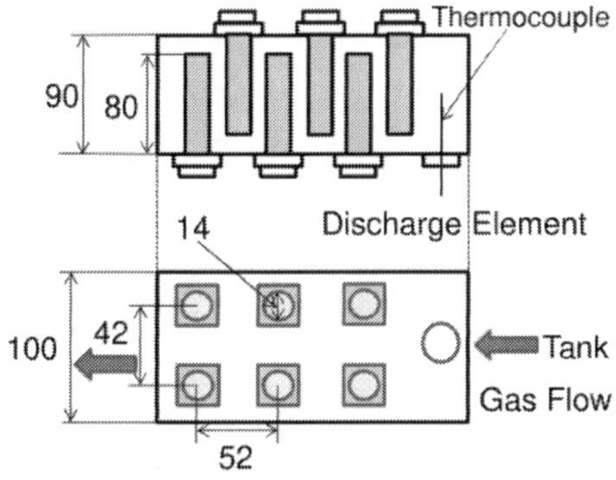

Figure 2. Schematic diagram of the plasma reactor.

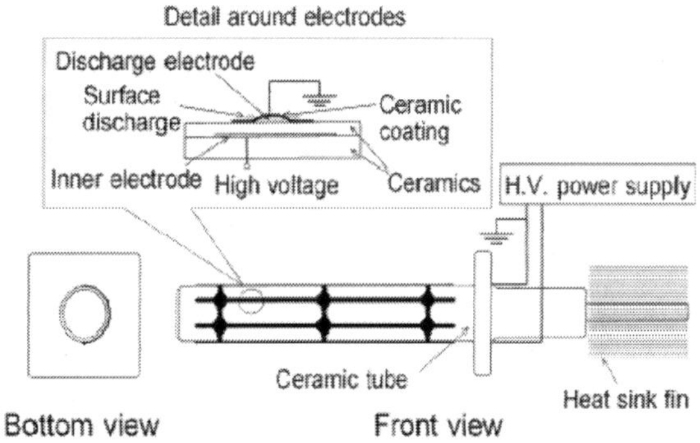

Figure 3. Schematic diagram of the discharged element.

completed when the upstream and downstream CO_2 concentrations reach a steady state. After the physical adsorption process, the flow channel in the apparatus is purged with N_2. First, only valves C, D, E, and H are opened, and 99.5% N_2 gas with a flow rate of 10 L/min is prepared by a 99.5% N_2 cylinder. Then, N_2 is introduced into the experimental apparatus by opening valve L for 60 min. After that, only valves E, G, and H are opened. Then, 99.5% N_2 gas is introduced into the experimental apparatus by opening valve L for 10 min. The O_2 concentration in the flow channel is measured with an O_2 analyzer. When the N_2 purge is completed, the O_2 concentration in the flow channel becomes less than 1.0%.

In the N_2 plasma reduction process, the circulation flow channel is set by opening valve C, D and G, and the N_2 plasma flow is then generated by a plasma reactor and a blower with a flow rate of 0.54 m^3/min. Owing to the N_2 plasma flow, the CO_2 adsorbed by the adsorbent is desorbed and reduced to CO with plasma chemical reactions shown in reference [2]. The CO, CO_2, and O_3 concentrations are measured at a tank with a gas-detecting tube every 10 min. O_2 concentration is also measured every 10 min. After the discharge is finished, only valves C, D, E, H, and K are opened; then, the gas in the flow channel is discharged to the air through a MnO_2 catalyst by a blower.

In both the physical adsorption process and N_2 plasma process, the temperatures of the adsorbent and upstream temperatures of the plasma reactor are measured by thermocouples every 30 s. Upstream temperatures of the plasma reactor in the adsorption and reduction processes are approximately 25°C and 40°C, respectively.

Figures 2 and 3 show a schematic diagram of the plasma reactor (90 mm × 425 mm × 100 mm) and a discharge element, respectively. The plasma reactor consists of twelve surface discharge elements that are energized by a 10-kHz bipolar pulsed high-voltage power supply. The power consumption is 300 W, and the electrodes face toward upstream gas.

The conversion efficiency of CO_2 to CO (η_{CO}) is defined as

$$\eta_{CO}(\%) = \frac{V_{CO}}{(V_{adsCO2} + V_{CO2})} \times 100 \quad (1)$$

where V_{CO} and V_{CO2} are the volume of the CO generation and initial volume of CO_2 in the N_2 plasma reduction process, respectively. V_{adsCO2} is the volume of the adsorbed CO_2 in the adsorption process.

3. Experimental results and discussion

Figure 4 shows the CO_2 concentrations as a function of the treatment time in the physical adsorption process. It is observed in figure 4 that the upstream concentration is stable at approximately 9.5% until 120 min from 10 min. On the other hand, the downstream concentration increases monotonically until 70 min from the start of treatment and is stable at 8.0% after that time. Therefore, the absorbent is almost saturated at 70 min. The volume of adsorbed CO_2 is 32.6 L during the process. The temperature in the adsorbent chamber increases rapidly until 20 min and decreases slowly after that time. Therefore, CO_2 is actively adsorbed until 20 min from the start of treatment.

Figure 5 shows the CO, CO_2, and O_2 concentrations as a function of the treatment time for the N_2 plasma reduction process. The CO concentration reaches approximately 1% at 70 min and is stable after that time. The CO_2 concentration increases monotonically until 30 min from the start of treatment and is stable at 4.0% after that time. The O_2 concentration increases monotonically. Therefore, the treatment time of 70 min is the suitable condition in this experiment because this time is required to reduce CO_2 to CO at the lowest power possible. In addition, the conversion efficiency is at most 2.2% when the treatment time is 70 min. The O_3 concentration is less than 5 ppm during the treatment.

The results when the experiment is repeated with the same adsorbent are shown in figures 6 and 7. The number of repetitions of the experiment is four. It is observed in figure 6 that the CO_2 concentrations of experiments 2–4 increase more rapidly and are stable earlier than experiment 1. The volume of adsorbed CO_2 in experiment 1 is 32.6 L, whereas those of experiments 2, 3, and 4 are between 10 L and 20 L. This is because part of the CO_2 adsorbed by the adsorbent is not desorbed during the N_2 plasma reduction process. It is assumed that CO generation increases by increasing the amount of CO_2 desorbed. Therefore, it is necessary to improve CO_2 desorption by heating the adsorbent using waste heat from CO_2 emission source such as boilers and engines.

It is observed in figure 7 that the CO concentration reaches approximately 1%, regardless of the number of experiments, whereas the CO_2 concentration varies between 3% and 5%. The CO concentrations of experiments 2 and 4 increase with the elapsed time until the treatment is finished. The conversion efficiency reaches at most 5.3%. Therefore, it is necessary that we increase the treatment time. Furthermore, it is necessary to change other parameters such as the flow rate of the blower, the volume of adsorbent, and the temperature of the adsorbent chamber to improve CO generation and the conversion efficiency.

4. Conclusion

In this study, a CO_2 reduction treatment system using an adsorbent and a nonthermal plasma flow is investigated. A volume of CO_2 of 32.6 L is adsorbed by the adsorbent with a volume of 1.22 L in 120 min during the physical adsorption process. In the N_2 plasma reduction process, it is observed that CO_2 is reduced to CO. We repeat this process four times with the same adsorbent, and it is observed that the CO concentration reaches approximately 1%, regardless of the number of experiments. The conversion efficiency reaches at most 5.3%.When the treatment is finished, The CO concentrations of experiments 2 and 4 increase with the elapsed time until the treatment is finished. Therefore, it is necessary to increase the treatment time, and change other parameters such as the flow rate of the blower, the volume of the adsorbent, and the temperature of the adsorbent chamber to improve CO generation and conversion efficiency.

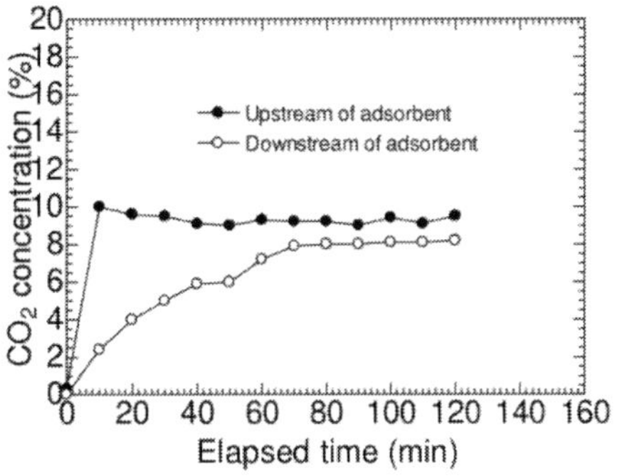

Figure 4. CO_2 concentrations during the physical adsorption process.

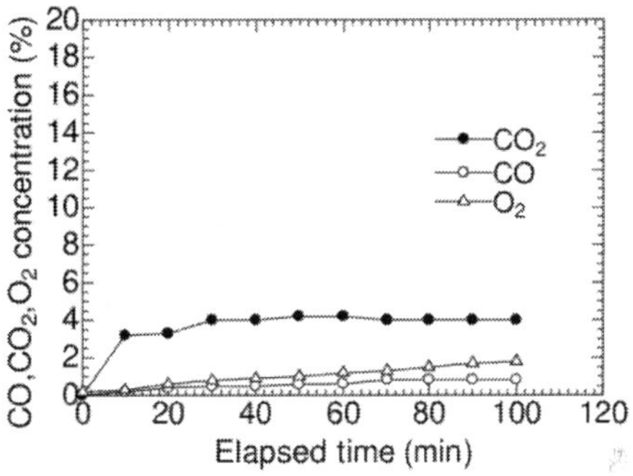

Figure 5. CO_2, CO, O_2 concentrations during the N_2 plasma reduction process.

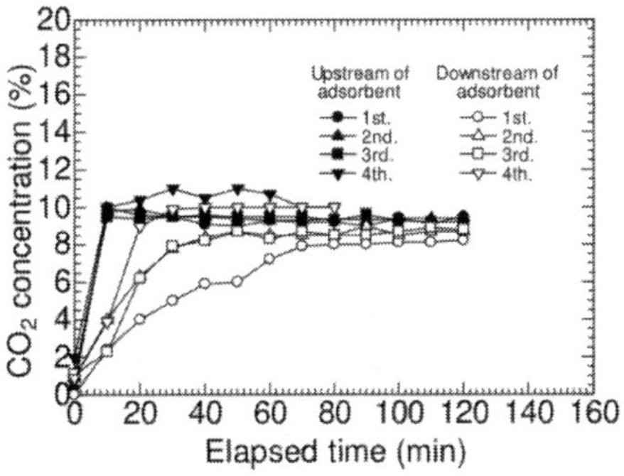

Figure 6. CO_2 concentrations during the physical adsorption process. The experiment is repeated four times.

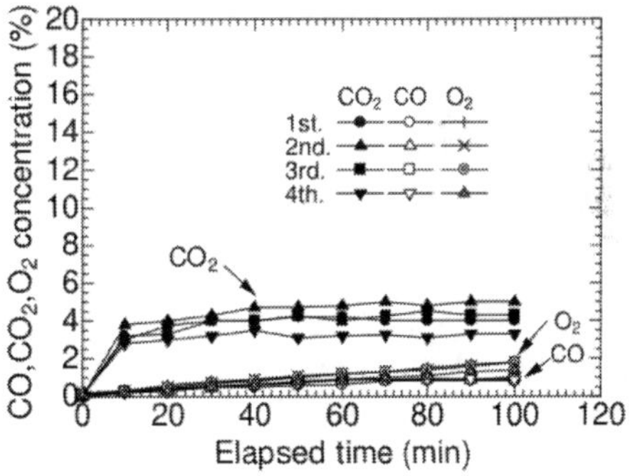

Figure 7. CO_2 concentrations during the N_2 plasma reduction process. The experiment is repeated four times.

Acknowledgments
This work is partially supported by JSPS KAKENHI, Grant Number 25630408, and a grant from the Steel Foundation for Environmental Protection Technology.

References

[1] Yu C-H, Huang C-H and Tan C-S 2012 *Aerosol Air Quality Res.* **12** 745
[2] Spencer L F and Gallimore A D 2011 *Plasma Chem. Plasma Process.* **31** 79
[3] UOP LLC, a Honeywell Company, UOP MOLSIV™ APG-III Adsorbent, catalog, http://www.uop.com/?document=uop-apg-iii-adsorbent-brochure&download=1

Observation of charge separation and gas discharge during sliding friction between metals and insulators

T Miura[1]
Electrical Safety Research Group, National Institute of Occupational Safety and Health, Japan, 1-4-6 Umezono, Kiyose-shi, Tokyo 204-0024, Japan

E-mail: miurat@s.jniosh.go.jp

Abstract. Charge generation due to friction between stainless steel and fused quartz in a vacuum was measured, and it was found that the density of the charge separation at the friction contact was 4×10^{-4} C/m^2. In experiments in ambient gas, reduction of the separated charge caused by microgap gas discharge was observed. The residual rate of the charge, which is the ratio of charge accumulation in an ambient gas to that in a vacuum, in argon ambient gas was small, and it seemed to be effective for the relaxation of generated static electricity due to friction between solids.

1. Introduction

Gas discharge is one of the well-known origins of triboluminescence. Previous microscopic studies of triboluminescence make it clear that the gas discharge occurs in a microscale gap between electrified surfaces around the real contact during friction between solid materials [1]. In other words, the gas discharge acts to reduce static electricity due to charge separation in friction.

To prevent accidents caused by static electricity, such as explosion and fire at an industrial chemical plant, suppression of the charge accumulation is essential. A part of the negative and positive charges separated due to friction is recombined as a result of the microgap gas discharge around contact. We need to investigate the mechanism from the charge generation to the gas discharge in detail.

In the present study, the charge accumulation during friction was measured in a vacuum and in several ambient gases and at several pressures. The efficiency of the charge reduction due to the gas discharge was evaluated by the residual rate of charge derived from the measurements.

2. Experimental setup

Figure 1 shows the experimental equipment of the present study. One purpose of this study is to measure the static electricity generated by sliding friction between metals and insulators under controlled ambient gas conditions. The mechanics of sliding friction between solids was based on a pin-on-disk technique. This type of equipment was also used in previous studies of triboluminescence [1].

The pin and the disk were settled in a vacuum chamber in order to control the ambient gas, such as the kind of gas, pressure, and humidity of the air. The gas was pumped out by using a turbomolecular pump combined with a scroll pump as a fore-vacuum pump; the pumping system was completely operative without oil pollution, and the samples in the vacuum chamber were kept clean. The pressure was measured with a diaphragm gauge and an ion gauge: typically, the base pressure of the vacuum is 1×10^{-3} Pa, and this state is defined as a vacuum condition in this study.

The chamber is made of stainless steel and grounded electrically, so the friction is performed inside the space shielded electromagnetically: this gives higher reliability to measurement of the static electricity by means of an electrometer.

A pin made of stainless steel was used in this study. The tip of the pin was a spherical surface with a radius of 0.5 mm. The pin was held at the end of an arm with a pair of plate springs fixed on a 3-axis manipulator, as shown in Figure 1. The contact force of the friction could be adjusted by the manipulator from outside the chamber without breaking the vacuum and gas condition. A strain gauge fixed on a plate-spring surface is sensitive to deformation, and then normal force at the contact is detected. It was fixed to 130 mN in the experiments.

The disk was made of synthetic fused quartz, which is optically transparent; its diameter was 50 mm, and its thickness was 1 mm. The contact area between the tip of the pin and the disk surface was directly observed with a microscope and a CCD camera from outside the chamber through the disk and a viewport as shown in Figure 1. The diameter of the circular contact, which corresponds to the width of the friction track, w, was 40 μm.

The rotational speed of the disk, n, is controlled with a micro-step stepping motor at a fixed speed. The sliding velocity of friction, v, can be calculated from n and the distance between the contact and the centre of the disk rotation, r; in the present experiments, $v = 0.75$ mm/s ($n = 0.005$ revolution per second (rps) and $r = 24$ mm).

The metal pin was directly connected to the electrometer input; the electrical charge generated on the metal side could be measured, and the charge accumulation was monitored during the sliding friction in real time.

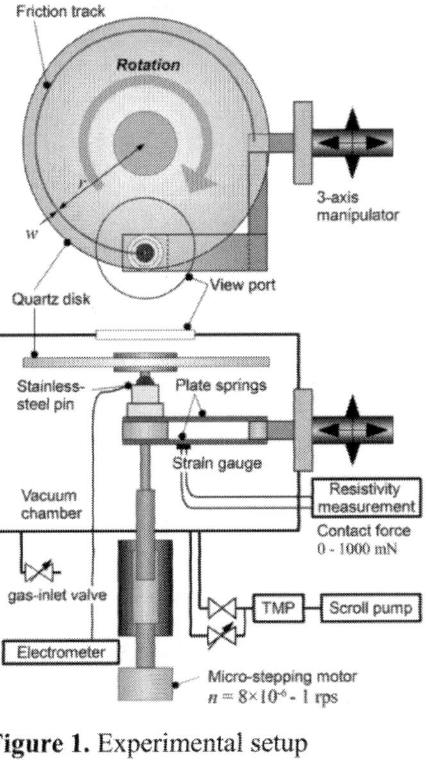

Figure 1. Experimental setup

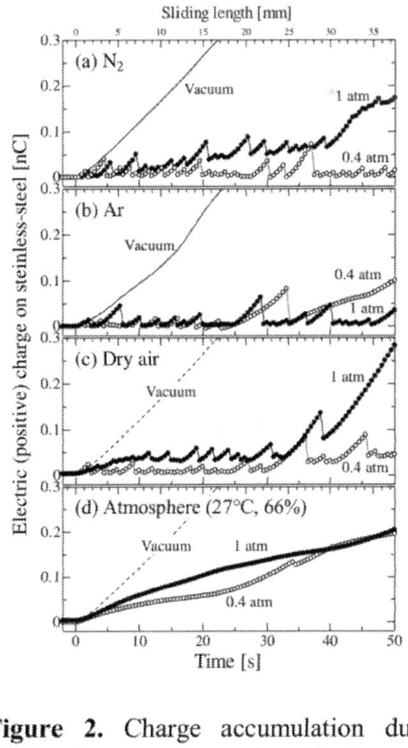

Figure 2. Charge accumulation during friction in a vacuum and ambient gases at different pressures

3. Results

Figure 2 shows the experimental results: the charge accumulation during the sliding friction between stainless steel and fused quartz in a vacuum and four kinds of ambient gases was measured. The results in a vacuum are shown in Figure 2(a) and 2(b). As a result of the present experiments, it was found that the stainless steel was electrified positively and the fused quartz was simultaneously negative. The charge accumulation developed at a fixed rate under the constant velocity of friction. In the process of charge accumulation in a vacuum, charge recombination caused by gas discharge did not occur; therefore, the fixed rate of the charge accumulation, dq/dt, indicates initial charge separation due to friction, i.e., essential triboelectrification without charge back-flow. The charge density of the initial charge separation at the contact, σ, could be estimated using an equation, $\sigma = (dq/dt)/vw$; the result of calculation is 4×10^{-4} C/m^2.

Experimental results for nitrogen gas at pressure of 1 and 0.4 atm are shown in Figure 2(a). It was confirmed that friction between solids in an ambient gas was accompanied by gas discharges [1]. The discharges were clearly observed as intermittently sudden reductions of the charge accumulation. At 0.4 atm of nitrogen gas, the amount of discharge was almost equal to all of the accumulated charge; this was always repeated during sliding friction. Therefore, the charge accumulation remained at a low value, while the rate of the charge separation, i.e., the slope of the accumulation process, was almost the same as that in a vacuum. In the case of measurement at 1 atm of nitrogen gas, the nature of the charge accumulation and the discharge process was almost the same as that of the measurement at 0.4 atm. The charge reduction was, however, small in comparison with the accumulated charge after the last discharge. Therefore, the charge cancelation was insufficient to erase the electrification on the fused quartz surface totally.

Figure 2(b) and 2(c) show measurements of argon gas and dry air, respectively. Results in the case of the dry air were similar to the case of the nitrogen gas, as shown in Figure 2(a). The value of the charge accumulation remained low at 0.4 atm due to the gas discharge; however, the cancelation was insufficient, and the "residual charge" accumulated slightly in the 1 atm pressure. In the case of the argon gas, the gas discharge was sufficient to erase the charge accumulation at both 1 atm and 0.4 atm.

Introducing room air to the vacuum chamber, the charge accumulation during sliding friction in atmospherically humid air was measured, and the results are shown in Figure 2(d). The temperature was 27 degrees centigrade, and the relative humidity was 66%. It was found that development of the charge accumulation at 0.4 and 1 atm was very smooth and lower than that in a vacuum. There is, however, no sudden falling of the charge accumulation.

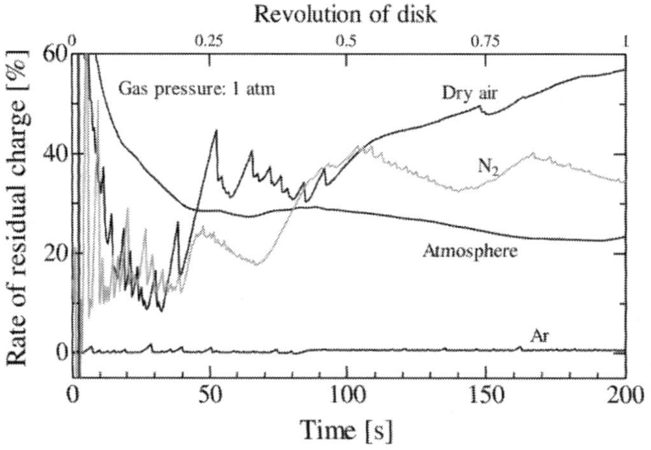

Figure 3. Residual rate of charge under the gas pressure of 1 atm

Table 1. Experimental results of residual rate and reference data of the corona starting gradient [2]

	Residual rate (%)		corona starting
	1 atm	0.4 atm	gradient (kV/cm)
N_2	32 ±7.4	0.9 ±1.3	38.0
Ar	0.5 ±0.2	1.8 ±0.5	7.2
Dry air	40 ±6.0	4.4 ±0.9	35.5
Atmosphere	28 ±1.3	22 ±1.6	-

4. Analysis and discussion

Figure 3 shows the residual rate of charge. These curves are produced from the ratio between the measurement of the charge accumulation in an ambient gas at 1 atm and the measurement in a vacuum at each time. For example, the residual rate in the case of friction under dry air is about 40% in the range between 50 and 150 seconds. This means that electrification of the fused quartz surface due to friction with a stainless steel pin under dry air was only 40% of the initial charge separation; in other words, 60% of the charge generated by friction is eliminated by gas discharge in a microgap. Remarkably, the residual rate of charge is very small in the case of the argon ambient gas.

The average and standard deviations of the residual rate of charge in the range between 50 and 150 seconds were calculated from the data. The results are shown in Table 1. The electric field needed to ignite the corona discharge in each gas at 1 atm [2] is also indicated in the table for reference. In nitrogen, argon, and dry air at 0.4 atm, the microgap discharge could effectively reduce the electrification on the fused quartz surface. However, nitrogen and dry air at 1 atm were not affected sufficiently to eliminate the charge by discharge. Argon gas worked strongly to eliminate electrification, even at 1 atm.

The residual rate of charge in the atmosphere at 1 atm was 28%. Although this is smaller than that in dry air (40%), microgap discharge, i.e., a sudden reduction of charge, was not observed. The effect of humidity on this charge relaxation has not been revealed.

The charge density before the microgap discharge was calculated to be 4×10^{-4} C/m^2 from the charge measurement in a vacuum; the electric field in a microgap should be 450 kV/cm. While it is much higher in comparison with the electric field for ignition of the corona discharge, the fact that the ignition field of argon is small seems to indicate that the residual rate was very low in the experiment in the case of argon gas.

5. Summary

Charge generation due to friction between stainless steel and fused quartz in a vacuum was measured, and it was found that the density of the charge separation at the friction contact was 4×10^{-4} C/m^2. In experiments in ambient gas, reduction of the separated charge caused by microgap gas discharge was observed. The residual rate of the charge in argon ambient gas was smallest, and it seems to be effective for relaxing the static electricity generated due to friction of solids.

Acknowledgments

This work was supported by JSPS KAKENHI Grant number 25350491, Grant-in-Aid for Scientific Research (C).

References

[1] Miura T and Arakawa I 2007 *IEEE Trans. Dielectrics and Electrical Insulation* **14** 560
[2] Mott N F and Bullard E 1953 *Electrical Breakdown of Gases* (London, Oxford University Press) p321

Laboratory bench for the characterization of triboelectric properties of polymers

Bogdan Neagoe, Yopa Prawatya, Thami Zeghloul, Dominique Souchet and Lucian Dascalescu

Institut P', UPR 3346 CNRS - Université de Poitiers - ENSMA, Angoulême, France

E-mail : marian.bogdan.neagoe@univ-poitiers.fr; lucian.dascalescu@univ-poitiers.fr

Abstract. The use of polymers as materials for sliding machine components is due to their low cost, ease of manufacturing, as well as appropriate mechanical and thermal properties. The aim of this paper is to present the experimental bench designed for the study of the triboelectric charge generated in sliding conformal contacts between flat polymer materials. The experiments were performed with 4-mm-thick samples of polystyrene and 5-mm-thick samples of poly-vinyl-chloride.The normal contact force can be adjusted using an appropriate control system and measured by a force sensor (± 50 N). The translational back-and-forth motion of the samples is produced by a crank-shaft system that generates a sinusoidal translational speed profile, with amplitudes between 12 and 50 mm/s, for strokes of 36 to 60 mm. The distribution of charge at the surface of the samples is measured by the capacitive probe of an electrostatic voltmeter (± 10 kV). The experiments pointed out that this bench enables the evaluation of the non-uniformity of the electric charge accumulated on the sliding bodies and the study of the correlations that might exists between this charge and the external forces applied to the contact.

1. Introduction

Polymers are more and more frequently used as materials for sliding machine components. Their choice is due to their low cost, ease of manufacturing, as well as appropriate mechanical and thermal properties. The design of such components should take into consideration the tribological processes in both lubricated and dry polymer-on-polymer contacts [1].

Generation of electric charges through contact and friction between insulating materials has already been thoroughly studied [2], using various test methodologies [3]. This so-called "triboelectrification" effect affects friction and wear [4]. It strongly depends on the load and the speed of the sliding motion, on the state of the surfaces in contact, and on the ambient conditions [5].

Most of the triboelectrification studies were performed for non-conformal contacts [4] – [10]. The area of contact between non-conformal surfaces is very small (almost point or line contact); as a result, the contact pressure is high and the wear of the surfaces is significant even under low loads [6]. The respective experimental benches typically provide uni-directional relative movement of the bodies in contact: a pin and a rotating disk or two rolling rings. The rotational movement is accompanied by charge generation due to the continuous friction between the two bodies [4] - [10].

Fewer studies were made on the tribocharging phenomena associated to conformal contacts [11],[12], where the surface of the contact area is comparable with the size of the bodies involved [6]. The benches described in the literature were typically making use of crank – shaft mechanisms to provide back-and-forth motion between the bodies in contact [11], [12].

The characterization of the charging state of the surfaces in contact is a difficult challenge when the process involves two insulating materials. Some researchers employed the scanning electron microscopy method [3][11], others took advantage of surface electric potential measurements [4][10].The latter is used in the present study, which is aimed at describing a laboratory bench designed to enable the repeatability of the tribocharging conditions in sliding conformal contacts between polymers.

2. Experimental bench

The design of the experimental set-up (figure 1) is such that a conformal contact is provided between two flat polymer samples (1) and (2). The top sample (1) is much smaller (80 mm x 15 mm) than the bottom sample (50 mm x 183 mm). The pressure is evenly distributed on the contact area, as the top sample holder (3) has a system by which the contact planarity between the two samples can be adjusted. The rail guide system (5) assures the horizontal displacement of the bottom sample holder (4). The vertical guiding bars system (6) maintains the normal force perpendicular to the surface of sample (2). This force can be adjusted using the mechanical force control system (7) and measured by the force sensor (N) (model STC 1205, JPRM, maximum force: ± 50 N resolution: 0.025 N).

The sliding bar (8) is parallel with the rail guide system (5) and used to transfer the back-and-forth motion (b) generated by the crank-shaft mechanism (9), which is entrained in rotational motion (c) by the electric motor (EM). The crank-shaft system generates translational motion with a sinusoidal speed profile, with amplitudes ranging between 12 and 50 mm/s, set by modifying the velocity of (EM). The stroke of the crank-shaft mechanism can be adjusted to 60 mm, 50 mm, 42 mm and 36 mm. A displacement sensor (D) (model LVDT L50R, TNC, resolution: 0.05 mm) is attached to the bottom sample holder. The tangential force is measured by a force sensor (T), similar to (N).

The three sensors (N), (T) and (D) are connected to a NI USB 6210 data acquisition board. The measurement data are transferred to a computer using LabView software. The same software and data acquisition board are used to trigger the start and stop of the back-and-forth motion, by controlling the power supply of the electric motor.

The electric potential at the surface of the bottom sample is measured by the capacitive probe (model 6300-7, TREK) of an electrostatic voltmeter (± 10 kV; resolution: 1 V; model P0865, TREK). The probe is mounted on translational stage that can adjust its position within a range of 25.4 mm (each turn of the control knob corresponds to a 1.2 mm step). The measurements are usually made from the center to the edge of the sample (figure 2). They enable the characterization of the charge distribution at the surface of the respective sample, and make possible the study of the correlations that exist between this distribution and the external forces applied to the contact.

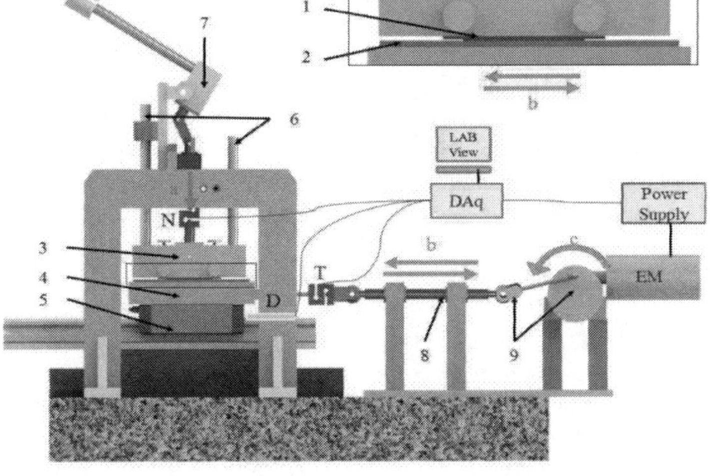

Figure 1. Experimental bench schema.
1- Top sample,
2- Bottom sample,
3-Top sample holder,
4- Bottom sample holder,
5- Horizontal rail guide system,
6- Vertical bar guide system,
7-Vertical force control system,
8- Sliding bar,
9- Crank-shaft system
T, N- Force sensor,
D- Displacement sensor,
EM- Electric motor;

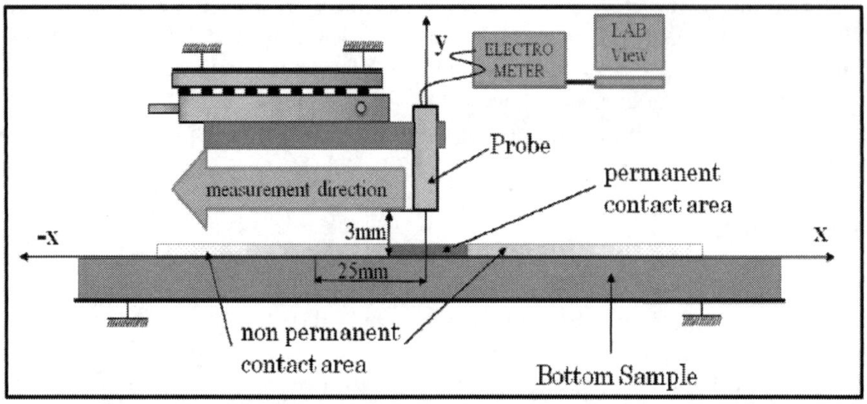

Figure 2. Surface potential measurements.

Because of the back-and-forth motion, the contact area of bottom sample is divided in two zones, one with permanent contact and one with non-permanent contact. The measurements are made after well-defined numbers of back-and-forth motions. All the data are collected from the electrostatic voltmeter trough an electrometer connected to a computer, using LabView software.

3. Materials

The experiments are performed with 4-mm-thick samples of polystyrene (PS) and 5-mm-thick samples of poly-vinyl-chloride (PVC). The samples are cut in two sizes, (1): 80 mm x 15 mm, and (2): 50 mm x 183 mm. The samples (1) are fixed to the top sample holder (3) (Figure 1), while the samples (2) are disposed on a grounded metallic electrode (200 mm x 80 mm), placed on the bottom sample holder (4), attached to the variable-speed sliding system described in the previous section of the paper.

4. Results and Discussion

A set of three experiments are made in ambient air (temperature: 17.9 – 18.7°C; relative humidity: 55 – 59%) using PS for top sample and PVC for bottom sample. In each experiment the normal force is 1 N, the amplitude of the translational speed is 24 mm/s, the stroke of the bottom sample holder is 60 mm, and the duration of a tribocharging (rubbing) cycle is 60 s. The samples are subjected to 12 such rubbing cycles. After each cycle, the electric potential is measured in 11 points, starting from the center of the probe and distanced from each other at 2.4 mm.

The results displayed in figures 3 and 4 are obtained for one sample, but are typical of the data obtained for the whole set of experiments. The measurements show that the charge reaches a maximum level after 9 cycles, than it decreases and begins to stabilize after 12 cycles (figure 3). This saturation phenomenon is likely to be correlated with the various factors of the tribocharging process: nature of the surfaces in contact, normal force, translational speed, etc.

The samples continue to preserve part of the charge acquired by tribocharging. However, the decay rate of this charge is not uniform on the surface of the sample, as pointed out by the electric potential distribution curves recorded at 8 and 24 hours after the 12th and last tribocharging cycle and displayed in figure 4. In the zone of permanent contact, the potential level is close to zero after 24 hours, while a certain amount of charge can still be detected in the non-permanent contact zone.

The more or less rapid decrease of the charge after reaching the saturation might be related to the degradation of the material surface caused by the wear. This is one of the hypotheses that should be confirmed by further studies. Another hypothesis that needs experimental confirmation is related to the different nature of the charge carriers in the permanent and non-permanent contact zones of the bottom electrode. The differences in the charge decay rates of the two zones may be explained by the fact that during the tribocharging process the latter zone is periodically affected by the ions in the ambient air (non-permanent contact with the top sample), while the former is not.

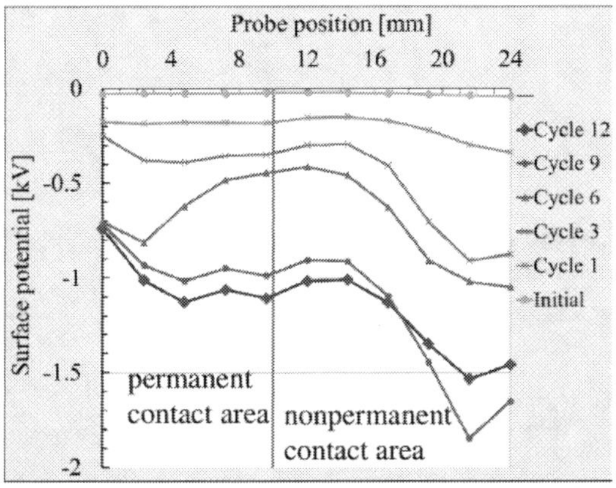

Figure 3. Distribution of the electric potential at the surface of the bottom sample after 0, 1, 3, 6, 9, and 12 tribocharging cycles.

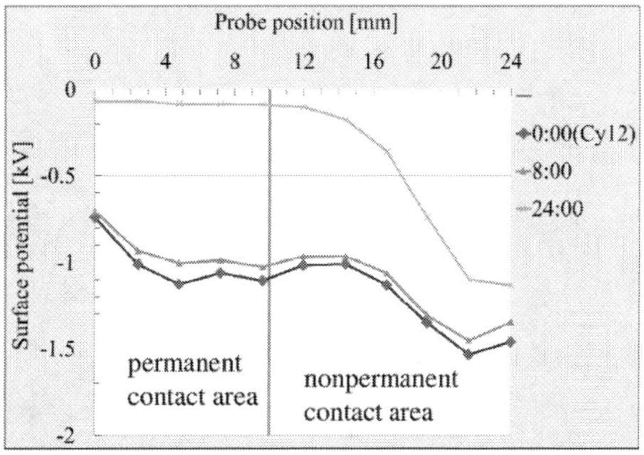

Figure 4. Distribution of the electric potential at the surface of the bottom sample at 0, 8 and 24 hours after the completion of 12 tribocharging cycles.

5. Conclusions

(i) The experimental bench described in this paper enables the study of the tribocharging phenomena induced by the conformal contact between two flat polymers. Its design facilitates the accurate adjustment and monitoring of the normal and tangential contact forces, as well as of the stroke and the relative sliding speed between the two bodies.
(ii) The distribution of the electric charge at the surface of the samples can be investigated by measurements performed with the capacitive probe of an electrostatic voltmeter.
(iii) The absolute value of the surface potential increases with the number of tribocharging cycles, then it saturates at levels related to the specific sliding conditions.
(iv) The charge acquired by triboelectric effect decreases faster in the zones of permanent contact between the samples than in the areas only periodically exposed to friction.

Acknowledgement
This work was partially funded by the French Government program "Investissements d'Avenir" (LABEX INTERACTIFS, reference ANR-11-LABX-0017-01).

References

[1] Zeghloul T, Dascalescu L, Rouagdia K, Fatihou A, Renoux P and Souchet D 2014 *Conf Rec. IEEE IAS Ann. Meet.* Vancouver, Canada, DOI 10.1109/IAS.2014.6978340
[2] Bailey A G 2001 *J. Electrost.* **51-52** 82-90
[3] Guerret-Piecourt C, Bec S and Tréheux D 2001 *C. R. Acad. Sci. Paris* **IV 2** 761–774
[4] Nakayama K 1996 *Wear* **194** 185-189
[5] Hiratsuka K and Hosotani K 2012 *Tribo. Int.* **55** 87–99
[6] Johnson K L 1985 Contact mechanics (Cambridge:Cambridge Univ. Press) p107-196
[7] Medevielle A, Thévenot F and Tréheux D 1995 *J. Euro. Ceram. Soci.* **15** 1193-99
[8] Harvey T J, Wood R J K, Denuault G and Powrie H E G 2002 *Tribo. Int.* **35** 605–614
[9] Lee D-Yo, Lee J, Hwang J and Choa S-H 2007 *Tribo. Int.* **40** 1253–57
[10] Ning L, Jian L, Yang S, Wang J, Ren J and Wang J 2010 *Tribo. Int.* **43** 568–576
[11] Changa Y-P, Yur J-P, Chou H-M and Chub H-M 2006 *Wear* **260** 1209–16
[12] Kchaoua B, Turki C, Salvia M, Fakhfakh Z and Tréheux D 2008 *Wear* **265** 763–771
[13] Antoniu A, Dascalescu L, Vacar I V, Plopeanu M C, Tabti B and Teodorescu H N 2011 *IEEE Trans. Ind. Appl.* **47** (2011) 1118-25

Experimental characterization of triboelectric charging of polyethylene powders

Simon Jantač, Ladislav Konopka, Juraj Kosek

University of Chemistry and Technology Prague, Department of Chemical Engineering, Technicka 5, 166 28 Prague 6, Czech Republic,
E-mail: ladislav.konopka@vscht.cz

Abstract. Triboelectric charging causes serious problems in the industrial processing of powders. We focus on the charging of polyethylene (PE) powder particles, whose agglomeration can cause serious economic problems in PE production in fluidized-bed reactors. The 'cascade method' apparatus, i.e., a slide followed by the Faraday's pail, was utilized to observe the particle-wall charging of PE particles in friction contact with various materials (glass, aluminium, PE) and allowed us to characterize the charging dynamics. Our results indicate that the evolution of the charge on the particles follows a saturation curve, where the saturated state is represented by maximum (outcome) charge. Such a trend can be conveniently fitted by a function representing the first-order dynamics. We determine the dependency of charging dynamics on various factors, e.g., the humidity, the slide surface roughness and the slide material. Our measurements imply that air humidity influences the charging process substantially more than the choice of the slide material. Moreover, we observe significant charging even in the case of the same materials being in contact. The work contributes to a better understanding of tribocharging and the estimation of charging-related parameters provides the input for the modelling of this complex process.

1 Introduction

The mechanism of triboelectrification resists our attempts for its full understanding and subsequent accurate description of the charging process. Triboelectrification is in practice often associated with the collisions of industrially important powders (e.g. polymers, drugs, flour). Powder particles undergo the repeated particle-particle and particle-wall friction contacts as they are manufactured and/or transported, which results in undesired charge accumulation. In this work, we focused on the qualitative and quantitative systematic description of the triboelectric charging of polyethylene (PE) particles.

2 Experiments

The granular material used in these experiments was PE (density 0.94 g/cm^3, diameter of the particles was 0.8-1 mm). PE particles are rough and porous, representing the nascent material leaving the polymerization reactor. The electrostatic charging of PE particles was characterized using the cascade method apparatus (described below), which was in some cases combined with the corona discharge unit.

The cascade method apparatus was constructed from a plate used as slide. The dimensions (length and width) of the slide were 50x5 cm. The slide inclination was fixed at 30°.Weighted PE sample was poured many times on the top of the slide and the gradual accumulation of the charge on the particle surface was measured in Faraday's pail after each cycle (Figure 1). Design of the apparatus was inspired by experiments associated with tribocharging of glass [1] and silicon [2], respectively.

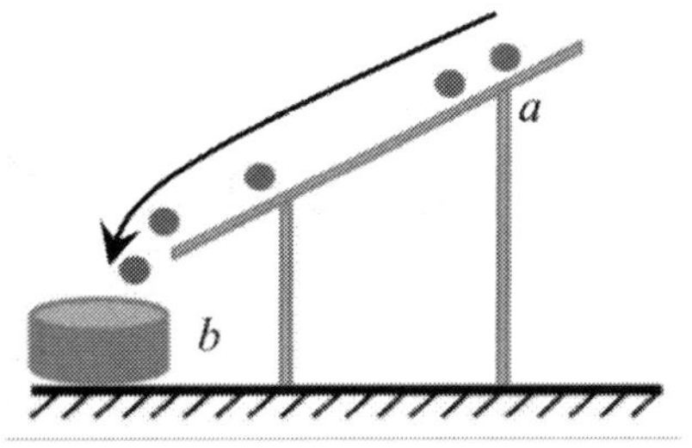

Figure 1. The Cascade method apparatus, constructed of the slide (*a*) and the Faraday's pail (*b*).

Experiments were carried out both in the humid air and in the air mixed with pure nitrogen in order to mimic inert atmosphere of the industrial reactor, respectively. Moreover, the sliding experiments were performed for different slide materials, namely PE, aluminium (Al) and glass.

The procedure of the experiment was the following. First, we measured the weight of the sample, then the sample was poured on the top of the slide. The charge of particles was then measured in the Faraday's pail. The particles were then extracted from the pail to the Petri dish and again poured on the top of the slide. This procedure was repeated until no significant additional charge on particles was generated.

The effect of corona discharge on the charging dynamics was tested using JCI 155v5 Charge Decay Test unit. Corona discharge was applied on the PE powder only after attaining the saturation charge.

3 Results and Discussion

Example of the charging dynamics is shown in Figure 2, where each point represents the measured value of specific charge accumulated on particle surface after one sliding cycle. Significant charging of the same materials in contact is also observed, which is in contradiction with some published results[3].

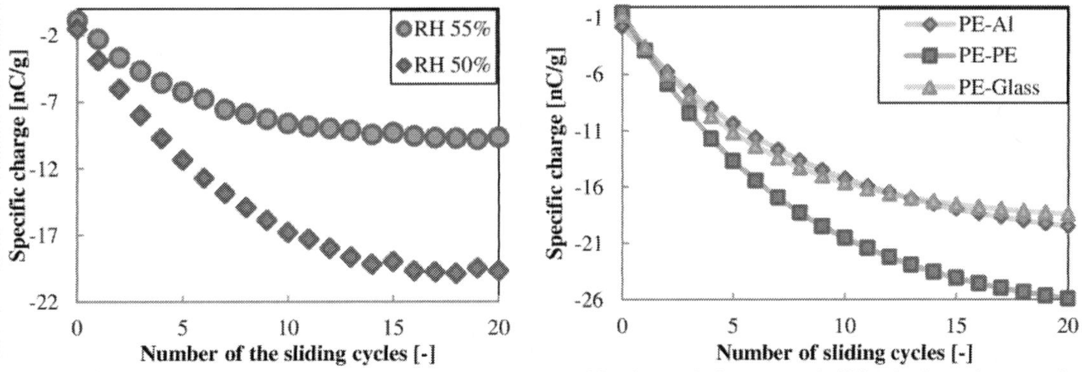

Figure 2. Measured evolution of the polyethylene powder specific charge during repeated sliding in dependence on the relative humidity (PE-Al contact) and on the choice of the material (relative humidity fixed at 50%).

Figure 2 shows the significant effect of the relative humidity and of the slide material on the triboelectric charging. Larger saturation charge was observed for lower humidity. The results indicate that the triboelectric charging is not trivially dependent on the choice of slide material. According to our observations, materials vary in their sensitivity to humidity with respect to charging. These two experimental sets were used to determine the charge of PE particles as the function of a charging period (number of the sliding cycles) for a given slide material and a given humidity.

Since the charge was recorded after each sliding cycle until the saturation value was reached, it is possible to estimate the dynamics of the charging. From Figure 2 it is obvious that the dynamic evolution of specific charge has a saturation character, so that it can be fitted by a combination of elementary functions. We use specifically the following equation

$$Q = A + B \cdot \exp\left(-\frac{x}{C}\right), \tag{1}$$

where Q is the specific electrostatic charge, x is the number of sliding cycles, parameter A represents the saturation charge, B is associated with the initial charge on the particles and C stands for the reciprocal value of the charging rate constant. Parameters A, B, C of equation (1) depend on the experimental conditions (e.g. humidity, slide material). Figure 3 illustrates that equation (1) fits the experimental data accurately.

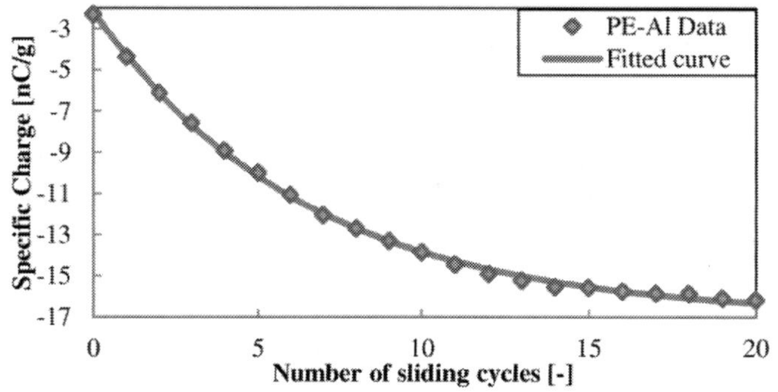

Figure 3. Dynamic evolution of the specific charge in PE – Al system. Measured data (blue points) and their fit by Equation (1) (red curve).

Equation (1) is the analytical solution of differential equation (2):

$$\frac{dQ}{dx} = \frac{1}{C} \cdot (A - Q). \tag{2}$$

Since the number of sliding cycles x is only a specific interpretation of charging time, the left-hand side of equation (2) represents the charge accumulation. As the charge accumulation is proportional to the first power of the charge Q, we state that the charging process follows the first order charging kinetics.

Table 1. Table of parameters of Eq. (1) relevant to Figure 3

A	B	C
-16.99	14.79	6.45

In experiments using the corona discharge unit (Figure 4), the sample was first charged by two corona discharges, and then the sample was slid more than once until the saturation charge was achieved. This procedure was repeated several times.

It is evident from Figure 4 that the sample treated with the corona discharge did repeatedly charge approx. 20% less than the original sample. The value of the saturation charge seems to be independent on the number of the corona discharges if the corona discharge is applied at least once. The corona treated sample was examined by Confocal Raman Microscope and no change of the particle surface in the comparison to the original sample was observed. Even after 20 discharges the Raman spectra remains unchanged. This suggests that the corona discharge might deplete only part of high energy electrons on the particle surface.

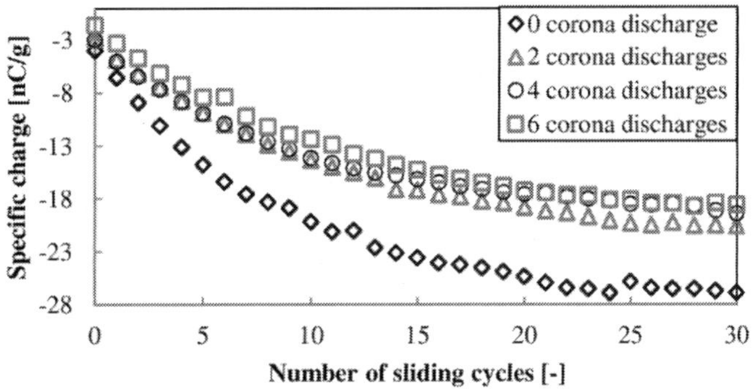

Figure 4. The charging curves after repeated corona discharges on the sample (at RH 46%).

4 Conclusions

Polyethylene represents the most commonly produced polymer worldwide. Agglomeration of PE particles during their production caused by the accumulation of the electrostatic charge on their surface can lead to significant economic problems. We measured the triboelectric charge generated by the particle-wall contact using the in-house designed cascade apparatus. Systematic experiments with a good reproducibility were conducted in dependence on the slide material and the air humidity). We demonstrated that evolution of the charge can be fitted by the formal first order kinetics. The air humidity affects the charging process substantially more than the choice of the slide material. However, materials appeared to vary in their sensitivity to humidity with respect to charging. The significant charge accumulation was measured also in the case of the same materials being in the friction contact. The corona discharge applied on the sample reduced the saturation charge approximately about 20%, therefore it might be a fast applicable method for the reduction of the charging during the processing and transportation of powders.

References

[1] Hogue MD, Calle CI, Weitzman PS, Curry DR 2008 Calculating the trajectories of triboelectrically charged particles using Discrete Element Modeling (DEM) *J. Electrostat.* **66** pp 32-38
[2] Ireland PM 2010 Triboelectrification of particulate flows on surfaces: Part I — *Experiments Powder. Technol.* **198** pp 189-198
[3] Park CH, Park JK, Jeon HS 2008 Triboelectric series and charging properties of plastics using the designed vertical-reciprocation charger *J. Electrostat.* **66** pp 578-583

Electrospun materials for affinity-based engineering and drug delivery

T J Sill and H A von Recum[1]
Department of Biomedical Engineering, Case Western Reserve University, Cleveland, OH USA

E-mail: horst.vonrecum@case.edu

Abstract. Electrospinning is a process which can quickly and cheaply create materials of high surface to volume and aspect ratios from many materials, however in application toward drug delivery this can be a strong disadvantage as well. Diffusion of drug is proportional to the thickness of that device. In moving from macro to micro to nano-sized electrospun materials drug release rates change to profiles that are too fast to be therapeutically beneficial. In this work we use molecular interactions to further control the rate of release beyond that capable of diffusion alone. To do this we create materials with molecular pockets, which can "hold" therapeutic drugs through a reversible interaction such as a host/guest complexation. Through these complexes we show we are able to impact delivery of drug from electrospun materials, and also apply them in tissue engineering for the reversible presentation of biomolecules on a fiber surface

1. Introduction

Electrospinning and spraying are processes which have been known for more than 100 years, however have only recently be "rediscovered" and applied toward applications in medicine including drug delivery and tissue engineering. The process uses a high voltage course to inject charge on a polymer solution or melt. As this solution is accelerated toward a collector of opposite polarity, electrostatic and solution forces drive the formation of structures ranging from droplets (electrospraying) to micro and nanofibers (electrospinnning) (Figure 1).

Electrospinning has been exponentially increasing in its usage for tissue engineering and drug delivery over the last decade [1]. Recent advances using polymer fibers with composite materials has even allowed for new directions to take these materials [2]. Some of the major advantages of electrospun materials are that they can be easily manufactured using a fairly inexpensive benchtop apparatus, and simple polymer solvents or melts. Many different polymers can be electrospun from natural materials to synthetic. Also the resulting fibers are easily within the size range of biological matrix molecules (100 nm – 1 μm).

However one of the major limitations that these materials, particularly in regards to drug delivery, is that small fiber diameter. Fick's First Law of Diffusion $J = D\, dc/dx$ which relates drug diffusive flux J to concentration dc/dx , or drug amount per unit distance; and as such the smaller the distance the drug has to travel, the more rapidly it will diffuse across that distance (x) [3]. D is the empirically determined diffusivity. Acceleration of drug release corresponding to a decrease in device size is a phenomenon that had been well-known for the micro- and nano-particulate world but had to be re-learned for electrospun micro- and nano-fibers. Basically this means that drug which would take weeks to months to get out of a macro-sized polymer slab, would get out of an electrospun fiber of similar mass within minutes to days. Typically drug delivery applications require therapeutic rates

[1] To whom any correspondence should be addressed.

across biological time frames such as cell proliferation cycles (days) to tissue remodeling (weeks). So delivery that is completed within hours is not much better than a bolus injection of that drug.

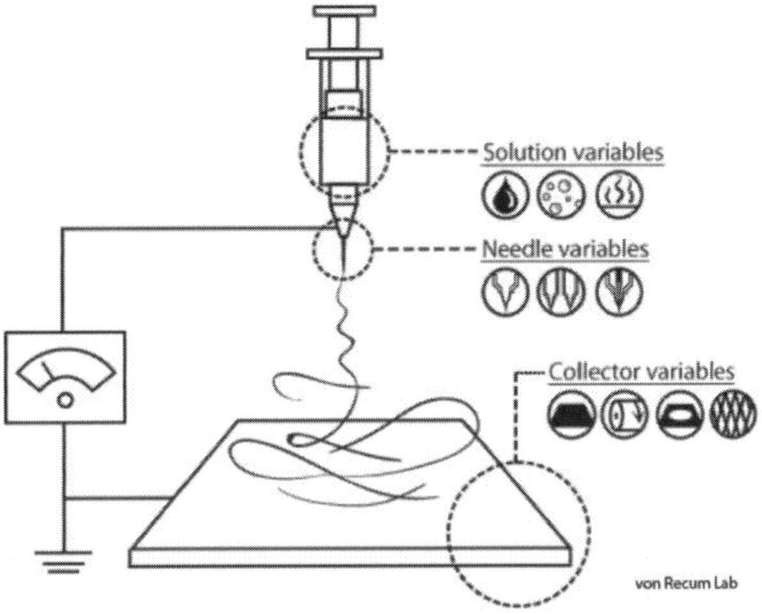

Figure 5. Electrospinning Device Setup. Variables altered are the solution, the needle arrangement, and collector layout (plate, rotating drum, metal gap plate, biomedical device).

In our laboratory we have been exploring the use of molecular interactions to control the rate of drug release. These interactions are specific affinities between device polymer and drug, which if strong enough can slow down the rate of drug release beyond that capable of diffusion alone. Essentially drug- decomplexing from the material interactions becoming the rate-limiting step. This new research field is termed "affinity-based drug delivery" [4]. In this paper we explore whether through affinity-based drug delivery strategies we can overcome this major limitation of utilizing electrospun fibers for drug delivery.

Another recently popular application for electrospun materials is as tissue engineering or regenerative medicine scaffolds. Often these materials are electrospun from natural polymers such as collagen to improve biocompatibility; however, the most commonly used solvent for electrospinning collagen has previously been hexafluoroisopropanol (HFIP), which is extremely cytotoxic, even with the miniscule residual amounts found in well-washed electrospun collagen mats. Only recently have investigators begun to electrospin collagen from more biologically friendly solvents [5].

One way to gain the benefits of biological integration and cell attachment, but also keep the benefits of the high flexibility of electrospinning, is to decorate electrospun fibers with cell attachment peptides such as the ubiquitous Arginine-Glycine-Aspartic Acid (RGD). With this strategy one can encourage endothelial cell attachment and ingrowth of new blood vessels to form a vascular tissue bed [6]. Decorating an electrospun fiber with RGD is done either after the spinning is complete, or as polymer modification before the electrospinning process. The former method is limited in its chemical conjugation strategies, and the latter method limits electrospinning since the peptide is molecularly more fragile than most synthetic polymers, and as a charged molecule it changes spinning dynamics.

Further another limitation of RGD decorated fibers is that they are simultaneously both chemically unstable and biologically unchangeable. The instability comes from general peptide instability, which can either spontaneously hydrolyze or specifically be degraded by a number of nonspecific proteases. Once degraded, the precious cell attachment signal that the investigator has laboriously added to the electrospun fiber is now lost. The unchangability is of concern due to the highly dynamic environment of tissue engineering and regenerative medicine in terms of tissue remodeling. One cell signal might be needed at one time point while a different cell signal might be needed at another time point. As an example for bone tissue engineering, initially a signal to recruit endothelial cells and generate blood vessels is required. While at later time points no more new blood vessels are needed, but rather osteogenic cells are needed. A chemically attached signal such as RGD cannot change post-implantation.

In this work we also wish to show how the use of these same affinity groups can begin to address some of these problems found in electrospun materials for tissue engineering, namely a) how to introduce a biofunctional molecule easily and stably, b) how to replace that molecule if needed after natural degradation; or c) how to replace that molecule if its signal needs to be replaced by that of a different biomolecule.

2. Methods

Poly(ethylene vinyl alcohol) (pEVOH) is modified with pendant β-cyclodextrins (CD), due to CD's well known capacity to form inclusions with small molecular weight drugs. This is done in a two step process. pEVOH is first modified with the bi-functional crosslinker N-(p-Maleimidophenyl) isocyanate (PMPI). The resulting polymer is then electrospun using conventional techniques [7].

In parallel a CD-thiol is made by converting 6-Tosyl-β-CD by heating for 2 days 100°C in thiourea and methanol. Trichloroethylene is added to precipitate the product which is then washed and recrystalization from water. CD-thiol is reduced in basic conditions prior to use, and then added to pEVOH-PMPI to form pEVOH-CD. Conjugation is verified by NMR analysis, and conjugation efficiency by FTIR. Feed ratios were calculated to produce polymers with either 2.5 or 10% CD substitutions on polymer side chains. pEVOH which underwent the same processing steps but no CD conjugation was used as a non-affinity diffusion-only polymer control.

Fibers were hydrodynamically loaded with a model drug, Rhodamine B, in an aqueous solution, extensively washed, and then evaluated for drug release in an infinite sink of phosphate buffered saline agitated at 37°C, removing release aliquots every 12 h.

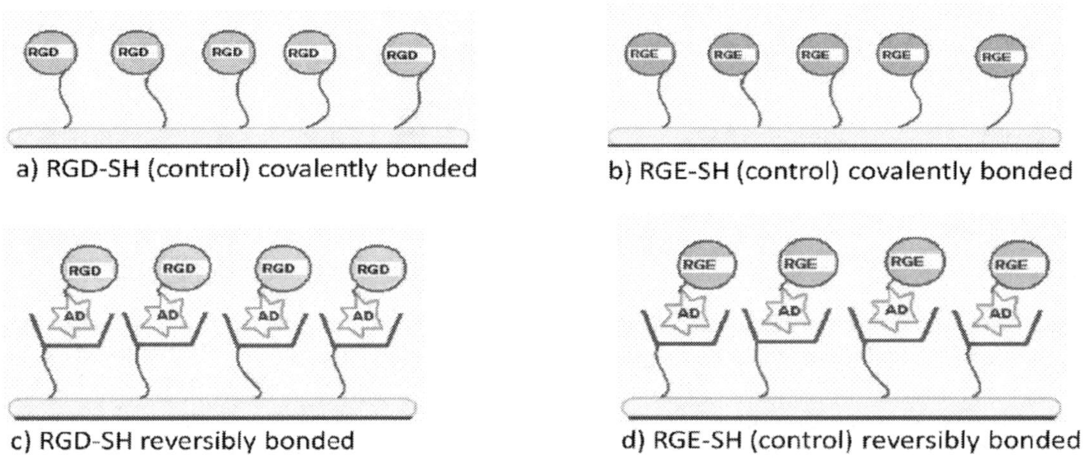

Figure 2. Scheme for reversible and non-reversible attachment of peptides. Gold surfaces were conjugated either with peptide directly (a & b), or indirection through association with conjugated cyclodextrin (b & c).

For the tissue engineering studies, CD-thiol was directly conjugated to a gold surface, complexed with either the cell attachment peptide RGD, or the non-attaching control peptide RGE. To cause the association between the peptide and CD coated surface, by using a bifunctional crosslinker with N-hydroxysuccinimide at one end reacting to adamantine methylamine; and maleimide at the other end reacting to an extra cysteine residue; with a poly(ethylene glycol) spacer in between. The adamantine forms a strong but reversible complex with CD. In parallel RGD and RGE were directly attached to different gold substrates through use of the thiol on the extra cysteine residue. Green fluorescent protein (GFP) labeled fibroblast cells were added to the surfaces and evaluated by reflectance microscopy (Figure 2).

The pEVOH-CD was also evaluated for reversible modification with association with adhesion peptides. In this case an interpenetrating network was made between the pEVOH-CD polymer manufactured above, and tissue culture polysytrene, loaded with the various peptide constructs, and then incubated with fibroblast cells. Lastly the pEVOH-CD was electrospun, loaded with peptide constructs, incubated with GFP cells, and the resulting mat examined by fluorescence microscopy both top-down, as well as in cross-section after serial-sectioning by cryomicrotome.

3. Results

The drug delivery experiments show that addition of affinity through use of cyclodextrin substantially delays release of drug out of electrospun nanofibers, with the majority of the drug released after three days from the non-affinity pEVOH control, however at that same time point only ¼ of the drug was released from the affinity-based pEVOH-CD polymer (Figure 3).

In the tissue engineering studies reflectance microscopy shows that GFP labeled cells do not adhere to the control gold surface conjugated only with CD. However once an adamantane modified cell attachment peptide RGD is added there is rapid cell attachment with many focal adhesions and cell motility. When this peptide is exchanged with an adamantane (Ad) modified RGE cell attachment is rapidly diminished with few attached cells (Figure 4).

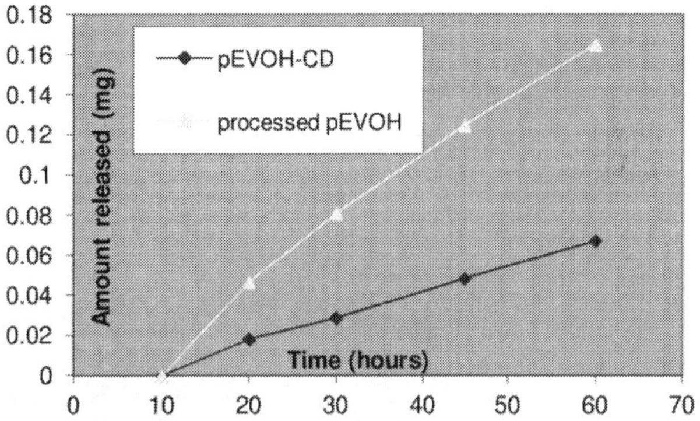

Figure 1. Rhodamine B Release from Affinity-Based Polymers. Affinity-based polymers made from CD-modified pEVOH (blue) shows substantial reduction in release rate when compared to non-affinity controls of unmodified pEVOH (yellow).

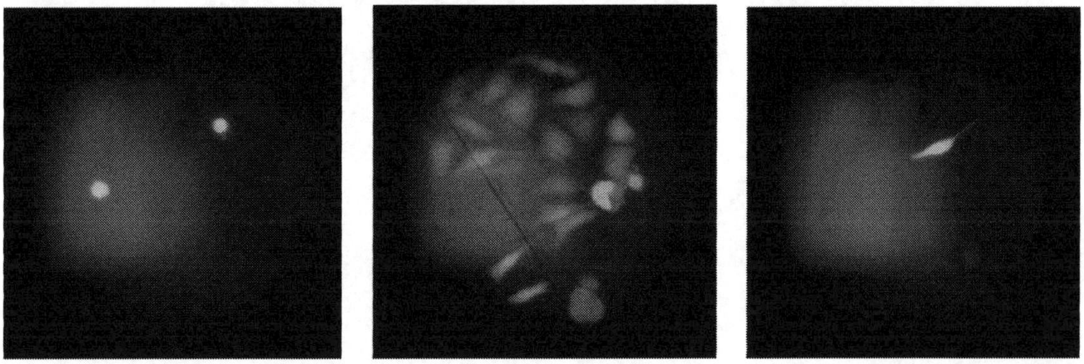

Figure 2. Reflectance Microcopy of GFP cells on gold surfaces reversibly decorated with RGD. Left) CD-SH alone. Middle) Ad-RGD on CD-SH. Right) Surface replaced with Ad-RGE on CD-SH.

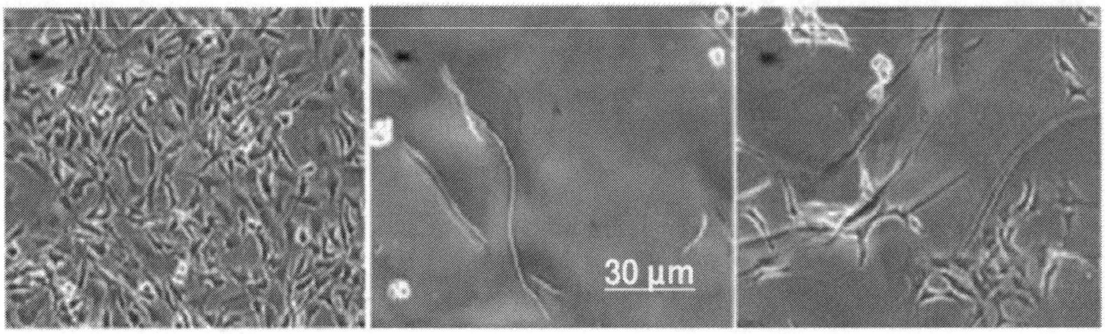

Figure 3. Cells on pEVOH-CD surfaces reversibly decorated with RGD. Left) TCPS control shows good cell attachment. Middle) PEVOH-CD face of interpenetrating network shows no cell attachment. Right) PEVOH-CD with Ad-RGD shows cell attachment and spreading.

Fibroblasts the pEVOH-CD surfaces show similar behaviour. Cells do not attach to the pEVOH-CD face of the interpenetrating network modified surface. However upon incubation with an Ad-modifed RGD cells begin to attach and spread similar to tissue culture styrene controls (Figure 5).

Lastly, electrospun mats were seeded with GFP labeled fibroblasts. From top-down microscopy individual cells can be seen which are approximately 10 times bigger than individual fibers estimated at 1µm in diameter. In cross-section many individual cells can be seen adhering to the outermost fibers (Figure 6).

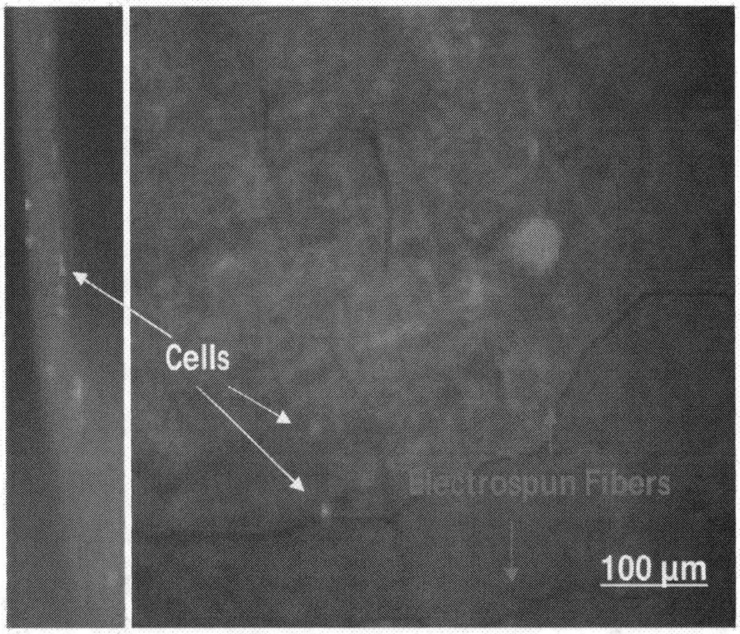

Figure 4. GFP labeled cells on electrospun pEVOH-CD fibers with associated RGD peptides. Left) Cross-section showing cells adhered and spread (yellow arrows). Right) Top-down with individual (dark) fibers visible (red arrows).

4. Discussion

In this work we have shown that through affinity-based molecular interactions with cyclodextrin we can control the rate of drug release out of electrospun materials, reducing the disadvantage caused by their small diameter, and bringing release rates into therapeutically relevant timescales. Further we found that we could use the molecular interactions to cause reversible associate of affinity-modified cell-attachment peptides RGD, promoting cell adhesion on simple gold surfaces, affinity-polymer modified tissue culture surfaces, and electrospun affinity-based materials.

References

[1] Sill T J and von Recum H A 2008 Electrospinning: applications in drug delivery and tissue engineering. *Biomaterials* **29** 1989-2006
[2] Merritt S R, Exner A E, Lee Z and von Recum H A 2012 Electrospinning and Imaging *Adv. Engin. Mater.* **14** B266-B278
[3] Fu A S, Thatiparti T R, Saidel G M and von Recum H A 2011 Experimental studies and modeling of drug release from a tunable affinity-based drug delivery platform *Ann. Biomed. Eng.* **39** 2466-2475
[4] Wang N X and von Recum H A 2011 Affinity-based drug delivery *Macromol. Biosci.* **11** 321-332
[5] Dong B, Arnoult O, Smith M E and Wnek G E 2009 Electrospinning of collagen nanofiber scaffolds from benign solvents *Macromol. Rapid Commun.* **30** 539-542
[6] Choi W S, Bae J W, Lim H R, Joung Y K, Park J-C, Kwon I K and Park K D 2008 RGD peptide-immobilized electrospun matrix of polyurethane for enhanced endothelial cell affinity. *Biomed. Mater.* **3** 044108
[7] Kenawy el R, Layman J M, Watkins J R, Bowlin G L, Matthews J A, Simpson D G and Wnek G E 2003 Electrospinning of poly(ethylene-co-vinyl alcohol) fibers. *Biomaterials* **24** 907-913